Advances in Additive Manufacturing for Metallic Materials and Their Applications (3rd Edition)

Advances in Additive Manufacturing for Metallic Materials and Their Applications (3rd Edition)

Guest Editors

Petru Berce
Răzvan Păcurar

MDPI

Basel • Beijing • Wuhan • Barcelona • Belgrade • Novi Sad • Cluj • Manchester

Guest Editors
Petru Berce
Department of
Manufacturing Engineering
Technical University
of Cluj-Napoca
Cluj-Napoca
Romania

Răzvan Păcurar
Department of
Manufacturing Engineering
Technical University
of Cluj-Napoca
Cluj-Napoca
Romania

Editorial Office
MDPI AG
Grosspeteranlage 5
4052 Basel, Switzerland

This is a reprint of the Special Issue, published open access by the journal *Metals* (ISSN 2075-4701), freely accessible at: https://www.mdpi.com/journal/metals/special_issues/C98J60W836.

For citation purposes, cite each article independently as indicated on the article page online and as indicated below:

Lastname, A.A.; Lastname, B.B. Article Title. *Journal Name* **Year**, *Volume Number*, Page Range.

ISBN 978-3-7258-7234-3 (Hbk)
ISBN 978-3-7258-7235-0 (PDF)
https://doi.org/10.3390/books978-3-7258-7235-0

Contents

Preface vii

Răzvan Păcurar and Petru Berce
Advanced Insights into Laser-Based Metal Additive Manufacturing: From Microstructural Control to Functional Performance
Reprinted from: *Metals* **2026**, *16*, 69, https://doi.org/10.3390/met16010069 1

Simone Paggetti, Enrico Bedogni and Paolo Veronesi
Factors Affecting the Surface Roughness of the As-Built Additively Manufactured Metal Parts: A Review
Reprinted from: *Metals* **2025**, *15*, 1069, https://doi.org/10.3390/met15101069 7

Chen Liu, Junxiao Liu, Xiuyuan Yin, Xiaoyu Zhang, Shuo Shang and Changsheng Liu
Multi-Model Collaborative Optimization of Inconel 690 Deposited Geometry in Laser-Directed Energy Deposition: Machine Learning Prediction and NSGA-II Decision Framework
Reprinted from: *Metals* **2025**, *15*, 905, https://doi.org/10.3390/met15080905 35

Gianluca Pirro, Alessandro Morri, Alessandra Martucci, Mariangela Lombardi and Lorella Ceschini
Effect of Heat Treatments and Related Microstructural Modifications on High-Cycle Fatigue Behavior of Powder Bed Fusion–Laser Beam-Fabricated Ti-6Al-2Sn-4Zr-6Mo Alloy
Reprinted from: *Metals* **2025**, *15*, 849, https://doi.org/10.3390/met15080849 50

Yu Qiao, Marius Grad and Aida Nonn
Toward an Efficient and Robust Process–Structure Prediction Framework for Filigree L-PBF 316L Stainless Steel Structures
Reprinted from: *Metals* **2025**, *15*, 812, https://doi.org/10.3390/met15070812 67

Conall Kirk, Weijie Xie, Shubhangi Das, Ben Ferguson, Chenliang Wu, Hau-Chung Man and Chi-Wai Chan
Microstructure, Porosity, and Bending Fatigue Behaviour of PBF-LB/M SS316L for Biomedical Applications
Reprinted from: *Metals* **2025**, *15*, 650, https://doi.org/10.3390/met15060650 89

Cosmin Cosma, Martin Melichar, Stelian Libu, Alexandru Popan, Glad Contiu, Cristina Teusan, et al.
The Effect of Drag Finishing on Additively Manufactured Customized Dental Crowns
Reprinted from: *Metals* **2025**, *15*, 471, https://doi.org/10.3390/met15050471 113

Ozgur Poyraz, Daniel Tomlinson, Anthony Molyneux, Marie E. Baxter, Evren Yasa and James Hughes
Optimized and Additively Manufactured Face Mills for Enhanced Cutting Performance
Reprinted from: *Metals* **2025**, *15*, 376, https://doi.org/10.3390/met15040376 137

Zerui Li, Jingyuan Xu, Jincheng Tang, Zhuo Sang and Ming Yan
Additively Manufactured Biomedical Ti-15Mo Alloy with Triply Periodical Minimal Surfaces and Functional Surface Modification
Reprinted from: *Metals* **2025**, *15*, 355, https://doi.org/0.3390/met15040355 153

Zhengjie Guo, Yuting Ma, Tayyeb Ali, Yi Yang, Juan Hou, Shujun Li and Hao Wang
Enhanced Compressive Properties of Additively Manufactured Ti-6Al-4V Gradient Lattice Structures
Reprinted from: *Metals* **2025**, *15*, 230, https://doi.org/10.3390/met15030230 177

Preface

This Special Issue presents eight original research articles and one comprehensive review that explore advanced additive manufacturing of metallic materials, including data-driven process optimization, process–structure prediction, surface integrity, and fatigue behavior, as well as customized biomedical and industrial components and topology-optimized designs. The contributions presented in this Special Issue reflect the ongoing advancement toward more reliable, performance-driven, and application-oriented metal additive manufacturing technologies.

Petru Berce and Răzvan Păcurar
Guest Editors

metals

Editorial

Advanced Insights into Laser-Based Metal Additive Manufacturing: From Microstructural Control to Functional Performance

Răzvan Păcurar and Petru Berce *

Department of Manufacturing Engineering, Faculty of Industrial Engineering, Robotics and Production Management, Technical University of Cluj-Napoca, Blv. Muncii No. 103-105, 400641 Cluj-Napoca, Romania
* Correspondence: petru.berce@tcm.utcluj.ro

1. Introduction and Scope

Laser based metal additive manufacturing has emerged as a transformative solution for producing next-generation metallic components, from architected lattices [1] and biofunctional implants [2] to wear-resistant coatings [3] and optimized industrial tools [4], bridging computational design [5], process modeling [6] and experimental validation to deliver lightweight [7], durable [8], and performance-tailored structures [9] for biomedical, aerospace and engineering applications [10]. In particular, laser-based AM methods, including Laser Powder Bed Fusion (L-PBF) [11] and Laser-Directed Energy Deposition (L-DED) [12], have enabled unprecedented design freedom (e.g., intricate lattices and triply periodic minimal surface (TPMS) structures) [13] alongside the fabrication of fully dense metallic parts [14]. This Special Issue focuses on advances in L-PBF and DED technologies that push the boundaries of microstructural control, mechanical performance, and functional integration. Key themes of the nine articles of this Special Issue include innovations in L-PBF process optimization and topology-optimized designs, functional coatings and surface modifications for enhanced bio-performance, investigations of fatigue behavior in critical alloys (Ti-6Al-4V, Ti-15Mo, Ti-6Al-2Sn-4Zr-6Mo, and SS316L), as well as microstructural modeling and simulation frameworks that bridge process parameters to material structure. The Special Issue also highlights novel applications of triply periodic minimal surface (TPMS) lattices in biomedical implants, hybrid material approaches (combining AM-built structures with advanced coatings) and post-processing techniques like drag finishing to improve surface quality of AM parts. Finally, a comprehensive review of factors affecting as-built surface roughness in metal AM provides a unifying perspective on how process parameters, such as layer thickness, energy input and build orientation influence surface finish and how this can be modeled, optimized, and complemented by post-processing methods for improved part quality.

Collectively, the contributions in this Special Issue underscore a shift toward integrated and intelligent AM approaches that couple materials science insights with computational optimization and post-processing strategies. Innovations such as machine learning-driven process optimization and cellular automated microstructure simulation reflect the growing convergence of data-driven modeling with experimental AM process development. By addressing challenges from lattice design trade-offs and fatigue life enhancement to surface quality control, this collection not only demonstrates improved functional performance of additively manufactured metal components but also emphasizes the efficiency and reliability of the manufacturing processes themselves. Looking forward, continued research

 https://doi.org/10.3390/met15101069

is expected in standardizing process–structure–property prediction tools, developing environmentally conscious post-processing methods, and exploring hybrid manufacturing routes that ensure consistent quality and performance in critical metallic applications.

2. Contributions

This Special Issue presents nine original research articles that address critical challenges in laser-based metal additive manufacturing, including lattice design optimization, functional coatings for biomedical applications, fatigue resistance of titanium and stainless steel alloys, advanced microstructural modeling, process parameter optimization by machine learning, showcasing the latest strategies to enhance performance, reliability, and manufacturability in metallic AM components.

The first contribution, by Guo, Z. et al., titled "Enhanced Compressive Properties of Additively Manufactured Ti-6Al-4V Gradient Lattice Structures" explores how graded lattice architectures can improve the mechanical efficiency of L-PBF Ti-6Al-4V structures. Through a combination of finite element design optimization and experimental testing, the authors compare a uniform lattice to a functionally graded lattice (both with ~84% porosity) under compression. The graded lattice showed a layer-by-layer collapse mechanism under load, resulting in superior energy absorption and load-bearing capacity. Notably, the gradient design achieved approximately 20% higher cumulative energy absorption than the one of the uniform lattices. Under longitudinal compression the graded lattice outperformed the uniform design, whereas under transverse loading both lattices behaved similarly. These findings provide a foundation for designing lightweight Ti-6Al-4V lattice components with optimized stiffness and crashworthiness for aerospace and biomedical applications.

The second contribution, by Li, Z. et al., titled "Additively Manufactured Biomedical Ti-15Mo Alloy with Triply Periodical Minimal Surfaces and Functional Surface Modification" presents a novel β-Ti alloy implant combining a low-modulus TPMS lattice structure with bioactive surface coatings. A porous Ti-15Mo structure was fabricated via PBF-LB with an elastic modulus tailored between 15 and 50 GPa (closer to bone) and ~100 MPa yield strength. To address biomedical functionality, the lattice surfaces were enhanced using a microporous TiO_2 layer by micro-arc oxidation, which was subsequently coated with silver nanoparticles (Ag) for broad antibacterial effect, tannic acid (TA) for antioxidant properties, and fluorinated hydroxyapatite (FHA) to promote osteogenesis. The result was a Ti-15Mo implant with a biofunctional multi-layer coating that demonstrated good cytocompatibility, hemocompatibility, and bactericidal performance in vitro. This work offers a comprehensive approach to aligning mechanical and biological performance in AM implants, showing that TPMS-based porous designs combined with advanced surface modification can mitigate stress-shielding while imparting infection resistance and improved osseointegration.

The third contribution, by Poyraz, O. et al., titled "Optimized and Additively Manufactured Face Mills for Enhanced Cutting Performance" applies L-PBF and topology optimization to the development of metal cutting tools. A set of indexable face milling cutters was re-engineered to reduce vibrational amplitudes during machining without sacrificing stiffness. Using M300 maraging steel and L-PBF, three face mill designs with varying weight-reduction targets were produced, leveraging the geometric freedom of AM to integrate internal damping features. Experimental modal analysis and cutting trials showed that the optimized AM face mills exhibited significantly lower vibration magnitudes and reduced tool wear compared to a conventional design. In fact, the AM-optimized tools achieved more stable cutting with diminished chatter and extended insert life, highlighting the potential of PBF-manufactured tooling to outperform traditional mills. This study demonstrates how coupling topology optimization with metal AM can yield

https://doi.org/10.3390/met15101069

cutting tools with improved dynamic performance, pointing toward broader adoption of AM in the tooling industry.

The fourth contribution, by Cosma, C. et al., titled "The Effect of Drag Finishing on Additively Manufactured Customized Dental Crowns" investigates a post-processing technique to enhance the surface finish of CoCr dental crowns produced by L-PBF. Drag finishing (DF), a mechanical polishing method, was applied to as-printed CoCr crowns to overcome the inherently high roughness of AM surfaces. The authors report that DF treatment on the external crown surfaces achieved a 70–90% reduction in surface roughness, bringing the roughness (Ra) down to about 0.6 μm in the best case. This smoothness is comparable to that of conventionally milled surfaces and is achieved in a fraction of the time required by manual polishing. While inner (concave) surfaces saw only limited improvement due to restricted media access, 3D surface texture analysis confirmed that drag finishing effectively uniformizes the surface by leveling microscopic peaks. In addition, coordinate measurements showed that DF improved dimensional accuracy of the crowns' exterior (deviations reduced to +0.01–0.05 mm) and the process, coupled with a proper heat treatment, increased surface hardness to ~520 HV. This contribution underlines the importance of post-processing in AM for dental applications, demonstrating that automated finishing can yield patient-specific metal restorations with clinically acceptable surface quality and fit.

The fifth contribution, by Kirk, C. et al., titled "Microstructure, Porosity, and Bending Fatigue behavior of PBF-LB/M SS316L for Biomedical Applications" examines how inherent AM defects impact the fatigue life of 316L stainless steel, a common implant material. SS316L samples fabricated by L-PBF were subjected to microstructural analysis and four-point bending fatigue tests to simulate cyclic loading conditions of biomedical implants. The study finds that process-induced defects, specifically micro-pores and unmelted powder particles within the as-built metal, act as critical stress concentrators that facilitate early crack initiation under cyclic bending. Even when loading is below the yield strength, these internal flaws markedly reduce fatigue life, as real-time microscopy of the tests captured crack nucleation at defect sites. The authors conclude that the fatigue performance of L-PBF 316L is strongly governed by its microstructural integrity, directly linking higher porosity to shorter lifespan. This insight is crucial for the design of reliable patient-specific implants, suggesting that improved process parameters or post-build hot isostatic pressing (HIP) may be necessary to decrease the level of the defects. The work supports ongoing efforts to ensure AM-built biomedical components meet stringent durability requirements by controlling porosity and microstructure.

The sixth contribution, by Qiao, Y. et al., titled "Toward an Efficient and Robust Process–Structure Prediction Framework for Filigree L-PBF 316L Stainless Steel Structures" introduces a computational modeling approach to predict microstructural outcomes in thin-walled AM parts. The authors develop a cellular automata (CA) simulation framework to model grain growth and texture formation during the solidification of 316L stainless steel in L-PBF. Filigree geometries (with a wall thickness approaching the scale of individual grains) pose a particular challenge for prediction, as even minor thermal fluctuations can influence grain structure. By calibrating the CA model with relevant process parameters (like laser power and scan speed) and validating against EBSD measurements, the study demonstrates that the simulation can effectively reproduce key microstructural features observed experimentally. The CA model captured the grain size distribution, aspect ratio, and crystallographic texture in the thin walls, proving its robustness in reflecting how process conditions translate into microstructure. This work highlights the value of physics-based modeling in AM: reliable process–structure prediction tools can help optimize scan strategies and parameters virtually, ensuring desired grain structures (and

https://doi.org/10.3390/met15101069

thus consistent mechanical properties) in complex L-PBF components. It marks a step toward integrating microstructural simulation into the design loop of high-performance, safety-critical AM parts.

The seventh contribution, by Pirro, G. et al., titled "Effect of Heat Treatments and Related Microstructural Modifications on High-Cycle Fatigue Behavior of PBF-LB-Fabricated Ti-6Al-2Sn-4Zr-6Mo Alloy", investigates how tailored post-build heat treatments can enhance the fatigue life of a high-strength titanium alloy produced by L-PBF. The alloy Ti-6Al-2Sn-4Zr-6Mo (Ti6246) was studied under two conditions: a standard $\alpha + \beta$ anneal at 875 °C (AN875) and a solution treatment at 825 °C followed by aging at 500 °C (STA825). High-cycle fatigue tests revealed that the STA825 heat treatment delivers approximately 25% greater fatigue strength than the conventional anneal. Microstructural analysis explains this improvement: STA825 yields a finer bi-lamellar α/β microstructure with thinner primary α lamellae and a more homogeneous dispersion of secondary α within the β matrix. Such a refined microstructure effectively delays crack nucleation and slows crack propagation, as confirmed by defect-sensitivity models, showing a higher fatigue limit and threshold stress intensity for STA825-treated samples. Notably, the optimized STA825-treated Ti6246 outperformed even the well-established Ti-6Al-4V alloy in fatigue endurance, underscoring Ti6246's potential as a superior material for load-bearing AM components. This contribution emphasizes that post-processing heat treatments are vital for unlocking the full fatigue performance of AM Ti alloys, enabling their use in demanding aerospace and automotive applications where fatigue failure is a primary concern.

The eighth contribution, by Liu, C. et al., titled "Multi-Model Collaborative Optimization of Inconel 690 Deposited Geometry in Laser-Directed Energy Deposition: Machine Learning Prediction and NSGA-II Decision Framework" addresses the challenge of achieving precise geometric control in L-DED processes through a data-driven approach. Focusing on Inconel 690 (a nickel alloy relevant for nuclear industry repairs), the authors built a large experimental dataset relating process parameters to single-bead geometry (width, height, depth) via full-factorial laser deposition trials. By applying statistical analysis, they identified strong correlations (e.g., laser power positively correlating with bead width $r \approx 0.82$ and depth $r \approx 0.85$, and powder feed rate correlating with bead height $r \approx 0.70$). Using this knowledge, a hybrid machine learning model based on a backpropagation neural network (BPNN) was trained to predict bead dimensions with high accuracy (R^2 up to ~0.96). The model was then coupled with a multi-objective optimization using the NSGA-II genetic algorithm, which generated a Pareto set of 100 optimal parameter combinations for controlling bead geometry. Remarkably, when tested, the ML-driven framework could predict and achieve target bead widths and depths within $\leq$5% error, and heights within ~5.3%. This confirms the effectiveness of the approach for reliable, first-try process setup in L-DED, reducing trial-and-error and ensuring dimensional accuracy in deposited features. The work showcases how combining machine learning prediction with evolutionary optimization can guide process parameter selection for complex AM processes, opening new directions to adaptive control in critical applications like nuclear component repair.

The ninth contribution, by Paggetti, S. et al., titled "Factors Affecting the Surface Roughness of As-Built Additively Manufactured Metal Parts: A Review," provides a comprehensive synthesis of the main factors influencing surface quality across metal AM technologies. It covers all major metal AM processes, including powder bed fusion (laser-based and electron-beam), direct energy deposition (both laser and arc-based), binder jetting, material extrusion, and sheet lamination, outlining the typical as-built roughness levels and the mechanisms behind them. The authors systematically identify how factors such as layer thickness, scan strategy, energy input, feedstock form (powder or wire), and build orientation/inclination angle affect the resulting surface texture. For example, inclined or

https://doi.org/10.3390/met15101069

curved surfaces built in PBF tend to exhibit higher roughness due to the stair-step effect, which can be mitigated by applying in situ re-melting scans. Likewise, in DED processes, increased laser power or powder flow can inadvertently increase roughness by enlarging melt pools and spatter, whereas optimized spot size and travel speed yield smoother tracks. The review presents these correlations in a series of tables and charts, serving as practical guidelines for engineers to predict or control surface finish. By providing a concise summary of roughness drivers across technologies, this work helps practitioners design parts and select process parameters that meet surface quality requirements or determine when post-processing (like machining or polishing) will be necessary. In doing so, it bridges the knowledge gap for achieving application-specific surface roughness in metal AM in close correlation with better process control.

3. Acknowledgments

As Guest Editors of this Special Issue, "Advances in Additive Manufacturing for Metallic Materials and Their Applications (3rd Edition)," we would like to express our sincere gratitude to all the contributing authors for sharing their innovative research and insights into laser-based metal AM. We extend our thanks to the expert reviewers for their thorough evaluations and constructive feedback, which have greatly enhanced the quality of these articles. We are also grateful to the editorial team of Metals and the MDPI staff for their professional support throughout the publication process. We trust that the diverse advancements compiled in this Special Issue will serve as a valuable resource for researchers, engineers, and practitioners in the additive manufacturing community and inspire further developments in this rapidly evolving field in the future.

Conflicts of Interest: The author declares no conflicts of interest.

List of Contributions:

1. Guo, Z.; Ma, Y.; Ali, T.; Yang, Y.; Hou, J.; Li, S.; Wang, H. Enhanced Compressive Properties of Additively Manufactured Ti-6Al-4V Gradient Lattice Structures. *Metals* **2025**, *15*, 230. https://doi.org/10.3390/met15030230.
2. Li, Z.; Xu, J.; Tang, J.; Sang, Z.; Yan, M. Additively Manufactured Biomedical Ti-15Mo Alloy with Triply Periodical Minimal Surfaces and Functional Surface Modification. *Metals* **2025**, *15*, 355. https://doi.org/10.3390/met15040355.
3. Poyraz, O.; Tomlinson, D.; Molyneux, A.; Baxter, M.E.; Yasa, E.; Hughes, J. Optimized and Additively Manufactured Face Mills for Enhanced Cutting Performance. *Metals* **2025**, *15*, 376. https://doi.org/10.3390/met15040376.
4. Cosma, C.; Melichar, M.; Libu, S.; Popan, A.; Contiu, G.; Teusan, C.; Berce, P.; Balc, N. The Effect of Drag Finishing on Additively Manufactured Customized Dental Crowns. *Metals* **2025**, *15*, 471. https://doi.org/10.3390/met15050471.
5. Kirk, C.; Xie, W.; Das, S.; Ferguson, B.; Wu, C.; Man, H.-C.; Chan, C.-W. Microstructure, Porosity, and Bending Fatigue Behaviour of PBF-LB/M SS316L for Biomedical Applications. *Metals* **2025**, *15*, 650. https://doi.org/10.3390/met15060650.
6. Qiao, Y.; Grad, M.; Nonn, A. Toward an Efficient and Robust Process–Structure Prediction Framework for Filigree L-PBF 316L Stainless Steel Structures. *Metals* **2025**, *15*, 812. https://doi.org/10.3390/met15070812.
7. Pirro, G.; Morri, A.; Martucci, A.; Lombardi, M.; Ceschini, L. Effect of Heat Treatments and Related Microstructural Modifications on High-Cycle Fatigue Behavior of Powder Bed Fusion–Laser Beam-Fabricated Ti-6Al-2Sn-4Zr-6Mo Alloy. *Metals* **2025**, *15*, 849. https://doi.org/10.3390/met15080849.
8. Liu, C.; Liu, J.; Yin, X.; Zhang, X.; Shang, S.; Liu, C. Multi-Model Collaborative Optimization of Inconel 690 Deposited Geometry in Laser-Directed Energy Deposition: Machine Learning

https://doi.org/10.3390/met15101069

Prediction and NSGA-II Decision Framework. *Metals* **2025**, *15*, 905. https://doi.org/10.3390/met15080905.
9. Paggetti, S.; Bedogni, E.; Veronesi, P. Factors Affecting the Surface Roughness of the As-Built Additively Manufactured Metal Parts: A Review. *Metals* **2025**, *15*, 1069. https://doi.org/10.3390/met15101069.

References

1. Dar, J.; Ponsot, A.G.; Jolma, C.J.; Lin, D. A Review on Scan Strategies in Laser-Based Metal Additive Manufacturing. *J. Mater. Res. Technol.* **2025**, *36*, 5425–5467. [CrossRef]
2. Wei, C.; Zhang, Z.; Cheng, D.; Sun, Z.; Zhu, M.; Li, L. An Overview of Laser-Based Multiple Metallic Material Additive Manufacturing: From Macro- to Micro-Scales. *Int. J. Extrem. Manuf.* **2021**, *3*, 012003. [CrossRef]
3. He, B.; Deng, C.; Yao, H.; Wang, S.; Huang, Y.; Zhu, H.; Cui, L. Microstructure and Mechanical Properties of Wear-Resistant SiCp/Al Composite Layers on 6061 Aluminum Alloy Fabricated by Additive Friction Stir Deposition. *Addit. Manuf.* **2025**, *109*, 104880. [CrossRef]
4. Li, X.; Pan, Z.; Smolej, L.; Nadimpalli, V.K.; Moshiri, M. Towards Manufacturing Intra-Layer Multi-Material Mould Tools with Vertical Interfaces Using Laser-Based Powder Bed Fusion. *Mater. Des.* **2024**, *243*, 113056. [CrossRef]
5. Sanchez-Poncela, M.; Jin, Q.; Hebert, R.; Aindow, M.; Amigó-Borrás, V.; Rementeria, R. Computational Design of Grain-Refined Isotropic Triplex Steels for Laser-Powder Bed Fusion. *Prog. Addit. Manuf.* **2025**, *10*, 10563–10580. [CrossRef]
6. Nath, A. Process Modeling of Laser-Based Metal Additive Manufacturing of Metal Alloys. In *Advances in Processing of Lightweight Metal Alloys and Composites*; Springer: Cham, Switzerland, 2022; pp. 443–455. [CrossRef]
7. Paul, C.; Jinoop, A.N.; Kumar, A.; Bindra, K. Laser-Based Metal Additive Manufacturing: Technology, Global Scenario and Our Experiences. *Trans. Indian Inst. Met.* **2021**, *74*, 2581–2595. [CrossRef]
8. Taghizadeh, M.; Zhu, Z.H. A Comprehensive Review on Metal Laser Additive Manufacturing in Space: Modeling and Perspectives. *Acta Astronaut.* **2024**, *222*, 403–421. [CrossRef]
9. Altun, F.; Bayar, A.; Hamzat, A.K.; Asmatulu, R.; Ali, Z.; Asmatulu, E. AI-Driven Innovations in 3D Printing: Optimization, Automation, and Intelligent Control. *J. Manuf. Mater. Process.* **2025**, *9*, 329. [CrossRef]
10. Sibisi, T.H.; Shongwe, M.B.; Tshabalala, L.C.; Mathoho, I. LAM Additive Manufacturing: A Fundamental Review on Mechanical Properties, Common Defects, Dominant Processing Variables, and Its Applications. *Int. J. Adv. Manuf. Technol.* **2023**, *128*, 2847–2861. [CrossRef]
11. Scarpellini, A.; Finazzi, V.; Schito, P.; Bionda, A.; Ratti, A.; Demir, A.G. Laser Powder Bed Fusion of a Topology Optimized and Surface Textured Rudder Bulb with Lightweight and Drag-Reducing Design. *J. Mar. Sci. Eng.* **2021**, *9*, 1032. [CrossRef]
12. Ferreira, N.M.; Vila Pouca, M.; Fernandes, C.; Seabra, J.; Lesiuk, G.; Parente, M.; Jesus, A. Direct Energy Deposition Parametric Simulation Investigation in Gear Repair Applications. *Materials* **2023**, *16*, 3549. [CrossRef] [PubMed]
13. Alkunte, S.; Fidan, I.; Naikwadi, V.; Gudavasov, S.; Ali, M.A.; Mahmudov, M.; Hasanov, S.; Cheepu, M. Advancements and Challenges in Additively Manufactured Functionally Graded Materials: A Comprehensive Review. *J. Manuf. Mater. Process.* **2024**, *8*, 23. [CrossRef]
14. Chowdhury, S.; Yadaiah, N.; Prakash, C.; Ramakrishna, S.; Dixit, S.; Gupta, L.R.; Buddhi, D. Laser Powder Bed Fusion: A State-of-the-Art Review of the Technology, Materials, Properties, Defects, and Numerical Modelling. *J. Mater. Res. Technol.* **2022**, *20*, 2109–2172. [CrossRef]

metals

MDPI

Review

Factors Affecting the Surface Roughness of the As-Built Additively Manufactured Metal Parts: A Review

Simone Paggetti [1,*], Enrico Bedogni [1,2] and Paolo Veronesi [1]

[1] Department of Engineering 'Enzo Ferrari', Universita degli Studi di Modena e Reggio Emilia, 41125 Modena, MO, Italy; enrico.bedogni@kosme.it (E.B.); paolov@unimore.it (P.V.)
[2] Kosme s.r.l. Unipersonale, Via dell'Artigianato, 5, 46048 Roverbella, MN, Italy
* Correspondence: simone.paggetti@unimore.it

Abstract

Nowadays, additive manufacturing technologies continue to increase in number, and with them, the various challenges they bring for the optimal design of components. However, many relevant applications require that a certain surface finishing level is reached; in particular, surface roughness should stay below a certain threshold. The aim of this work is to provide, for each of the most used metal additive manufacturing technologies, a short review of parameters affecting as-built surface roughness, indicating possible correlations with process parameters. The identified correlations, summarized visually as tables, could serve as starting guidelines for the design and production of parts with controlled surface roughness or having a surface roughness suitable for the application of possible surface finishing post-processes.

Keywords: additive manufacturing; surface roughness; power bed fusion; direct energy deposition; binder jetting; material extrusion; layer laminate manufacturing; surface finishing

1. Introduction

Additive Manufacturing (AM) is a production process that creates three-dimensional objects by adding material layer by layer, unlike subtractive manufacturing, where material is removed from a solid block to achieve the desired shape. It is commonly used for prototyping, given its flexibility in creating a wide variety of geometries, but also for standard parts production. Parts made using AM are particularly recognizable because the production method is clearly reflected on the 'as-built' surface, which usually shows the construction layers used. The roughness of surfaces produced by metal AM is influenced by many parameters, among which [1]:

- Technology.
- Raw materials: type, shape (wire, sheets, or powders), and dimensions or particle size.
- Process-related parameters such as energy density, layer thickness, and building speed.
- Parts-related parameters, such as the presence of inclined surfaces, supports, etc.

The ones listed above are the ones with the greater impact, but many more have a smaller role, such as building strategies within each layer, orientation in the building plate, inert gas flow, etc.

Surface roughness plays a critical role in determining the functional performance, reliability, and quality of components across a wide range of industrial sectors. In mechanical engineering, rough surfaces can lead to increased friction, wear, and premature fatigue failure of moving parts. This is particularly problematic in precision assemblies, where tight tolerances and smooth interactions are essential for optimal operation [2–4].

In the aerospace industry, surface roughness directly influences aerodynamic performance. Components such as turbine blades, airframes, and nozzles require extremely smooth surfaces to minimize drag and improve fuel efficiency. Even minor deviations in surface texture can lead to turbulent airflow, increased energy consumption, and reduced operational lifespan [5–7].

The food industry also faces challenges related to surface roughness, especially in equipment used for processing, packaging, and handling. Rough surfaces can harbor contaminants and microorganisms, making cleaning and sterilization more difficult. This not only compromises hygiene standards but also increases the risk of product contamination and regulatory non-compliance. A direct example is the requirement for a specific surface roughness value when food products come into direct contact with metallic components [8–10].

In the medical field, implants and surgical tools must exhibit controlled surface roughness to ensure biocompatibility and proper integration with biological tissues. Excessive roughness can lead to inflammation, poor healing, or rejection, while insufficient roughness may hinder cell adhesion and osseointegration [11–13].

Surface roughness also affects coating adhesion, corrosion resistance, and fatigue strength in automotive and marine applications. For instance, rough substrates may lead to uneven coating thickness, resulting in weak spots and accelerated degradation under environmental stress [14–16].

Moreover, in additive manufacturing, surface roughness is a key parameter influencing post-processing requirements, dimensional accuracy, and mechanical properties. Various technologies exhibit different degrees of surface roughness depending on process parameters, material properties, layer resolution, etc.

Despite its importance, surface roughness is often under-characterized in industrial studies. More comprehensive investigations are needed to understand its behavior under different processing conditions and its long-term impact on component performance.

In conclusion, surface roughness is a multifaceted issue that affects product quality, safety, and efficiency across diverse industries. Addressing it through improved design, material selection, and process control is essential for advancing manufacturing technologies and meeting increasingly stringent performance standards.

Concerning technologies, the most widely used for metal parts are Powder Bed Fusion (PBF), Direct Energy Deposition (DED), Binder Jetting (BJ) and Material Extrusion (ME).

PBF technologies are a group of additive manufacturing techniques that produce 3D parts from powder materials. The laser source is fixed, and the beam is directed across the printing bed using a system of lenses [17]. Among the PBF technologies, the most relevant are:

- Selective Laser Melting (SLM). This technology uses one or more lasers to melt layers of metal powder. The laser focuses on a single spot, melting the powder with an average layer thickness between 30 and 60 μm or larger [18]. The laser spot is circular and can vary in diameter from 20 to 200 μm.
- Area Wise Laser Powder Bed Fusion (AWLPBF)/Diode-based Additive Manufacturing (DiAM). Also called area printing LPBF, this technology uses a wide pulsed beam composed of multiple lasers to create an image of approximately 1 cm^2 that melts the powder. The layer thickness ranges between 50 and 100 μm [19].
- Electron Beam Melting (EBM) uses an electron beam as a heat source instead of a laser. The beam focuses on a single point, similarly to SLM. Unlike the typical argon-shielded

chamber, EBM operates in a vacuum. Common layer thickness ranges between 50 and 100 μm [20].

Selective Laser Sintering (SLS) technologies, while architecturally similar to SLM, and thus typically listed among the PBF methods, SLS does not fully melt the metal powders. There are many ways to realize metal components by sintering. In Solid State Sintering, the part is formed at the sintering temperature of the printing metal, and the raw material is just made of the metal powder in which the components are printed. In Chemically Induced Binding, chemical reactions occurring between the material powders or between the powders and the atmosphere of the printing chamber are used to impart form to the powders. The products of these reactions act as binders for the primary powders. In Liquid Phase Sintering/Partial Melting, there are three further subclasses according to the type of powder grains used: separate grains, composite grains, and coated. These three categories explain how the binder and powder material are mixed or prepared before printing. In this case, the binder can be polymeric or metallic [21].

Direct Energy Deposition (DED) technologies are a group of additive manufacturing techniques that build 3D objects using either powders or wire as raw materials, usually assisted by a laser as a heat source. Unlike PBF, in DED the material is deposited and melted simultaneously, and the heat source is mobile, moving across the printing area. The layer thickness in DED processes is typically in the millimeter or sub-millimeter range, which is significantly higher than in PBF. These technologies are particularly efficient for adding material only where needed, making them well-suited for repair applications or for building large, near-net-shape components [17,22]. Among the DED techniques, the most widely used are:

- Laser Metal Deposition (LMD)/Laser-engineered net shaping (LENS)/Powder Laser Energy Deposition (PLED): in this process, the raw material—typically metal powder—is melted by a laser and simultaneously deposited onto a substrate to build the part layer by layer [23].
- Wire Laser Energy Deposition (WLED)/Wire Laser Additive Manufacturing (WLAM): This technique operates similarly to LMD but uses metal wire instead of powder as the feedstock. The wire is melted by a laser during deposition.
- Wire Electron Beam Energy Deposition (WEBED)/Wire Electron Beam Additive Manufacturing (WEBAM)/Electron Beam Freeform Fabrication (EBF^3)/Electron Beam Wire-Feed Additive Manufacturing (EBWFAM): A wire feedstock is melted using an electron beam rather than a laser. Due to the need for a vacuum environment to operate the electron beam, powder-based deposition (as in LMD) is not feasible [24].
- Wire Arc Additive Manufacturing (WAAM): This process employs a Gas Metal Arc Welding (GMAW) torch mounted on a robotic arm. Similar to conventional welding, the electric arc is shielded using inert gas [22].
- Submerged Arc Additive Manufacturing (SAAM): while the hardware setup is similar to WAAM, this method replaces the inert gas shielding with a layer of sand that submerges the melt pool, thereby protecting it from oxidation [25].
- Rapid Plasma Deposition (RPD)/Wire Plasma Arc Energy Deposition (WPAED)/Plasma Metal Deposition (PMD): this variant uses a Gas Tungsten Arc Welding (GTAW) torch instead of GMAW, while maintaining the same general configuration as WAAM [26].
- Metal Powder Application (MPA), also known as Cold Spray Additive Manufacturing (CSAM): A variant of DED, using kinetic energy instead of a heat source alone. It employs a high-velocity gas stream to accelerate metal powder particles toward a substrate, where they accumulate to form the desired component. Upon impact, the particles undergo severe plastic deformation, converting kinetic energy into localized thermal energy, which facilitates bonding without melting. Originally developed for

coating applications, this solid-state process has since been adapted for additive manufacturing, particularly for materials that are difficult to process using conventional fusion-based techniques [27,28].

Binder technologies utilize metal powders combined with a binder, which imparts sufficient mechanical strength to the green parts prior to sintering. On average, the layer thickness used in these processes ranges between 50 and 100 μm. In most cases, a debinding step is required either before or during the sintering phase to remove the binder and ensure proper densification of the final component, or impregnation with a lower melting point metal is performed alternatively to sintering. Among these technologies, the most relevant are:

- Metal Binder Jetting (MBJ): This process spreads a layer of metal powder and selectively deposits binder only where the part is to be formed. The binder is then activated using a heat source, typically a lamp. MBJ does not require support structures [29].
- Mold Slurry Deposition (MSD)/MoldJet (MJ)/Layerwise Slurry Deposition (LSD): In this method, metal powder and binder are pre-mixed into a slurry. A wax mold is first created additively using a drop-on-demand technique and then filled with slurry. The wax mold provides integrated support during printing [30].
- Lithography-based Metal Manufacturing (LMM)/Area Wise Vat Polymerization (AWVP)/Digital Light Processing (DLP)/Vat Photo Polymerization (VPP): Even if not commonly listed among the BJ technologies, it combines metal powders and binders, and it is here included for this reason. This technology uses a photopolymerization process similar to Stereolithography (SLA). Metal powder is suspended in a photopolymer binder, which is selectively cured by light exposure. LMM is the only technology among these that requires support structures [31].
- Material Jetting (MJ)/Metal Material Jetting (MMJ). This technique is very similar to Metal Binder Jetting (MBJ). Instead of using a recoater that deposits only metallic powder, which is subsequently impregnated with a binder, MMJ employs a slurry in which the binder and metallic powder are pre-mixed. This mixture is deposited through a series of nozzles, like MBJ. Additionally, a second set of nozzles is used to deposit support material. The powders used in this technology are typically very fine, often in the nanometric range, as they must be able to pass through the narrow nozzles required by jetting systems [32,33].

Material Extrusion (ME) technologies are polymer additive manufacturing processes, such as extrusion and jetting-based methods, which have been adapted for metal printing. These technologies share the need for post-processing steps such as debinding and sintering to remove the binder and densify the final part or, alternatively, impregnation. Among them:

- Fused Deposition Modeling (FDM)/Filament Material Extrusion (FME), also known as Fused Filament Fabrication (FFF): It is one of the most widely used 3D printing technologies. It operates by extruding a thermoplastic filament through a heated nozzle, depositing material layer by layer onto a build platform. When applied to metal printing, the filament is loaded with metal powder [34].
- Pellet Material Extrusion (PME): follows the same principle as FDM, but the feedstock is in the form of pellets rather than filament, offering potential advantages in terms of material cost and feed flexibility.
- Dispersion Material Extrusion (DME): Follows the same architecture of the previous two technologies; instead of using a polymer as a binder during extrusion, a liquid binder is employed, resulting in a metallic powder dispersed in the form of a slurry.

This slurry is then deposited onto the build platform via an extruder, similarly to a conventional FME process [35].

Layer Laminate Manufacturing (LLM), also referred to as Sheet Lamination (SL), is an additive manufacturing process used to fabricate three-dimensional objects by stacking and bonding successive layers of material. The process begins with sheets of material-typically paper, plastic, or metal foil-which are layered and bonded using techniques such as adhesive application, ultrasonic welding, heat, or pressure. Each sheet is precisely cut to the desired geometry either before or after the bonding step. This method is considered a hybrid additive-subtractive technique, as it involves both the addition of material through layering and the removal of excess material during the cutting phase [17], and it will be excluded from this review.

Friction Energy Deposition (FED), Friction Stir Additive Manufacturing (FSAM), Friction Stir Powder Additive Manufacturing (FSPAM), or Additive Friction Stir Deposition (AFSD): This technology utilizes the energy generated by friction resulting from the rotation of the deposition head against the build platform initially, and subsequently against the underlying layers [36]. It operates either with a feedstock in the form of rods or powder, or in combination with LLM as a method for welding metallic sheets [37,38].

Liquid Metal Printing (LMP), Direct Metal Writing (DMW), Drop-On-Demand Liquid Metal Jetting (DOD-LMJ) or Molten Metal Jetting (MMJ): This technology resembles Drop-on-Demand (DoD) printing where the feedstock was replaced by molten metal [39]. This makes it challenging to work with metals that have high melting points. As a result, one of the most commonly used materials is aluminum and its alloys [40,41].

Localized Electrodeposition in Liquid (LEL), Localized Electrochemical Deposition (LECD): It utilizes a conductive substrate immersed in a liquid electrolyte. The extrusion head injects a metal solution that is electrochemically deposited onto the substrate, which acts as the cathode, while the anode is positioned around the perimeter of the build area, separated by an insulating layer. This submicrometric technique is patented by Exaddon [42]. Although its industrial application is currently limited due to the small build volume and low production speed, this technology has been included in the overview of metal additive manufacturing methods for completeness. Nevertheless, it finds application in the medical field [43,44].

Figure 1 presents all the technologies mentioned above.

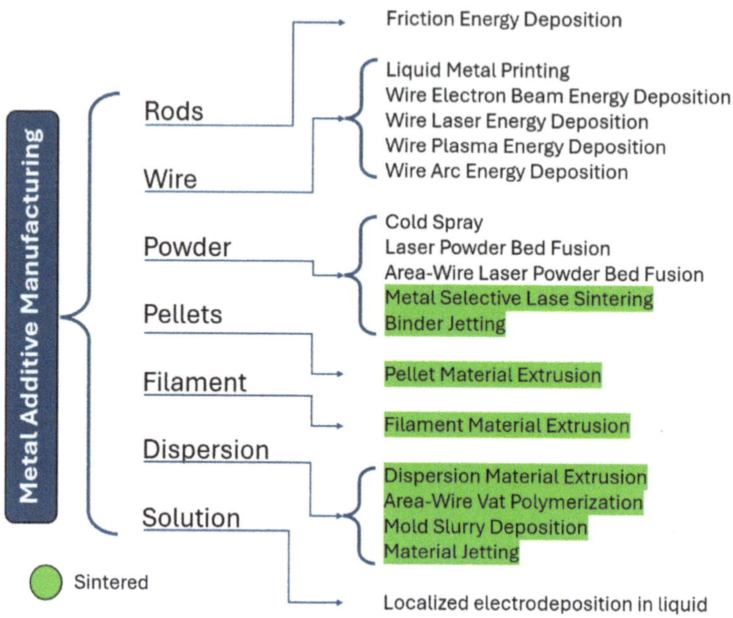

Figure 1. Overview map of AM technologies.

2. As-Built Surface Roughness

2.1. Selective Laser Melting

Starting with SLM technologies, the effect of processing parameters on surface roughness has been widely investigated. Hatching spacing, scan speed, and laser power effects were recently studied by Mansoura et al. by analyzing the effect of two parameters at a time, while keeping the layer thickness constant at 40 μm [18]. Measurements were taken on the 0° up-skin surface. The reported values by Mansoura et al. for 316L stainless steel powder show an average roughness (Ra) ranging from 4.30 to 6.47 μm. The analysis of the three contour plots reported in that study highlights the influence of process parameters on surface roughness, showing that increasing both hatching spacing and scan speed is responsible for an increase in the as-built surface roughness. The same study explores the relationship between laser power and hatch spacing, and it demonstrates that roughness decreases with increasing laser power but increases with increasing hatch spacing. In the case of the combined effects of laser power and scan speed, as expected, roughness increases when laser power increases and scan speed increases.

The combination of the three plots provides useful information for the combined optimization of these parameters, providing a more detailed insight into the process with respect to the energy density parameter alone. Overall, the interaction plots of this study serve as a fundamental tool for identifying optimal operating conditions aimed at minimizing surface roughness, thereby contributing to the improvement of additive manufacturing or laser processing surface quality [45]. However, these studies contrast with the findings of Correa-Gómez et al. and Yildiz et al., which report opposite effects in response to changes in energy density, despite similar trends in scan speed, hatch spacing, and laser power. In this case, a reduction in surface roughness was observed with decreasing laser power and increasing scan speed [46,47].

This shift in the effect of process variables at different energy densities can be attributed to changes in melt pool behavior, as demonstrated in Jiang et al. studies. It is shown that, at both high and low energy densities, fusion defects can occur—ranging from unmelted powder to open surface porosity—which contribute to the final surface roughness of the as-built component [48].

This encourages the search for an optimal intermediate energy density that can yield the best possible as-built surface roughness.

As previously mentioned, another variable that influences surface roughness is the geometry of the component and, specifically, the inclination of the surface relative to the printing plate.

This correlation was studied by Wang et al. [19], who demonstrated that surface roughness can increase by up to a factor of 10 when comparing the best and worst surface orientations. For up-skin surfaces, the average roughness (Ra) measured for processing 316L stainless steel powders ranges from 5.03 μm to 24.2 μm, while the Rz values range from 12.7 μm to 55.90 μm. Similar trends were observed for down-skin surfaces, where Ra values range from 6.25 μm to 65.10 μm, and Rz values range from 15.80 μm to 230.00 μm [49].

The last variable considered is the metal powder particle size, which also plays a significant role in surface roughness, not only because it practically sets a lower limit to PBF layer thickness.

For SLM, the commonly used particle size distribution has a D10 of 15 μm and a D90 of 45 μm, often referred to as 15–45. In their study, Sinico et al. [20] analyzed three different Maraging 300 powder size distributions: 15–45 μm, 10–30 μm, and 5–15 μm. The finest distribution, 5–15 μm, was excluded from further analysis due to its poor flowability, which negatively affects the powder spreading process during printing [50]. Considering all tested combinations of energy density and laser power, specimens fabricated with smaller particles consistently exhibited higher surface quality, with a reduction in Ra ranging from approximately 20% to over 50% with respect to the coarser particles.

The following table (Table 1) summarizes the main findings of the aforementioned studies:

Table 1. Effect of SLM processing parameters of steel powders on the as-built surface.

Variables	Action	Effect	Physical Mechanisms Involved
Hatching spacing	Increase	Increase roughness	Melt pool behavior
Scan speed (Low J/mm^3)	Increase	Increase roughness	Melt pool behavior
Laser power (Low J/mm^3)	Increase	Decrease roughness	Melt pool behavior
Scan speed (High J/mm^3)	Increase	Decrease roughness	Melt pool behavior
Laser power (High J/mm^3)	Increase	Increase roughness	Melt pool behavior
Surface angles	Approach 0° and 90°	Decrease roughness	Gravity on melt pool
Particle size	Decrease	Decrease roughness	Powder packing
Particle size	Too small	Poor flowability	Powder rheology

2.2. Area Wise Laser Powder Bed Fusion

This technology was developed and patented by Seurat [19,51]. Currently, there are no published studies specifically analyzing surface roughness for this process.

However, since it employs a layer-by-layer approach with a layer thickness range similar to other LPBF techniques, it is plausible that the surface roughness for up-skin and down-skin surfaces at various angles is of the same order of magnitude, except for the 0° up-skin and 90° vertical surfaces. These two orientations are significantly influenced by the laser scanning pattern and the melt pool dynamics, which differ substantially from those in conventional L-PBF systems.

Few papers have been published on this technology, and even fewer have focused on surface roughness. However, the works by Fotovvati et al. and Matthews et al. report roughness values for 0° up-skin surfaces, varying with energy density, ranging from 1.7 to 10 μm [52] and the Spearman correlation matrix for the technology's parameter [51].

2.3. Electron Beam Melting

Compared to SLM technologies, the greater layer thickness used in EBM has a direct impact on surface roughness. While EBM is typically better suited for mass production thanks to its higher build rate, it also tends to result in increased roughness [53].

Due to the ferromagnetic nature of steels, they are rarely used in Electron Beam Melting (EBM). As a result, only a limited number of studies have been conducted on this material. For this reason, Ti-6Al-4V alloy was selected for investigation in the context of EBM.

The influence of various process parameters on surface roughness in EBM of Ti-6Al-4V was investigated using powders with a particle size distribution of 50–100 μm. According to Safdar et al. and Wang et al. [54,55], the average surface roughness (Ra) ranged from 2 to 24 μm. No Rz values were reported, and the study includes only vertical surfaces. Nevertheless, the study illustrates the variation of surface roughness (Ra) as a function of different process conditions. The influence of material thickness was investigated by comparing two configurations: 5.5 mm and 7.5 mm thick samples. Results demonstrate that lowering the thickness reduces roughness. The same study addresses the effect of electrical current, comparing the use of 6 mA and of 4 mA on samples having the same thickness. Results show that roughness is decreased when reducing the current. Moreover, results show that a faster scan speed improves surface quality. Finally, the influence of focus offset is investigated, resulting in a surface roughness reduction when using a larger offset focus. Collectively, these findings provide a comprehensive overview of how variations in thickness, current, scan speed, and focus offset affect surface roughness. The primary factor contributing to the increase in surface roughness is the thermal load on the material, which leads to the sintering of loose powder particles on the surface [55].

The study conducted by Galati et al. [53] focused on the influence of part geometry on surface roughness for Ti-6Al-4V. In their experiments, the average particle size was 75 μm, and the layer thickness was fixed at 50 μm. The analyses were conducted on surfaces with varying inclinations, ranging from the horizontal up-skin surface up to an angle of 60° and from 80° to 50° for down-skin surfaces.

For upward-facing surfaces, the average roughness (Ra) was 14.922 μm, with a standard deviation of 0.657 μm. In contrast, downward-facing surfaces exhibited a higher average Ra of 19.133 μm, with a standard deviation of 0.306 μm. The main factor responsible for the increase in surface roughness is the staircase effect [53].

The following table (Table 2) summarizes the main findings of the aforementioned studies:

Table 2. Effect of EBM processing parameters on the as-built surface of Ti6Al4V samples.

Variables	Action	Effect	Physical Mechanisms Involved
Sample thickness	Increase	Increase roughness	Component core sintering surroundings
Current	Increase	Increase roughness	Melt pool sintering surroundings
Scan speed	Increase	Decrease roughness	Melt pool sintering surroundings

Table 2. *Cont.*

Variables	Action	Effect	Physical Mechanisms Involved
Offset focus	Increase	Decrease roughness	Melt pool sintering surroundings
Surface angles (Up-skin)	From 0° to 90°	Increase roughness	Staircase effect
Surface angles (Down-skin)	From 90° to 180°	Roughness: Maximum at the extremes and central minimum	Staircase effect

2.4. Selective Laser Sintering

Due to the specific sub-category of binding methods that do not fully melt the metal powder—or only melt the binder component—Selective Laser Sintering (SLS) is considered separately from other Laser Powder Bed Fusion (L-PBF) techniques in this study.

Among the various bonding systems, this work focuses on those most commonly used for the manufacturing of metal components, starting with metal powders, either coated with a binder or not [21].

Like L-PBF, the main parameters influencing surface roughness in SLS are energy density and its related variables, such as laser power, layer thickness, scan speed, hatch spacing, and spot size. Additional factors include part bed temperature and scan strategy.

Wang et al. investigated an iron-based powder blend containing 29% Nickel, 8.3% Copper, and 1.35% Phosphorus. Their study reported surface roughness values for 0° up-skin surfaces and those perpendicular to the build direction. The average roughness was 18.2 μm for the top view and 12.6 μm for the side view. These values were primarily attributed to thermal stresses that caused surface cracking. Pores formed during sintering were correlated with the build direction, with areas ranging from approximately 120 ± 280 μm^2 on the top surface and 57 ± 114 μm^2 on the side surface. Pore formation was most prevalent between layers [56].

A comprehensive study by Sachdeva et al. [57] focused on the roughness of 0° up-skin surfaces in SLS, analyzing additional parameters such as bed temperature, hatch length, and scan count. Their reference energy density was 28 J/mm^2, significantly higher than that used in Song and Koenig's work [58]. Sachdeva's findings illustrate the trends of Ra and Rz roughness values as functions of laser power, hatch spacing, bed temperature, and hatch length. The variation of the aforementioned process parameters affected both Ra and Rz similarly. The reference values were 28 W laser power (ranging from 24 to 32 W), 0.2 mm hatch spacing (ranging from 0.1 to 0.3 mm), 175 °C bed temperature (ranging from 172 to 178 °C), and 100 mm hatch length (ranging from 40 to 120 mm). Ra and Rz both showed a maximum at the reference laser power and a minimum with lower scan speed and bed temperature. Hatch length exhibited a progressive reduction in roughness with increasing values. The only notable difference was that the Rz maximum was slightly shifted to a higher laser power compared to Ra [57].

The roughness values reported in both studies ranged from 5.46 to 18.2 μm for Ra and from 30.87 to approximately 160 μm for Rz [57,58].

From the work of Attarzadeh et al., it is possible to derive the trends in surface roughness for surfaces parallel to the build plate, under even higher energy densities—effectively continuing the investigation initiated by Song and Koenig. Across this entire energy range, a reduction in surface roughness is observed, attributed to the removal of defects [59].

Zhu et al. conducted extensive research on the sintering of coated metal powders. Their findings indicate that varying binder contents have a direct impact on the maximum sintering temperature, along with the particle size distribution of the powders used. Powders with an average size of 14.2 μm do not exhibit lower thermal stability compared to those of 22 μm, yet they produce components with higher density at lower temperatures and offer a superior surface finish, 1.72 μm [60]. This technique does not directly modify the metal powders through laser interaction; therefore, the printing parameters are of secondary importance.

The following table (Table 3) summarizes the main trends that emerged from the literature search:

Table 3. Effect of SLS processing parameters on the as-built surface.

Variables	Action	Effect	Physical Mechanisms Involved
Laser power	Increase	Roughness: Maximum at reference point parameters (28 J/mm^2)	Low and high energy create defects through unsintered or stick powder
Hatch spacing	Increase	Roughness: Minimum at reference point parameters (28 J/mm^2)	Low and high energy create defects through unsintered or stick powder
Hatch length	Increase	Decrease roughness	Melt pool behavior
Bed temperature	Increase	Roughness: Minimum at reference point parameters (28 J/mm^2)	Low and high energy create defects through unsintered or stick powder

2.5. Laser Metal Deposition

In blown 316L stainless steel and Inconel 718 powder DED components, surface roughness (Ra) values have been reported to range between 17 and 110 μm [61].

The main DED process parameters influencing surface roughness, as analyzed by Jardon et al. [62], include:

- $\dot{m}_P$: powder mass flow
- $\dot{V}_{CG}$: carrying gas flow
- $\dot{V}_{SG}$: shielding gas flow
- Particle size distribution
- Presence of residual unmelted powder

These factors collectively affect the consistency of material deposition and the quality of the melt pool, which in turn influences the final surface finish.

An important factor influencing surface roughness in DED is the ratio between the carrying gas flow rate and the shielding gas flow rate. When $\dot{V}_{CG}/\dot{V}_{SG}$ ratio is high, the melt pool tends to be more uniform, with a circular front boundary, indicating stable powder delivery and consistent melting. Conversely, when the ratio is low, powder convergence decreases, resulting in less powder reaching the melt pool, which can lead to irregular deposition and increased surface roughness [62].

Powder particles can also adhere to the surface of the melt pool, contributing to an increase in overall surface roughness. This phenomenon is more pronounced with larger particles, which are more likely to remain partially melted or unmelted on the surface. In contrast, smaller particles tend to have a lower impact, as they are more easily absorbed into the melt pool and less likely to stick to the surface [62–64].

As studied by Jardon et al., increasing the powder mass flow rate ($\dot{m}_P$) correlates with an increase in surface roughness. At low mass flow rates, particles behave more like individual entities rather than a continuous stream, making them more likely to remain attached to the melt pool surface. At higher mass flow rates, although the force required to overcome the melt pool's surface tension is lower, the greater number of particles increases the likelihood of adhesion, thereby contributing to higher roughness [61,62].

The influence of part geometry and surface remelting on surface roughness is clearly demonstrated in the work of Rombouts et al. [65]. Their study highlights how both the shape of the component and the application of remelting strategies can significantly affect the final surface finish. The analyzed surface was the 0° up-skin, where the melt pool plays a key role in smoothing the surface. Their findings confirm that, under certain process parameters, the melt pool naturally contributes to reducing surface roughness. Moreover, applying a remelting strategy further enhances this effect, leading to even smoother surfaces.

Different up-skin surface angles, ranging from 15° to 75°, were analyzed in the same study. The results showed that as-built surface roughness increases with the inclination angle. However, applying a remelting strategy allows for reducing the roughness to values as low as 1.5 μm [65].

Layer thickness also contributes to the increase in surface roughness, due both to a larger melt pool—which increases the likelihood of melting loose powder on the surface—and to the staircase effect [63]. This is further influenced by the scan pattern, as studied by Ribeiro et al. on SS316L, who identified the Contour method as the most effective in achieving superior surface roughness [64].

The layer height is directly linked to the melt pool, which is in turn influenced by the laser power. As with other additive manufacturing technologies, excessively high or low energy densities can lead to the formation of defects in the deposited material track. Therefore, identifying an optimal intermediate point is essential to achieve desirable surface roughness [66].

The following table (Table 4) summarizes the main trends that emerged from the aforementioned literature search:

Table 4. Effect of LMD processing parameters on the as-built surface.

Variables	Action	Effect	Physical Mechanisms Involved
$\dot{V}_{CG}/\dot{V}_{SG}$	Increase	Increase track stability	Melt pool behavior
Particle size	Increase	Increase roughness	Melt pool behavior
$\dot{m}_p$	Increase	Increase roughness	Melt pool behavior
$\dot{V}_{SG}$ small particles	Increase	Not relevant	Not relevant
$\dot{V}_{SG}$ large particles	Increase	Roughness: Minimum at the extremes and central maximum	Dispersive powder
Surface remelting	Increase energy density	Decrease roughness	Reduce defects
Surface angles (Up-skin)	From 0° to 90°	Increase roughness	Staircase effect
Layer height	Increase	Increase roughness	Staircase effect

2.6. Wire Laser Energy Deposition

As with LMD technology, surface roughness in WLED varies depending on multiple factors and can be categorized into two types:

- Overall roughness, which accounts for the valleys between layers and the scan pattern tracks.
- Single-track roughness, which refers to the roughness measured only on the top surface of an individual track, is shaped by the melt pool after solidification.

Overall roughness is primarily influenced by geometric variables such as surface angle, layer thickness, and wire diameter [67].

The key process parameters affecting roughness include spot diameter, laser pulse rate and duration, wire feed angle, wire feed direction [68], wire feed rate, average laser power, and stage speed.

Shaikh et al. [67] conducted their analysis on 304 and 316L steel wire by keeping all other parameters constant while varying three key factors for single track: average laser power, pulse duration, and stage speed, and just two for multilayer deposition, fixing stage speed. The influence of average power, pulse duration, and stage speed on the width and height of the deposition track is illustrated in the work of Shaikh et al., where the relevant variables are, respectively, the overlap of adjacent tracks and the geometry of the deposition of different surface angles. An increase in width and a decrease in height is recorded with an increase in laser power and pulse duration, while an increase in stage speed reduces the height of the welding bead but almost does not affect the width.

Concerning overlap, ranging from 30% to 80% in 10% increments, it affects the continuity and uniformity of the surface, thereby influencing the final roughness. Results show that overlapping multiple tracks shows an improvement of surface finishing, reaching a maximum for 70% overlap, but then a drastic increase in roughness at 80% is observed. The effect of the deposition angles is also analyzed, with a maximum in roughness at 30° and lower values for 0° and 90° [24,67].

The feeding direction has a significant impact on surface roughness. Findings by Syed and Li indicate that when the wire approaches the melt pool from the front (front feeding), the resulting roughness is considerably lower compared to rear feeding. Furthermore, the angle at which the wire enters the melt pool also plays a role: in front feeding, roughness increases with the angle, whereas in rear feeding, the effect is reversed. These phenomena are primarily governed by the behavior of the melt pool and its cooling dynamics [68].

By comparing the studies conducted by Shaiki et al., Li et al., and Syed and Li, it becomes evident that the wire diameter directly affects the height of the weld bead and, consequently, the final surface roughness of the part. Specifically, an increase in wire diameter leads to a corresponding increase in roughness.

In the best cases, average surface roughness (Ra) values of approximately one micron were measured.

A study on the effect of auxiliary heating conducted by Sang et al. on stainless steel 304. The study analyzes the influence of both print bed and wire preheating on surface roughness. The roughness measured under auxiliary heating conditions was lower than that observed without heating. The investigated effect appears to contribute significantly to the reduction in defects and control of the melt pool cooling. The results obtained in this study were based on a lower laser power compared to other works, which may lead to defects that are subsequently mitigated by substrate and wire heating, providing additional energy to the material through alternative mechanisms [69].

The following table (Table 5) summarizes the main trends that emerged from the literature search:

Table 5. Effect of WLED processing parameters on the as-built surface.

Variables	Action	Effect	Physical Mechanisms Involved
Average Laser power	Increase	Increase width, decrease height of track	Melt pool behavior
Pulse duration	Increase	Increase width, decrease height of track	Melt pool behavior
Stage speed	Increase	Decrease width, decrease height of track	Melt pool behavior
Overlap	Increase	Roughness: Decrease with minimum at 70% then drastically increase at 80%	Partial remelting, remove defects
Deposition angles	From 0° to 90°	Roughness: Minimum at the extremes and central maximum	Melt pool behavior with gravity
Wire diameter	Increase	Increase roughness	Melt pool dimension
Alternative heating	Increase	Decrease roughness	Melt pool behavior

2.7. Wire Electron Beam Energy Deposition

To date, this technology has not been extensively investigated regarding surface roughness. Given the similarity of its machine architecture to that of WLED and WLAM systems [70], it is reasonable to expect that the roughness behavior may follow comparable trends. Parameters such as overlap and average power are likely to influence surface roughness in a similar manner, at least in the context of a single-track scan.

Electron beam power has been shown to have a similar effect to laser power on the width and height of welding beads [24].

However, surface roughness has scarcely been investigated in this context. To date, only the study by Kovalchuk reports roughness measurements, with values ranging between 0.3 and 1 mm [71].

2.8. Wire Arc Additive Manufacturing

Among all Arc-based DED technologies, Wire Arc Additive Manufacturing (WAAM) is the most widely adopted. It stands out for its simplicity and ranks among the fastest in terms of deposition rate. Similar to WLED and WEBAM, the surface topography in WAAM is influenced by the height and width of the single-track bead. The use of an electric arc introduces current (amperage) as an additional process parameter affecting surface roughness, alongside scan speed and wire feed rate, both of which also influence the geometry of the deposited bead.

For instance, in the case of stainless steel 316L, roughness values measured on single-track beads range from 1.00 to 8.32 μm [61].

When considering a complete component manufactured via WAAM, surface roughness tends to increase significantly due to the deposition pattern and the inherent layer-by-layer nature of the process. The accumulation of successive weld beads and the resulting stair-step effect contribute to a more pronounced surface irregularity compared to single-track measurements [72].

The overall surface topography is not determined anymore by the roughness alone [73]. The overall surface profile is predominantly influenced by waviness rather than micro-scale roughness. The maximum amplitude of these surface undulations ranges between 1.12 mm and 2.13 mm. These values are attributed to the presence of waves, bulges, and randomly occurring overlapping weld metal, which collectively contribute to the irregular surface morphology.

In WAAM technologies, inter-layer temperature is a key factor influencing surface roughness. Studies by Xiong et al. and Belhadj et al. have demonstrated a correlation whereby surface roughness increases with higher temperatures and decreases with increased travel speed. Consequently, longer deposition intervals between successive layers allow the previously deposited layer to cool down, resulting in reduced surface roughness [74,75].

Whenever an increase in a process variable leads to a rise in layer height, a corresponding increase in surface roughness is also observed. Modifying either the wire feed speed or the travel speed—while keeping the other constant—affects the resulting layer height. Specifically, increasing the travel speed reduces the layer height and consequently the surface roughness. Conversely, increasing the wire feed speed results in a higher layer height, which is associated with increased surface roughness [74].

The following table (Table 6) summarizes the main trends that emerged from the literature search:

Table 6. Effect of WAAM processing parameters on the as-built surface.

Variables	Action	Effect	Physical Mechanisms Involved
Wire feed rate	Increase	Increase bead height, decrease width, decrease wetting angle	Melt pool behavior
Wire feed rate	Increase	Increase roughness	Melt pool size
Travel speed	Increase	Decrease bead width, decrease melt through depth	Melt pool behavior
Travel speed	Increase	Decrease roughness	Melt pool size
Current	Increase	Roughness: Slight increase and then decrease	Melt pool behavior
Current	Increase	Increase bead width, increase melt through depth	Melt pool behavior
Layer height	Increase	Increase roughness	Melt pool size

2.9. Submerged Arc Additive Manufacturing

Like WAAM, the overall surface topography of components produced via SAAM is not solely defined by surface roughness but also significantly influenced by waviness. Waviness is closely related to the height and width of the single-track bead.

As SAAM is still an emerging technology, there is currently a lack of direct data regarding its surface topography. However, based on similarities with other arc-based processes—such as comparable deposition rates, layer thicknesses, and materials—it is reasonable to assume that the resulting waviness is of the same order of magnitude as that observed in WAAM. Therefore, the expected surface undulations can be estimated to fall within the millimeter range.

Current research on this technology continues to focus primarily on its mechanical properties, while surface finish remains largely unexplored. One advantage supporting these studies is that the deposition rate of this technology is even higher than that of WAAM. Moreover, components are rarely used in their as-built condition and are typically subjected to post-processing through machining [76,77].

2.10. Rapid Plasma Deposition

Like WAAM and SAAM, this technology is characterized by pronounced surface waviness, primarily due to its high layer thickness (3–4 mm) and elevated deposition rate [78].

As a relatively recent and patented process, there is currently a limited number of publications available on this research specifically addressing the 'as-built' surface characteristics. The same considerations and conclusions drawn for SAAM—regarding the influence of process parameters on surface topography—can be reasonably extended to Rapid Plasma Deposition (RPD) as well [79].

Nonetheless, the study conducted by Kumar and Jain provides a comprehensive overview of the process parameters and their influence on surface roughness. This study investigates the effect of power supply, travel speed, and metal feed rate on surface roughness for both powder- and wire-based technologies, specifically in the deposition of Stellite-6 onto an AISI 4130 steel substrate. Surface roughness exhibits a similar trend with respect to changes in process variables for both powder- and wire-based feedstock materials. As observed in previous technologies, increasing the travel speed and decreasing the material feed rate led to a reduction in surface roughness, primarily due to changes in layer height; meanwhile, an increase in the supplied power leads to a corresponding increase in surface roughness. Within the examined range of values, no plateau or maximum/minimum point is observed. The measured roughness values of 118 and 227 μm may still increase or decrease depending on the influencing variables [80]. The following table (Table 7) summarizes the main trends that emerged from the above-mentioned literature search:

Table 7. Effect of RPD processing parameters on the as-built surface.

Variables	Action	Effect	Physical Mechanisms Involved
Laser power	Increase	Increase roughness	Melt pool behavior
Metal feed rate	Increase	Increase roughness	Melt pool size
Travel speed	Increase	Decrease roughness	Melt pool size

2.11. Cold Spray Additive Manufacturing

Since CSAM is a "cold" technology, the most suitable materials are ductile ones, which can deform and adhere effectively to the deposition surface, such as aluminum and copper [81,82]; nevertheless, steels, titanium, and nickel alloys can also be employed under suitable conditions [83,84].

The surface roughness is primarily influenced by particle diameter, gas temperature, and gas pressure. These parameters directly affect the particle velocity and deformation behavior upon impact, which in turn determine the quality of the deposited surface. Shao et al. [85] used a D10 = 9.11 μm, D50 = 20.9 μm, and D90 = 44.0 μm copper powder for their study. The two variables that have been investigated are pressure and temperature. These two ranged between 600 to 900 °C and from 4 to 5.5 MPa.

The average surface roughness (Ra) of cold-sprayed coatings was observed to decrease progressively with increasing gas pressure and gas temperature. In the study by Shao et al., Ra values ranged from 12.69 μm to 7.05 μm under varying process conditions [85].

The influence of particle size on surface roughness becomes evident when comparing these results with those reported by Sirvent et al. and similarly to Prashar and Vasudev [84,86], who employed larger particles and observed different roughness characteristics. Aluminum powder (indicated as Al-) used for their study has a D10 = 21.4 μm, D50 = 51.1 μm, and D90 = 100.3 μm, while the Titanium powder (indicated as Ti-) has a D10 = 8.5 μm, D50 = 24.8 μm, and D90 = 81.4 μm.

The sidewalls of components manufactured via Cold Spray exhibit surface waviness that significantly affects the overall topography, similarly to what has been observed in other additive technologies. These undulations are typically the result of cumulative deposition irregularities and process dynamics, and they contribute to a more complex surface morphology beyond micro-scale roughness alone.

Two scan strategies have been studied to analyze the effect on waviness: -T, traditional scan strategy, and -MK, metal knitting. The waviness parameter resulted in 531 $\pm$ 70 μm for the Al-MK sample and 296 $\pm$ 85 μm for the Ti-MK sample, whereas the -T samples did not exhibit such a marked difference. Similarly, surface roughness (Ra) values were comparable between aluminum and titanium samples, with approximately 70 μm for the MK samples and around 20 μm for the T samples [86].

The deposition of thick layers, as described by Prashar and Vasudev, leads to an increase in surface roughness, which correlates with waviness. Furthermore, deposits exceeding 1.5 mm in thickness exhibit low mechanical strength, necessitating both thermal and mechanical post-processing treatments [84].

The following table (Table 8) summarizes the main trends that emerged from the above-mentioned literature search:

Table 8. Effect of CSAM processing parameters on the as-built surface.

Variables	Action	Effect	Physical Mechanisms Involved
Particle diameter	Increase	Increase surface waviness	Interparticle defects
Gas temperature	Increase	Decrease roughness	Thermal energy involved in particle deformation
Gas pressure	Increase	Decrease roughness	Kinetic energy involved in particle deformation
Particle distribution-T	Increase	Not relevant	Not relevant
Particle distribution-MK	Increase	Increase surface waviness	Deposition strategy

2.12. Metal Binder Jetting

Like other additive manufacturing technologies, Binder Jetting is influenced by a variety of process parameters. Key variables include [29,87,88]:

- Layer thickness directly affects the overall surface topography.
- Printing speed of recoat is higher, where higher speeds may compromise dimensional accuracy and resolution.
- Roller rotation speed affects the powder flowability.
- Binder drops in size, as larger droplets tend to reduce surface resolution and detail.
- Binder saturation.
- Binder drying time.
- Powder characteristics, including particle size and spreading behavior, influence not only surface roughness but also internal porosity.

The interplay of these parameters plays a critical role in determining the final quality and performance of the printed component [29,87,88].

Powder size in Metal Binder Jetting (MBJ) can vary significantly depending on the application and desired resolution. While MBJ is compatible with fine powders typically used in Metal Injection Molding (MIM), it can also accommodate coarser powders. For instance, the study by Ziaee and Crane reports the use of powders with particle sizes exceeding 70 μm of many materials and alloys, such as K418 Nickel Superalloy and Copper. Such variations in powder size can influence not only the surface finish and resolution but also the packing density and final porosity of the printed parts [29,89].

Powder flowability can be managed with the Roller Rotation Speed, which help with powder flow and distribution, as Lores et al. findings show that, although to a lesser extent, it reduces surface roughness by improving powder packing [90].

There is currently limited research specifically addressing the influence of process parameters on surface roughness in Metal Binder Jetting (MBJ). While some studies suggest correlations, quantitative data remain scarce. One of the few detailed investigations is provided by Mostafaei et al. [91], who examined Inconel 625 using powder sizes ranging from 16 to 63 μm. The process parameters included a layer thickness of 100 μm, a recoat speed of 130 mm/s, a roller speed of 250 rpm, a roller traverse speed of 15 mm/s, and a drying speed of 17 mm/s. After sintering, the surface roughness resulted in having a root-mean-square height (Hrms) of 24.77 $\pm$ 5.13 μm, an Rq of 1.39 $\pm$ 0.2 μm, and an Ra of 1.10 $\pm$ 0.16 μm. Noticeably, the sintering process applied also affects such results.

The drying time of the binder is also a critical variable, as it not only contributes to increased surface roughness but also negatively affects the material's density. This is due to crack formation, which after prolonged drying periods, grows, and as a result, voids may form within the material, hindering the sintering process. A similar effect is caused by binder saturation. Insufficient binder leads to powder detachment, which in turn increases both surface roughness and porosity [90].

The following table (Table 9) summarizes the main trends that emerged from the above-mentioned literature search:

Table 9. Effect of MBJ processing parameters on the as-built surface.

Variables	Action	Effect	Physical Mechanisms Involved
Layer thickness	Increase	Increase roughness	Staircase effect
Particle size	Increase	Increase roughness	Lose resolution
Binder drops size	Increase	Increase roughness	Lose resolution
Printing speed/Recoating	Increase	Increase roughness	Binder drops drag
Surface angles	From 0° to 45°	Increase roughness	Staircase effect
Dring time	Increase	Increase roughness	Powder detachment
Binder saturation	Increase	Decrease roughness	Proper powder binding
Roller rotation speed	Increase	Decrease roughness	Proper powder flowability

2.13. *Mold Slurry Deposition*

Mold Slurry Deposition is an additive manufacturing technology patented by Tritone [30]. As a relatively recent development, it has not yet been extensively studied in terms of surface roughness on metallic materials. Due to its slurry-based process, it is widely used for ceramics, precious metals, and oxides [92–95]. The technique requires debinding and sintering, but the binder is highly robust, allowing for machining of the 'green' part prior to sintering. The wax mold not only defines the geometry of each individual layer but also serves as a support structure. As this is also layer-by-layer technology, it can be expected that surface roughness will depend on both layer thickness and the inclination of the surface relative to the build platform, but due to the lack of available references, the table reporting the effects of printing parameters on surface roughness will not be compiled.

2.14. *Lithography-Based Metal Manufacturing*

Like Metal Binder Jetting, the roughness in Lithography-based Metal Manufacturing (LMM) is influenced by key parameters that shape the component's topography. These include layer thickness, powder particle size, and the resolution of the Digital Light Processing (DLP) system—the component of the machine responsible for projecting the UV image that initiates polymerization of the powder—resin mixture [96]. Working with powders dispersed in a slurry/binder allows the use of finer particles and enables a reduction in layer thickness, with lower risks compared to the use of loose powders as in PBF technologies [31]. This technology typically employs MIM-grade 316L powders, such as those used in the study by Melentiev et al., with particle sizes ranging from 5 to 22 μm. The resulting surface roughness, measured across the entire surface, yielded a Sa value of 5.6 μm. High precision applications show an average surface roughness Ra of 1.6 μm [97,98].

The study by Mitteramskogler et al. highlighted how the inclination of the surface relative to the build plate affects surface roughness. A minimum roughness was observed for surfaces oriented either perpendicular or parallel to the build plate, while roughness significantly increased at small angles and then decreased again as the orientation approached vertical [99]. The following table (Table 10) summarizes the main trends that emerged from the above-mentioned literature search:

Table 10. Effect of LMM processing parameters on the as-built surface.

Variables	Action	Effect	Physical Mechanisms Involved
Layer thickness	Increase	Increase roughness	Staircase effect
Particle size	Increase	Increase roughness	Lose resolution
Surface angle	Minimum at 0° al 90°	Increase roughness	Staircase effect

2.15. Metal Material Jetting

MMJ can employ sub-micrometric powders suspended in the binder, allowing the printed layer thickness to be reduced to below 10 microns. Although no specific studies on surface roughness for metallic material are currently available, it can be reasonably stated—from studies conducted on polymers—that reducing layer thickness improves surface approximation, thereby minimizing the staircase effect. Furthermore, the use of ultra-fine powders contributes to enhanced sintering performance [33]. In addition to layer thickness, surface inclination also shows a direct correlation with roughness, as analyzed in the study by Kumar and Kumar [100]. More in-depth studies on the behavior of surface roughness in relation to changes in printing parameters remain to be explored.

2.16. Material Extrusion Additive Manufacturing

Filament and pellet extrusion technologies share the same printing architecture; the primary distinction lies in the form of the raw material, namely filament versus pellets. Several process parameters influence the resulting surface roughness, including layer thickness, surface angle, nozzle temperature, and printing speed. These variables affect the quality of interlayer bonding and the precision of material deposition, ultimately shaping the surface topography of the printed component [101].

The influence of process parameters on average surface roughness (Ra) in filament and pellet extrusion technologies has been investigated by Boschetto et al. [40], both in the green and sintered states, using a 316L powder-filled wire. Their findings indicate that Ra does not exhibit a consistent trend when varying nozzle temperature, printing speed, or layer thickness at a fixed surface angle. However, in both green and sintered parts, Ra shows a proportional increase with the surface angle, highlighting its dominant role in shaping surface topography. Figure 2 summarizes graphically the main results obtained by the reduced models obtained by Boschetto et al., where the lines indicate the layer thickness (L), in mm, the printing head temperature (T) in °C, and the speed (V) in mm/s.

This plot suggests that the primary factor influencing surface roughness is the surface angle, and therefore, the geometry of the printed component itself. As the surface angle increases, so does the average roughness (Ra), regardless of other process parameters, highlighting the critical role of part design in determining surface quality.

Consistent findings have been reported in other studies, showing similar roughness values under comparable surface analysis conditions [102–106].

Another variable that can affect surface roughness is the debinding and sintering process. As shown in the work of Obadimu and Kourousis, issues arising during sintering can lead to surface-related defects such as increased roughness, warping, and balling [106].

The following table (Table 11) summarizes the main trends that emerged from the above-mentioned literature search:

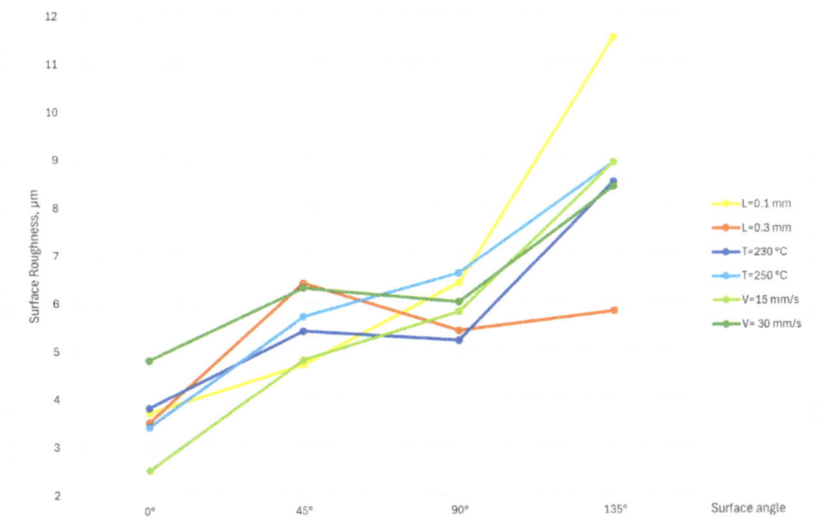

Figure 2. Ra value as a function of the surface angle for different printing parameters: L = layer thickness in mm, T = printing head temperature in °C, and V = head speed in mm/s.

Table 11. Effect of FDM processing parameters on the as-built surface.

Variables	Action	Effect	Physical Mechanisms Involved
Temperature	Increase	Different effects on different surface angles	Material deposition behavior
Deposition speed	Increase	Different effects on different surface angles	Material deposition behavior
Layer thickness	Increase	Different effects on different surface angles	Staircase effect
Surface angles	Increase	Increase roughness	Staircase effect

2.17. Dispersion Material Extrusion

The parameters that influence surface roughness in Direct Material Extrusion (DME) include extrusion thickness, extrusion width, printing speed, extrusion multiplier (a factor that adjusts the amount of extruded material, leading to over- or under-extrusion when this value is greater or less than 1), and finally, the particle size of the feedstock [35,107].

The parameter that most significantly affects surface roughness is the extrusion multiplier. When set too low, it leads to gaps between extrusion passes due to under-extrusion. Conversely, when set too high, it causes over-extrusion defects, resulting in a substantial increase in surface roughness. The Ra value measured in the worst case was 72 µm. Extrusion thickness ranks second in terms of its influence on surface roughness. Affecting the geometry of the deposited bead, it has a direct impact on the surface topography. This effect could potentially be amplified when considering inclined angles on the printed surface. Printing speed and bead width have the least influence on surface roughness. An increase in printing speed tends to raise roughness, as depositing a new layer onto a substrate where the binder has had time to dry results in fewer deformations. While increasing the bead width generally leads to a reduction in roughness, if the extruder is wider than the bead itself, positioning defects may occur [35,107].

An improvement in surface roughness was observed when sintering was performed under vacuum conditions. The best recorded Ra value was 18.59 microns [35,108].

The following table (Table 12) summarizes the main trends that emerged from the above-mentioned literature search:

Table 12. Effect of DME processing parameters on the as-built surface.

Variables	Action	Effect	Physical Mechanisms Involved
Extrusion thickness	Increase	Increase roughness when too high or too low	Material deposition behavior
Extrusion width	Increase	Decrease roughness	Decrease positioning defects
Printing speed	Increase	Increase roughness	Material deposition behavior
Extrusion multiplier	Increase	Increase roughness when too high or too low	Creating and filling void
Powder size	Increase	Decrease roughness	Increased sinterability

2.18. Laminate Layer Manufacturing

In LLM, surface roughness is influenced by both the layer thickness and the surface angle. This is due to the inherent roughness of the metal sheets used, which is typically characterized along the 0° and 90° orientations. As illustrated in Ahn et al. [109], surface roughness increases with layer thickness, a trend attributable to the layer-by-layer nature of the process. The roughness values measured in this study ranged from 2.09 μm to 75 μm, highlighting the significant impact of material and process parameters on the final surface quality. These values increase with a thicker layer. Due to the way the samples are made, in 0° and 180°, the roughness is the same as the stock material, and in 90°, the roughness is given by the cutting finishing [41]. The following table (Table 13) summarizes the main trends that emerged from the above-mentioned literature search:

Table 13. Effect of LLM processing parameters on the as-built surface.

Variables	Action	Effect	Physical Mechanisms Involved
Layer thickness	Increase	Increase roughness	Staircase effect
Base material roughness	Increase	Increase roughness	Direct exposure

2.19. Friction Energy Deposition

In FED, surface roughness can only be measured on the top surface parallel to the build platform, as the vertical walls formed during the process are significantly jagged. As a result, any functional application is not feasible without any kind of post-processing. This technology is therefore not suitable for producing near-net-shaped components [110].

The variables influencing surface roughness include the rotational speed of the tool, the feed rate, and the geometry of the tool head—specifically, the part responsible for generating heat through friction [111].

An increase in the rotational speed of the tool results in higher temperatures, which promotes the plasticization of the deposited material, improving its flow and thereby reducing surface roughness. Conversely, increasing the feed rate while maintaining the same rotational speed leads to lower temperatures, which in turn cause an increase in surface roughness [36,111]. The grooves on the tool head are believed to provide greater friction compared to a flat head, thereby generating more heat under identical movement and rotational speeds. However, the study does not offer a definitive conclusion regarding the optimal geometry of the tool [111]. The following table (Table 14) summarizes the main trends that emerged from the above-mentioned literature search:

Table 14. Effect of FED processing parameters on the as-built surface.

Variables	Action	Effect	Physical Mechanisms Involved
Tool geometry	Change	Grooves and rings on the surface	Increase heat generated by friction
Tool rotational speed	Increase	Decrease roughness	Increase temperature
Tool travel speed	Increase	Increase roughness	Decrease temperature

2.20. Drop-on-Demand Liquid Metal Jetting

In DOD-LMJ, the deposition process is characterized by a high-frequency release of discrete molten metal droplets, enabling precise control over material placement. The spatial arrangement of deposited droplets significantly affects the resulting surface topography. Gao et al. conducted a study analyzing its impact under four distinct droplet placement methodologies [112]. Even without changing the droplet distribution method, the spacing between droplets significantly affects surface roughness. For smaller droplets, roughness decreases as the spacing is reduced. In contrast, larger droplets may lead to flow inconsistencies, resulting in the formation of discontinuous lines and surface defects [113,114].

Another variable that significantly affects surface roughness is the substrate temperature. A higher substrate temperature prolongs the cooling time of the droplets, allowing the material to flow more effectively into surface pores and resulting in greater material densification and lower surface roughness [112,113].

However, the variable that most strongly influences surface roughness and morphology is the size of the individual droplets. By comparing the studies conducted by Gao et al., Lass et al., Simonelli et al., Traxel et al., and Watkins et al., it becomes evident that larger droplets lead to increased roughness. Due to the fact that, once solidified, larger droplets tend to form deeper valleys between them compared to smaller ones [40,112–115].

The average surface roughness, Ra, reported in these studies ranges from an optimal value of 4.62 µm up to 140 µm. The following table (Table 15) summarizes the main trends that emerged from the above-mentioned literature search:

Table 15. Effect of DOD-LMJ processing parameters on the as-built surface.

Variables	Action	Effect	Physical Mechanisms Involved
Drop arrangement	Change	Decrease roughness	Better drop positioning
Drop distance	Increase	Increase roughness	Deep valley formation
Substrate temperature	Increase	Decrease roughness	Slower cooling
Drop size	Increase	Increase roughness	Deep valley formation

3. Conclusions

Nearly all the cited studies on surface roughness have evaluated it using the average roughness parameter 'Ra'. Only a few works have reported additional values such as 'Rz', 'Sa', and 'Sz', and even fewer have included other roughness metrics. Given the aim of this review as a guideline for global analysis and machine selection based on surface roughness, and considering the industry's predominant demand for Ra, the conclusions will refer exclusively to this parameter. This choice is also due to the impracticality of providing inclusive results for all technologies discussed.

Metal Additive Manufacturing offers different processes, each one presenting its peculiarities and capabilities. This review addressed the process capabilities of the most used Metal AM techniques, in terms of surface quality in the as-built state, estimated by the average surface roughness, Ra. Even though the usefulness of Ra to assess the effective surface quality is highly debated, it is here taken as a reference to provide useful guidelines on how to modify process parameters to control the average surface roughness. All the previous Tables describe the trends reported in literature for certain classes of materials and technologies, suggesting the effect of a single parameter on the resulting Ra. However, the literature search indicated synergistic effects of the parameters many times, as described in the text for the available examples; hence, the trends indicated in the tables should be taken only as a starting point for a multi-parameter optimization. In conclusion, Figure 3 reports the Ra values associated with each technology, highlighting that some encompass multiple effects and measurement conditions, while others refer to roughness values obtained solely on surfaces parallel to the build plane—that is, under more favorable conditions compared to the former. The blue arrows in Figure 3 highlight regions exhibiting surface roughness levels that exceed the scale represented in the graph. The reported maximum roughness values may be exceeded under more extreme printing conditions.

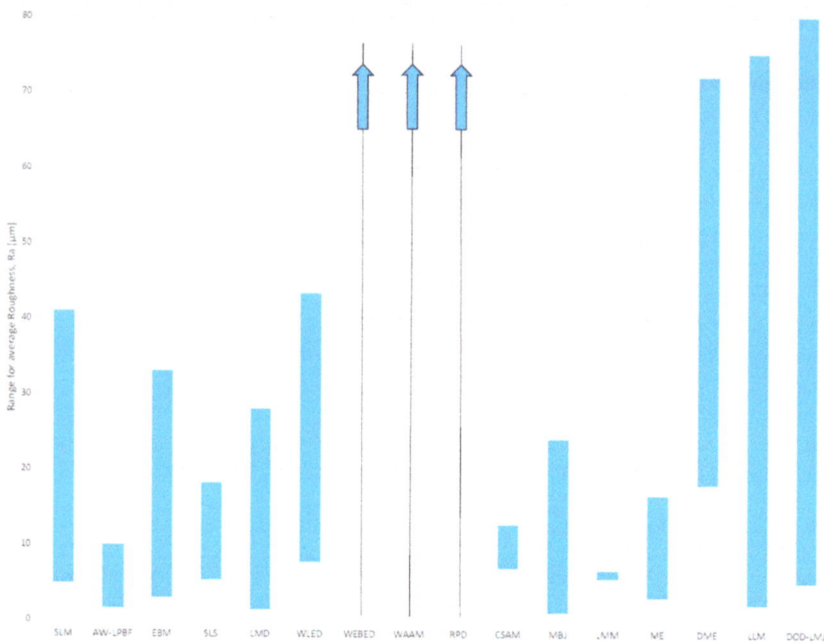

Figure 3. Range of the average roughness (Ra) for each of the reviewed technologies, presenting available data.

This review has also highlighted how printing parameters can lead to contrasting results across different studies, even when not limited to variations in materials. The conditions related to energy density proved to be decisive in this regard, showing that even when printing parameters are varied in the same way, the resulting surface roughness can differ significantly depending on the energy density levels.

Future research should aim at a more in-depth analysis of additional surface roughness parameters, while also expanding the scope to include all available printing materials and considering the influence of energy density. Moreover, some studies have focused solely on the roughness of the top surface parallel to the build plane. Given that additive manufacturing is inherently a layer-by-layer process and considering the strong influence of surface inclination, the use of test specimens printed with inclined surfaces, both in up-skin and down-skin orientations, to identify the critical angle beyond which the technology requires support structures, should be standardized and included in every research.

Author Contributions: Conceptualization, S.P. and E.B.; methodology, S.P. and P.V.; software, S.P.; formal analysis, P.V.; investigation, S.P.; resources, E.B.; data curation, S.P.; writing-original draft preparation, S.P.; writing-review and editing, P.V.; visualization, S.P.; supervision, P.V.; project administration, E.B.; funding acquisition, E.B. All authors have read and agreed to the published version of the manuscript.

Funding: This research received no external funding.

Data Availability Statement: No new data were created or analyzed in this study. Data sharing is not applicable to this article.

Conflicts of Interest: Author Enrico Bedogni was employed by the company Kosme s.r.l. Unipersonale. The remaining authors declare that the research was conducted in the absence of any commercial or financial relationships that could be construed as a potential conflict of interest.

Abbreviations

The following abbreviations are used in this manuscript:

AM	Additive Manufacturing
PBF	Powder Bed Fusion
DED	Direct Energy Deposition
BJ	Binder Jetting
ME	Material Extrusion
SLM	Selective Laser Melting
AWLPBF	Area Wise Laser Powder Bed Fusion
DiAM	Diode-based Additive Manufacturing
EBM	Electron Beam Melting
LMD	Laser Metal Deposition
LENS	Laser-Engineered Net Shaping
PLED	Powder Laser Energy Deposition
WLED	Wire Laser Energy Deposition
WLAM	Wire Laser Additive Manufacturing
WEBED	Wire Electron Beam Energy Deposition
WEBAM	Wire Electron Beam Additive Manufacturing
EBF^3	Electron Beam Freeform Fabrication
EBWFAM	Electron Beam Wire-Feed Additive Manufacturing
WAAM	Wire Arc Additive Manufacturing
GMAW	Gas Metal Arc Welding
SAAM	Submerged Arc Additive Manufacturing
RPD	Rapid Plasma Deposition
WPAED	Wire Plasma Arc Energy Deposition
PMD	Plasma Metal Deposition
GTAW	Gas Tungsten Arc Welding
MPA	Metal Powder Application
CSAM	Cold Spray Additive Manufacturing
MBJ	Metal Binder Jetting
MSD	Mold Slurry Deposition
MJ	MoldJet
LSD	Layerwise Slurry Deposition
LMM	Lithography-based Metal Manufacturing
DLP	Digital Light Processing
VPP	Vat Photo Polimerization
AWVP	Area Wise Vat Polimerization

UV	Ultraviolet
MJ	Material Jetting
MMJ	Metal Material Jetting
FDM	Fused Deposition Modeling
FME	Filament Material Extrusion
FFF	Fused Filament Fabrication
PME	Pellet Material Extrusion
DME	Dispersion Material Extrusion
LLM	Laminate Layer Manufacturing
SL	Sheet Lamination
FED	Friction Energy Deposition
FSAM	Friction Stir Additive Manufacturing
FSPAM	Friction Stir Powder Additive Manufacturing
AFSD	Additive Friction Stid Deposition
LMP	Liquid Metal Printing
DMW	Direct Metal Writing
DOD-LMJ	Drop-On-Demand Liquid Metal Jetting
MMJ	Molten Metal Jetting

References

1. Obilanade, D.; Dordlofva, C.; Törlind, P. Surface Roughness Considerations in Design for Additive Manufacturing-A Literature Review. *Proc. Des. Soc.* **2021**, *1*, 2841–2850. [CrossRef]
2. Afroz, L.; Qian, M.; Forsmark, J.; Li, Y.; Easton, M.; Das, R. Fatigue Life of Laser Powder Bed Fusion (L-PBF) AlSi10Mg Alloy: Effects of Surface Roughness and Porosity. *Prog. Addit. Manuf.* **2025**, *10*, 2423–2441. [CrossRef]
3. Cangussu, V.M.; Abrão, A.M.; Magalhães, F.d.C.; de Oliveira, D.A.; de Brito, P. The Influence of Deep Rolling on Surface Roughness, Corrosion Resistance and Fatigue Life of Annealed Aisi 1045 Steel. *Int J Adv Manuf Technol* **2025**, *139*, 4109–4124. [CrossRef]
4. Pagliari, L.; Gerosa, R.; Panzeri, D.; Fraccaroli, L.; Concli, F. High- and Low-Cycle Fatigue Behavior of Additively Manufactured Ti6Al4V and Influence of Surface Finish. *Eng. Fail. Anal.* **2025**, *180*, 109825. [CrossRef]
5. Perna, A.S.; Savio, L.; Coppola, M.; Scherillo, F. Investigating Fracture Behavior in Titanium Aluminides: Surface Roughness as an Indicator of Fracture Mechanisms in Ti-48Al-2Cr-2Nb Alloys. *Metals* **2025**, *15*, 49. [CrossRef]
6. Onafide, M.; Al Turk, A. Additive Manufacturing Techniques for Reducing Failure Modes and Enhancing Durability in Aerospace Components. *JMEI* **2025**, *18*, 42–52. [CrossRef]
7. Yin, T.; Miao, K.; Liu, H.; Fu, J. Contact and Wear Modelling of Rotary Seals under Mixed Lubrication in Piston Pumps for Aerospace Electro-Hydrostatic Actuators. *Chin. J. Aeronaut.* **2025**, *103575*. [CrossRef]
8. EHEDG: Guideline Catalogue. Available online: https://www.ehedg.org/guidelines-working-groups/guidelines/guidelines/guidelines/guidelines/detail/hygienic-design-principles (accessed on 14 September 2025).
9. Tian, X.; Li, T.; Liu, Y.; Tian, Y.; Li, Z.; Wang, Y.; Ma, Y.; Li, Y.; Wang, X.; Wang, W. Based on Surface Roughness and Hydrophobicity to Reduce the Bacterial Adhesion to Collagen Films for the Ability to Enhance the Shelf Life of Food. *Food Packag. Shelf Life* **2025**, *48*, 101473. [CrossRef]
10. Ihara, I.; Tokuda, H.; Schueller, J.K.; Mohamed, I.M.A.; Sakamoto, Y.; Toyoda, K.; Umetsu, K.; Yamaguchi, H. Surface Roughness and Cleanability: Evaluating the Impact of Magnetic Abrasive Finishing on Dairy Equipment. *J. Food Process Eng.* **2025**, *48*, e70106. [CrossRef]
11. Alsabti, M.A.; Ciuca, I.; Vasile, B.Ş.; Trusca, R.; Harb, A.A. The Effect of Surface Roughness and Hydroxyapatite Coating on the Biocompatibility of Co-Cr Alloys for Medical Implants. *Preprint* **2025**. [CrossRef]
12. Mathur, M.; Paliwal, N.; Kumar, S.; Kumar, K.; Gautam, S.K.; Jewariya, M. Texture Characterization of Bipolar Hip Joint Medical Implants Using Optical Methods. *MAPAN* **2025**, *40*, 471–476. [CrossRef]
13. Bharatish, A.; Kumar, A.; Siddhanth, K.S.; Manikant, V.; Jagdish, P.; Sharma, A.; Solaiachari, S. On Optimizing Wettability, Surface Roughness and Swelling Behaviour of Laser-Polished 3D-Printed PETG Polymer for Bio-Medical Implants. *Polymer* **2025**, *330*, 128482. [CrossRef]
14. Xu, R.; Jin, X.; Bi, H.; Zhang, Z.; Li, M. Effect of Surface Roughness on Contact Resistance and Electrochemical Corrosion Behavior of 446 Stainless Steel in Simulated Anode Environments for Proton Exchange Membrane Fuel Cell. *J. Solid State Electrochem.* **2024**, *28*, 3087–3098. [CrossRef]

15. Song, X.; Liu, F.; Yan, Y.; Wang, W.; Wang, Y.; Sun, W.; Cui, J. Influence of Surface Roughness of Thermal Barrier Coating on the Cooling Performance of a Film-Cooled Turbine Vane. *Case Stud. Therm. Eng.* **2025**, *65*, 105698. [CrossRef]
16. Heidarinejad, A.; Ashrafizadeh, F. Influence of Surface Texture and Coating Thickness on Adhesion of Nickel Plated Coatings to Aluminium Substrate. *J. Manuf. Process.* **2024**, *120*, 435–448. [CrossRef]
17. Ngo, T.D.; Kashani, A.; Imbalzano, G.; Nguyen, K.T.Q.; Hui, D. Additive Manufacturing (3D Printing): A Review of Materials, Methods, Applications and Challenges. *Compos. Part B Eng.* **2018**, *143*, 172–196. [CrossRef]
18. Sefene, E.M. State-of-the-Art of Selective Laser Melting Process: A Comprehensive Review. *J. Manuf. Syst.* **2022**, *63*, 250–274. [CrossRef]
19. Seurat | Area Printing | Metal Additive Manufacturing | 3D Printing. Available online: https://www.seurat.com/area-printing (accessed on 19 February 2025).
20. Murr, L.E.; Li, S. Electron-Beam Additive Manufacturing of High-Temperature Metals. *MRS Bull.* **2016**, *41*, 752–757. [CrossRef]
21. Kruth, J.; Mercelis, P.; Van Vaerenbergh, J.; Froyen, L.; Rombouts, M. Binding Mechanisms in Selective Laser Sintering and Selective Laser Melting. *Rapid Prototyp. J.* **2005**, *11*, 26–36. [CrossRef]
22. Ahn, D.G. Directed Energy Deposition (DED) Process: State of the Art. *Int. J. Precis. Eng. Manuf. Green Tech.* **2021**, *8*, 703–742. [CrossRef]
23. Graf, B.; Marko, A.; Petrat, T.; Gumenyuk, A.; Rethmeier, M. 3D Laser Metal Deposition: Process Steps for Additive Manufacturing. *Weld World* **2018**, *62*, 877–883. [CrossRef]
24. Ding, D.; Pan, Z.; Cuiuri, D.; Li, H. Wire-Feed Additive Manufacturing of Metal Components: Technologies, Developments and Future Interests. *Int. J. Adv. Manuf. Technol.* **2015**, *81*, 465–481. [CrossRef]
25. Treutler, K.; Wesling, V. The Current State of Research of Wire Arc Additive Manufacturing (WAAM): A Review. *Appl. Sci.* **2021**, *11*, 8619. [CrossRef]
26. Nickels, L. Additive Manufacturing: A User's Guide. *Met. Powder Rep.* **2016**, *71*, 100–105. [CrossRef]
27. Additive Manufacturing at HERMLE. Available online: https://www.hermle.de/en/additive-manufacturing/ (accessed on 25 February 2025).
28. Impact Innovations | Cold Spraying. Available online: https://impact-innovations.com/en/ (accessed on 3 March 2025).
29. Ziaee, M.; Crane, N.B. Binder Jetting: A Review of Process, Materials, and Methods. *Addit. Manuf.* **2019**, *28*, 781–801. [CrossRef]
30. Tritone Technologies. Available online: https://tritoneam.com (accessed on 21 February 2025).
31. Rieger, T.; Schubert, T.; Schurr, J.; Butschle, M.; Schwenkel, M.; Bernthaler, T.; Schneider, G. Slurry Development for Lithography-Based Additive Manufacturing of Cemented Carbide Components. *Powder Technol.* **2021**, *383*, 498–508. [CrossRef]
32. Gülcan, O.; Günaydın, K.; Tamer, A. The State of the Art of Material Jetting—A Critical Review. *Polymers* **2021**, *13*, 2829. [CrossRef]
33. Ibrahim, B.; Lopez, L.; Kulkarni, S.; Jobes, D.; Forgiarini, M.; Barber, J.R.; Gordon, J.V. Increasing Strength Properties in Sinter-Based Additive Manufacturing of SS316L via Metal Material Jetting of Sub-Micron Powders. *Addit. Manuf.* **2024**, *89*, 104268. [CrossRef]
34. Ramazani, H.; Kami, A. Metal FDM, a New Extrusion-Based Additive Manufacturing Technology for Manufacturing of Metallic Parts: A Review. *Prog. Addit. Manuf.* **2022**, *7*, 609–626. [CrossRef]
35. Hoffmann, M.; Elwany, A. Material Extrusion Additive Manufacturing of AISI 316L Pastes. *J. Manuf. Process.* **2023**, *108*, 238–251. [CrossRef]
36. Chen, L.; Lu, L.; Zhu, L.; Yang, Z.; Zhou, W.; Ren, X.; Zhang, X. Microstructure Evolution and Mechanical Properties of Multilayer AA6061 Alloy Fabricated by Additive Friction Stir Deposition. *Met. Mater. Trans. A* **2024**, *55*, 1049–1064. [CrossRef]
37. Hassan, A.; Pedapati, S.R.; Awang, M.; Soomro, I.A. A Comprehensive Review of Friction Stir Additive Manufacturing (FSAM) of Non-Ferrous Alloys. *Materials* **2023**, *16*, 2723. [CrossRef] [PubMed]
38. Abankar, M.; Lunetto, V.; De Maddis, M.; Russo Spena, P. Friction Stir Welding of Additively Manufactured A20X Aluminum Alloy: Welding Process, Mechanical Properties, and Microstructure. *Int. J. Adv. Manuf. Technol.* **2024**, *135*, 4635–4652. [CrossRef]
39. Elton, E.S.; Traxel, K.D.; Pascall, A.J.; Jeffries, J.R. Jet on Demand—A Pneumatically Driven Molten Metal Jetting Method for Printing Crack-Free Aluminum Components. *Addit. Manuf. Lett.* **2024**, *11*, 100240. [CrossRef]
40. Watkins, N.N.; Traxel, K.D.; Wilson-Heid, A.E.; Reeve, T.C.; Silva, C.M.; Jeffries, J.R.; Pascall, A.J. Process-Structure-Property Relationships for Droplet-on-Demand Liquid-Metal-Jetted Parts. *Addit. Manuf.* **2023**, *73*, 103709. [CrossRef]
41. Rice, C.S.; Mendez, P.F.; Brown, S.B. Metal Solid Freeform Fabrication Using Semi-Solid Slurries. *JOM* **2000**, *52*, 31–33. [CrossRef]
42. Ramachandramoorthy, R.; Kang, S.G. Process for the Manufacturing of Microscale Hollow Metal Bodies 2024. Available online: https://worldwide.espacenet.com/patent/search/family/083898424/publication/EP4357296A1?q=Exaddon%20ceres (accessed on 12 September 2025).
43. Microscale 3D Printing for Fundamental Research. Available online: https://www.exaddon.com/micro3d-research (accessed on 14 September 2025).
44. Ru, C.; Luo, J.; Xie, S.; Sun, Y. A Review of Non-Contact Micro- and Nano-Printing Technologies. *J. Micromech. Microeng.* **2014**, *24*, 053001. [CrossRef]

45. Mansoura, A.; Dehghan, S.; Barka, N.; Kangranroudi, S.S. Investigation into the Effect of Process Parameters on Density, Surface Roughness, and Mechanical Properties of 316L Stainless Steel Fabricated by Selective Laser Melting. *Int. J. Adv. Manuf. Technol.* **2024**, *130*, 2547–2562. [CrossRef]
46. Yildiz, R.A.; Ozdogan, C.; Malekan, M. On the Surface Roughness of 316L Stainless Steel Fabricated Using L-PBF Additive Manufacturing. *Finn. J. Trib.* **2024**, *41*, 54–56. [CrossRef]
47. Correa-Gómez, E.; Castro-Espinosa, H.; Caballero-Ruiz, A.; García-López, E.; Ruiz-Huerta, L. Effect of Process Parameters on the Roughness and Tensile Behavior of Parts Manufactured by the Metals LPBF Process. *Eng. Rep.* **2024**, *6*, e12904. [CrossRef]
48. Jiang, H.Z.; Li, Z.Y.; Feng, T.; Wu, P.Y.; Chen, Q.S.; Feng, Y.L.; Chen, L.F.; Hou, J.Y.; Xu, H.J. Effect of Process Parameters on Defects, Melt Pool Shape, Microstructure, and Tensile Behavior of 316L Stainless Steel Produced by Selective Laser Melting. *Acta Metall. Sin.* **2021**, *34*, 495–510. [CrossRef]
49. Wang, D.; Mai, S.; Xiao, D.; Yang, Y. Surface Quality of the Curved Overhanging Structure Manufactured from 316-L Stainless Steel by SLM. *Int. J. Adv. Manuf. Technol.* **2016**, *86*, 781–792. [CrossRef]
50. Sinico, M.; Witvrouw, A.; Dewulf, W. Influence of the Particle Size Distribution on Surface Quality of Maraging 300 Parts Produced by Laser Powder Bed Fusion. In Proceedings of the Special Interest Group meeting on Advancing Precision in Additive Manufacturing, Berlin, Germany, 6–9 October 2025; pp. 31–34.
51. Fotovvati, B.; Le, T.N.; Ferreri, N.; Shrestha, S.; Duanmu, N. Effects of Powder Characteristics of Different Alloy Powders on Part Quality in Area Printing® Additive Manufacturing. In *2024 International Solid Freeform Fabrication Symposium, Austin, TX, USA, 11-14 August 2024*; University of Texas at Austin: Austin, TX, USA, 2024.
52. Matthews, M.J.; Guss, G.; Drachenberg, D.R.; Demuth, J.A.; Heebner, J.E.; Duoss, E.B.; Kuntz, J.D.; Spadaccini, C.M. Diode-Based Additive Manufacturing of Metals Using an Optically-Addressable Light Valve. *Opt. Express* **2017**, *25*, 11788. [CrossRef]
53. Galati, M.; Minetola, P.; Rizza, G. Surface Roughness Characterisation and Analysis of the Electron Beam Melting (EBM) Process. *Materials* **2019**, *12*, 2211. [CrossRef]
54. Wang, P.; Sin, W.; Nai, M.; Wei, J. Effects of Processing Parameters on Surface Roughness of Additive Manufactured Ti-6Al-4V via Electron Beam Melting. *Materials* **2017**, *10*, 1121. [CrossRef] [PubMed]
55. Safdar, A.; He, H.Z.; Wei, L.; Snis, A.; Chavez De Paz, L.E. Effect of Process Parameters Settings and Thickness on Surface Roughness of EBM Produced Ti-6Al-4V. *Rapid Prototyp. J.* **2012**, *18*, 401–408. [CrossRef]
56. Wang, Y.; Bergström, J.; Burman, C. Characterization of an Iron-Based Laser Sintered Material. *J. Mater. Process. Technol.* **2006**, *172*, 77–87. [CrossRef]
57. Sachdeva, A.; Singh, S.; Sharma, V.S. Investigating Surface Roughness of Parts Produced by SLS Process. *Int. J. Adv. Manuf. Technol.* **2013**, *64*, 1505–1516. [CrossRef]
58. Song, Y.A.; Koenig, W. Experimental Study of the Basic Process Mechanism for Direct Selective Laser Sintering of Low-Melting Metallic Powder. *CIRP Ann.* **1997**, *46*, 127–130. [CrossRef]
59. Attarzadeh, F.; Fotovvati, B.; Fitzmire, M.; Asadi, E. Surface Roughness and Densification Correlation for Direct Metal Laser Sintering. *Int. J. Adv. Manuf. Technol.* **2020**, *107*, 2833–2842. [CrossRef]
60. Zhu, P.; He, X.; Guan, H.; Zhang, Z.; Zhang, T.; Qu, X. Study on Improving Properties of High-Density 316 L Stainless Steel Fabricated by Indirect Selective Laser Sintering. *Mater. Today Commun.* **2025**, *46*, 112496. [CrossRef]
61. Nuñez, L.; Downey, C.M.; Van Rooyen, I.J.; Charit, I.; Maughan, M.R. Analysis of Surface Roughness in Metal Directed Energy Deposition. *Int. J. Adv. Manuf. Technol.* **2024**. [CrossRef]
62. Jardon, Z.; Ertveldt, J.; Lecluyse, R.; Hinderdael, M.; Pyl, L. Directed Energy Deposition Roughness Mitigation through Laser Remelting. *Procedia CIRP* **2022**, *111*, 180–184. [CrossRef]
63. Gradl, P.R.; Cervone, A.; Gill, E. Surface Texture Characterization for Thin-Wall NASA HR-1 Fe–Ni–Cr Alloy Using Laser Powder Directed Energy Deposition (LP-DED). *Adv. Ind. Manuf. Eng.* **2022**, *4*, 100084. [CrossRef]
64. Ribeiro, K.S.B.; Mariani, F.E.; Coelho, R.T. A Study of Different Deposition Strategies in Direct Energy Deposition (DED) Processes. *Procedia Manuf.* **2020**, *48*, 663–670. [CrossRef]
65. Rombouts, M.; Maes, G.; Hendrix, W.; Delarbre, E.; Motmans, F. Surface Finish after Laser Metal Deposition. *Phys. Procedia* **2013**, *41*, 810–814. [CrossRef]
66. Yazar, K.U.; Pawar, S.; Park, K.S.; Choi, S.H. Effect of Process Parameters on the Clad Morphology, Microstructure, Microtexture, and Hardness of Single Layer 316 L Stainless Steel during Direct Energy Deposition. *Mater. Charact.* **2022**, *191*, 112148. [CrossRef]
67. Shaikh, M.O.; Chen, C.C.; Chiang, H.C.; Chen, J.R.; Chou, Y.C.; Kuo, T.Y.; Ameyama, K.; Chuang, C.H. Additive Manufacturing Using Fine Wire-Based Laser Metal Deposition. *RPJ* **2019**, *26*, 473–483. [CrossRef]
68. Syed, W.U.H.; Li, L. Effects of Wire Feeding Direction and Location in Multiple Layer Diode Laser Direct Metal Deposition. *Appl. Surf. Sci.* **2005**, *248*, 518–524. [CrossRef]
69. Sang, Y.X.; Xiao, M.Z.; Zhang, Z.J.; Fu, Q.Y. Effect of Auxiliary Heating Process on Low Power Pulsed Laser Wire Feeding Deposition. *Mater. Des.* **2022**, *218*, 110666. [CrossRef]

70. Metal 3D Printers | Metal 3D Printing Machines | Sciaky. Available online: https://www.sciaky.com/additive-manufacturing/industrial-metal-3d-printers (accessed on 8 September 2025).
71. Kovalchuk, D.; Melnyk, V.; Melnyk, I. A Coaxial Wire-Feed Additive Manufacturing of Metal Components Using a Profile Electron Beam in Space Application. *J. Mater. Eng. Perform.* **2022**, *31*, 6069–6082. [CrossRef]
72. Hensel, J.; Przyklenk, A.; Müller, J.; Köhler, M.; Dilger, K. Surface Quality Parameters for Structural Components Manufactured by DED-Arc Processes. *Mater. Des.* **2022**, *215*, 110438. [CrossRef]
73. Dinovitzer, M.; Chen, X.; Laliberte, J.; Huang, X.; Frei, H. Effect of Wire and Arc Additive Manufacturing (WAAM) Process Parameters on Bead Geometry and Microstructure. *Addit. Manuf.* **2019**, *26*, 138–146. [CrossRef]
74. Xiong, J.; Li, Y.; Li, R.; Yin, Z. Influences of Process Parameters on Surface Roughness of Multi-Layer Single-Pass Thin-Walled Parts in GMAW-Based Additive Manufacturing. *J. Mater. Process. Technol.* **2018**, *252*, 128–136. [CrossRef]
75. Belhadj, M.; Werda, S.; Kromer, R.; Darnis, P. Influence of Operating Parameters on the Mechanical and Geometric Properties of 316L Stainless Steel Structures Fabricated by WAAM-CMT. *Procedia CIRP* **2025**, *133*, 585–590. [CrossRef]
76. Li, Y.; Wu, S.; Li, H.; Dong, Y.; Cheng, F. Submerged Arc Additive Manufacturing (SAAM) of Low-Carbon Steel: Effect of in-Situ Intrinsic Heat Treatment (IHT) on Microstructure and Mechanical Properties. *Addit. Manuf.* **2021**, *46*, 102124. [CrossRef]
77. Cheng, F.; Xue, M.; Zhang, B.; Yan, S.; Wu, S. Submerged Arc Additive Manufacturing of Duplex Stainless Steel: Effect of Intrinsic Heat Treatment on Microstructure, Mechanical Properties and Pitting Resistance. *J. Mater. Process. Technol.* **2025**, *341*, 118885. [CrossRef]
78. Norsk Titanium | Technology. Available online: https://www.norsktitanium.com/technology#process (accessed on 5 May 2025).
79. Ariza-Galván, E.; Montealegre-Meléndez, I.; Pérez-Soriano, E.M.; Neubauer, E.; Kitzmantel, M.; Arévalo, C. Influence of Processing Conditions on the Mechanical Properties of 17-4PH Specimens Produced by Additive Manufacturing. *Machines* **2022**, *10*, 976. [CrossRef]
80. Kumar, P.; Jain, N.K. Surface Roughness Prediction in Micro-Plasma Transferred Arc Metal Additive Manufacturing Process Using K-Nearest Neighbors Algorithm. *Int. J. Adv. Manuf. Technol.* **2022**, *119*, 2985–2997. [CrossRef]
81. Wu, H.; Xie, X.; Liu, M.; Verdy, C.; Zhang, Y.; Liao, H.; Deng, S. Stable Layer-Building Strategy to Enhance Cold-Spray-Based Additive Manufacturing. *Addit. Manuf.* **2020**, *35*, 101356. [CrossRef]
82. Wang, X.; Feng, F.; Klecka, M.A.; Mordasky, M.D.; Garofano, J.K.; El-Wardany, T.; Nardi, A.; Champagne, V.K. Characterization and Modeling of the Bonding Process in Cold Spray Additive Manufacturing. *Addit. Manuf.* **2015**, *8*, 149–162. [CrossRef]
83. Vaz, R.; Garfias, A.; Albaladejo, V.; Sanchez, J.; Cano, I. A Review of Advances in Cold Spray Additive Manufacturing. *Coatings* **2023**, *13*, 267. [CrossRef]
84. Prashar, G.; Vasudev, H. A Comprehensive Review on Sustainable Cold Spray Additive Manufacturing: State of the Art, Challenges and Future Challenges. *J. Clean. Prod.* **2021**, *310*, 127606. [CrossRef]
85. Shao, L.; Xue, N.; Li, W.; Liu, S.; Tu, Z.; Chen, Y.; Zhang, J.; Dai, S.; Liu, Q.; Shi, X.; et al. Effect of Cold-Spray Parameters on Surface Roughness, Thickness and Adhesion of Copper-Based Composite Coating on Aluminum Alloy 6061 T6 Substrate. *Processes* **2023**, *11*, 959. [CrossRef]
86. Sirvent, P.; Lozano, A.; Garrido-Maneiro, M.A.; Poza, P.; Vaz, R.F.; Albaladejo-Fuentes, V.; Cano, I.G. Surface Topography Analysis in Cold Spray Additive Manufacturing. *Precis. Eng.* **2025**, *92*, 207–218. [CrossRef]
87. Mao, Y.; Li, J.; Li, W.; Cai, D.; Wei, Q. Binder Jetting Additive Manufacturing of 316L Stainless-Steel Green Parts with High Strength and Low Binder Content: Binder Preparation and Process Optimization. *J. Mater. Process. Technol.* **2021**, *291*, 117020. [CrossRef]
88. Myers, K.; Paterson, A.; Iizuka, T.; Klein, A. The Effect of Print Speed on Surface Roughness and Density Uniformity of Parts Produced Using Binder Jet 3D Printing 2021. *arXiv* **2021**, arXiv:2021010459. [CrossRef]
89. Fang, X.; Zu, Y.; Ma, Q.; Hu, J. State of the Art of Metal Powder Bonded Binder Jetting Printing Technology. *Discov. Mater.* **2023**, *3*, 15. [CrossRef]
90. Lores, A.; Azurmendi, N.; Agote, I.; Espinosa, E.; García-Blanco, M.B. A Study of Parameter and Post-Processing Effects on Surface Quality Improvement of Binder Jet 3D-Printed Invar36 Alloy Parts. *Prog. Addit. Manuf.* **2022**, *7*, 917–930. [CrossRef]
91. Mostafaei, A.; Neelapu, S.H.V.R.; Kisailus, C.; Nath, L.M.; Jacobs, T.D.B.; Chmielus, M. Characterizing Surface Finish and Fatigue Behavior in Binder-Jet 3D-Printed Nickel-Based Superalloy 625. *Addit. Manuf.* **2018**, *24*, 200–209. [CrossRef]
92. Vompe, T.; Goviazin, G.G.; Dolev, O.; Rudnik, L.; Katz-Demyanetz, A. Sintering of Additively Manufactured Zirconium by MoldJet Technology. *Powder Technol.* **2024**, *436*, 119494. [CrossRef]
93. Tian, X.; Mühler, C.; Günster, J.; Heinrich, J.G. Feasibility Study on Rapid Prototyping of Porcelain Products. *J. Ceram. Sci. Tech.* **2011**, *2*, 217–226. [CrossRef]
94. Addess, A.; Shilo, D.; Glass, B.; Paskovitch, Y.; Burkhardt, C.; Vogel, L.; Rabkin, E.; Faran, E. Additive Manufacturing of Ni–Ti SMA by the Sinter-Based MoldJet Method. *Shap. Mem. Superelasticity* **2025**, *11*, 89–99. [CrossRef]
95. Hoffmann, M.; Schubert, N.H.; Günster, J.; Stawarczyk, B.; Zocca, A. Additive Manufacturing of Glass-Ceramic Dental Restorations by Layerwise Slurry Deposition (LSD-Print). *J. Eur. Ceram. Soc.* **2025**, *45*, 117235. [CrossRef]

96. Harakaly, G.; Cano Cano, S.; Bosters, J.; Sperling, C.; Moedder, D.; Mitteramskogler, G. Recent Advances in the Biomedical Field with the Lithography-Based Metal Manufacturing Process. *J. Jpn. Soc. Powder Powder Metall.* **2025**, *72*, S1227–S1231. [CrossRef]
97. Melentiev, R.; Harakály, G.; Stögerer, J.; Mitteramskogler, G.; Wagih, A.; Lubineau, G.; Grande, C.A. High-Resolution Metal 3D Printing via Digital Light Processing. *Addit. Manuf.* **2024**, *85*, 104156. [CrossRef]
98. Melentiev, R.; Wagih, A.; Lagerweij, A.; Mitteramskogler, G.; Lubineau, G.; Grande, C.A. Unlocking Multiscale Metallic Metamaterials via Lithography Additive Manufacturing. *Virtual Phys. Prototyp.* **2024**, *19*, e2339368. [CrossRef]
99. Mitteramskogler, G.; Schwentenwein, M.; Burkhardt, C.; Cruchley, N. Lithography-Based Metal Manufacturing (LMM) of 316L Powder. Additive Manufacturing. Euro PM 2018 Congress and Exibition—Conference Paper—2020. Available online: https://sites.shocklogic.com/demos/portable_abstract_book/Euro%20PM2018%20Proceedings/EuroPM2018%20Proceedings/assets/3987667.pdf (accessed on 18 August 2025).
100. Kumar, K.; Kumar, G.S. A Study on Surface Roughness of Rapid Prototypes Fabricated Using Poly-Jet 3D Printing System. In Proceedings of the International Conference on Computer Aided Engineering (CAE 2013), Chennani, India, 19-21 December 2013; pp. 1–6.
101. Boschetto, A.; Bottini, L.; Miani, F.; Veniali, F. Roughness Investigation of Steel 316L Parts Fabricated by Metal Fused Filament Fabrication. *J. Manuf. Process.* **2022**, *81*, 261–280. [CrossRef]
102. Baligidad, S.M.; Chandra, B.T.; Girish, S.M. Fused Filament Fabrication of Stainless Steel 316L: Investigation on Effect of Printing Parameters on Microstructural and Mechanical Strength of Sintered Part. *IJSRP* **2024**, *14*, 242–255. [CrossRef]
103. Caminero, M.Á.; Romero Gutiérrez, A.; Chacón, J.M.; García-Plaza, E.; Núñez, P.J. Effects of Fused Filament Fabrication Parameters on the Manufacturing of 316L Stainless-Steel Components: Geometric and Mechanical Properties. *RPJ* **2022**, *28*, 2004–2026. [CrossRef]
104. Jacob, J.; Pejak Simunec, D.; Kandjani, A.E.Z.; Trinchi, A.; Sola, A. A Review of Fused Filament Fabrication of Metal Parts (Metal FFF): Current Developments and Future Challenges. *Technologies* **2024**, *12*, 267. [CrossRef]
105. Naim, M.; Chemkhi, M.; Alhussein, A.; Retraint, D. Effect of Post-Treatments on the Tribological and Corrosion Behavior of 17–4PH Stainless Steel Processed via Fused Filament Fabrication. *Addit. Manuf. Lett.* **2023**, *7*, 100158. [CrossRef]
106. Obadimu, S.O.; Kourousis, K.I. Shrinkage Behaviour of Material Extrusion Steel 316L: Influence of Primary 3D Printing Parameters. *RPJ* **2022**, *28*, 92–101. [CrossRef]
107. Utela, B.; Storti, D.; Anderson, R.; Ganter, M. A Review of Process Development Steps for New Material Systems in Three Dimensional Printing (3DP). *J. Manuf. Process.* **2008**, *10*, 96–104. [CrossRef]
108. Alaghmandfard, R.; Aghayar, Y.; Ester, D.; Mohammadi, M.; Borisoff, J.; Dotto, K. Sintering Process Optimization of the Additively Manufactured Pure Copper Parts through the Metal Paste Deposited Process. *J. Alloys Compd.* **2024**, *1009*, 176932. [CrossRef]
109. Ahn, D.; Kweon, J.H.; Choi, J.; Lee, S. Quantification of Surface Roughness of Parts Processed by Laminated Object Manufacturing. *J. Mater. Process. Technol.* **2012**, *212*, 339–346. [CrossRef]
110. Sharma, A.; Bandari, V.; Ito, K.; Kohama, K.; Ramji, M.; Himasekhar Sai, B.V. A New Process for Design and Manufacture of Tailor-Made Functionally Graded Composites through Friction Stir Additive Manufacturing. *J. Manuf. Process.* **2017**, *26*, 122–130. [CrossRef]
111. Chaudhary, B.; Jain, N.K.; Murugesan, J. Experimental Investigation and Parametric Optimization of Friction Stir Powder Additive Manufacturing Process for Aerospace-Grade Al Alloy. *Int. J. Adv. Manuf. Technol.* **2022**, *123*, 603–625. [CrossRef]
112. Gao, X.; Agyakwa, P.A.; Simonelli, M.; East, M.; Hague, R.J.M.; Gilani, N. Tailored Droplet Deposition Strategies for Direct Printing of Fully Functional Components via Molten Metal Jetting. *J. Manuf. Process.* **2025**, *151*, 206–213. [CrossRef]
113. Lass, N.; Gerdes, B.; Jehle, M.; Riegger, L.; Zengerle, R.; Koltay, P. Generation of High Aspect Ratio Metal Microstructures Exhibiting Low Surface Roughness by Drop-Wise Printing of Liquid Metal. *Procedia Eng.* **2015**, *120*, 1103–1106. [CrossRef]
114. Traxel, K.D.; Elton, E.S.; Petersen, A.M.; Silva, C.M.; Perron, A.; Jeffries, J.R.; Pascall, A.J. Processing and Microstructure of a Cu-Al-Fe-Mn Alloy via Droplet-on-Demand Additive Manufacturing. *Mater. Des.* **2024**, *237*, 112544. [CrossRef]
115. Simonelli, M.; Aboulkhair, N.; Rasa, M.; East, M.; Tuck, C.; Wildman, R.; Salomons, O.; Hague, R. Towards Digital Metal Additive Manufacturing via High-Temperature Drop-on-Demand Jetting. *Addit. Manuf.* **2019**, *30*, 100930. [CrossRef]

metals

MDPI

Article

Multi-Model Collaborative Optimization of Inconel 690 Deposited Geometry in Laser-Directed Energy Deposition: Machine Learning Prediction and NSGA-II Decision Framework

Chen Liu [1,2], Junxiao Liu [1,2], Xiuyuan Yin [1,2], Xiaoyu Zhang [1,2], Shuo Shang [1,2] and Changsheng Liu [1,2,*]

1 School of Materials Science and Engineering, Northeastern University, Shenyang 110819, China; liuchenneu@163.com (C.L.); 2300632@stu.neu.edu.cn (J.L.); 2110176@stu.neu.edu.cn (X.Y.); 2410250@stu.neu.edu.cn (X.Z.); shangs@mail.neu.edu.cn (S.S.)

2 Key Laboratory for Anisotropy and Texture of Materials Ministry of Education, Northeastern University, Shenyang 110819, China

* Correspondence: csliu@mail.neu.edu.cn

Abstract

The critical challenge of achieving precise geometric control in laser directed energy deposition (L-DED) of Inconel 690 for nuclear applications is addressed by this study. We established a data-driven optimization framework that reduces time-consuming trial-and-error experiments. A comprehensive process-geometry dataset was generated through full-factor experiments. Pearson correlation analysis revealed significant correlations: strong positive correlations between laser power and bead width (r = 0.82) and depth (r = 0.85), and between powder feed rate and height (r = 0.70). A hybrid machine learning model was subsequently developed. It used a Backpropagation Neural Network (BPNN) to achieve excellent prediction of width, height, and depth ($R^2 \leq 0.962$). It also generated 100 uniformly distributed Pareto optimal process parameter sets via the Non-dominated Sorting Genetic Algorithm II (NSGA-II). Experimental validation confirmed the model's high predictive accuracy—relative error $\leq$ 5% for width/depth, and a maximum relative error of 5.34% for height. This demonstrates the framework's effectiveness for reliable multi-objective process optimization in high-precision deposition tasks. It also highlights its potential for use in nuclear component repair and other material systems.

Keywords: Inconel 690; laser-directed energy deposition; machine learning; multi-objective optimization; bead geometry

1. Introduction

Laser-directed energy deposition (L-DED) is an additive manufacturing technology that employs high-energy laser beams to melt metallic feedstock, forming metallurgical bonds with substrates [1,2]. This process achieves near-net shape forming of complex structures through layer-wise deposition, offering superior material utilization (exceeding 95%) [3], enhanced design freedom (e.g., internal conformal channels) [4], and rapid prototyping capabilities compared with subtractive manufacturing. These advantages enable widespread L-DED adoption for high-performance metal components in aerospace, energy, and biomedical sectors [5–7]. However, complex coupling effects between process parameters and melt pool dynamics frequently cause geometric deviations, porosity, and microstructural inhomogeneity [8], limiting L-DED's application in critical load-bearing parts. Since single-bead deposition constitutes the fundamental unit determining final quality, investigating parameter–geometry relationships and establishing accurate bead dimension prediction models are essential for ensuring dimensional accuracy and process robustness.

The optimization of laser additive manufacturing processes is challenged by multi-parameter interactions, vast solution spaces, and complex physical-metallurgical phenomena, rendering trial-and-error approaches and numerical simulations prohibitively time-consuming and costly. Machine learning (ML) addresses these limitations by leveraging extensive process data to shorten experimental cycles, reduce computational expenses, and effectively predict/optimize process parameters. Veiga et al. [9] have systematically analyzed how critical parameters (current, wire feed speed) govern bead geometry (width/height) in Invar alloy wire-arc additive manufacturing through real-time melt pool monitoring, establishing process-geometry correlations via linear regression models. Afshari et al. [10] developed an integrated approach combining design of experiments, inverse analysis, and numerical simulation to evaluate bead geometry and cladding characteristics in Inconel 718 laser deposition, achieving precise prediction through multi-parameter optimization. Chigilipalli et al. [11] optimized Incoloy 825 deposition parameters via neuro-fuzzy modeling, creating an intelligent prediction framework for nickel-based alloys. Shen et al. [12] employed multiple ML algorithms (Random Forest, SVM, ANN) for dimension prediction in $CoCrFeNiMo_{0.2}$ high-entropy alloy deposition. Li et al. [13] proposed a novel differential evolution-optimized multi-output SVR (DE-MOSVR) model predicting bead morphology in 2319 aluminum wire-laser hybrid DED. Lee et al. [14] utilized Gaussian process regression (GPR) to forecast layer height and porosity in Fe-Ni alloy laser metal deposition, with Shapley Additive Explanations (SHAP) quantifying parameter significance.

Although machine learning (ML) can accurately predict single-target outputs, it lacks the ability to autonomously reconcile conflicts between parameters such as bead width, depth and height. Multi-objective optimization is the key tool for resolving such challenges in engineering decision-making. Wang et al. [15] developed an intelligent optimization method that integrates modified NSGA-II algorithms and multi-pass deposition profile prediction models to optimize parameters in aluminum alloy laser-arc hybrid additive manufacturing. Chen et al. [16] proposed an innovative, multi-objective optimization framework that considers multiple objectives, such as energy efficiency, material utilization and environmental impact in order to optimize process parameters systematically. Using a titanium alloy as a case study, the study verified the effectiveness of this framework in balancing processing performance and environmental benefits. Aghaei et al. [17] combined SVR prediction models with NSGA-II to map the relationship between different laser welding parameters and quality metrics (e.g., strength and penetration). This approach achieved multi-objective synergy through Pareto solutions, improving joint strength by 15% and reducing the width of the heat-affected zone by 20% compared with single-objective approaches. Panico et al. [18] used Taguchi methods and neural networks to model correlations between fused deposition modeling (FDM) 3D printing parameters and quality, utilizing NSGA-II to resolve conflicts between printing speed, strength, and accuracy. Yu et al. [19] created a hybrid framework that combines data augmentation, stacked ensemble learning, SHAP analysis and multi-objective optimization in order to design strong, tough, and ductile laser-melted titanium alloys. This framework simultaneously enhances the ultimate tensile strength (UTS) and elongation (EL) of LPBF Ti components. Together, these studies show that ML-multi-objective synergy is an innovative way to allocate weight precisely across competing objectives and identify optimal parameter sets in laser additive manufacturing.

Inconel 690 alloy, with its compatible filler ER NiCrFe-7A, has been shown to possess exceptionally high-temperature stability and corrosion resistance properties, thus making it a critical material for the manufacturing of components for nuclear power plants. The utilization of this alloy has been demonstrated to result in a substantial increase in service lifetimes. While the potential of additive manufacturing (AM) in the repair and fabrication of nuclear safety-critical components, including reactor penetration nozzles, large-curvature piping nozzles, and fuel assembly bottom grids (BG)/nozzles (BN), has been demonstrated [20–22], current research on Inconel 690 AM remains considerably limited. Existing studies have primarily focused on mitigating high-temperature impregnation cracks and enhancing dissimilar-weld mechanical properties, with insufficient attention being paid to predicting and optimizing single-bead geometry. It is imperative to acknowledge that precise dimensional control of individual beads is of the essence in ensuring the quality and stability of fabricated components. Consequently, the advancement of predictive modelling and process optimization for Inconel 690 deposition assumes a pivotal role in nuclear safety assurance.

Therefore, this study combines machine learning and multi-objective optimization to create a predictive model for the geometry of beads in the laser-directed energy deposition (L-DED) of Inconel 690 alloy. A full-factorial experimental design was implemented, using laser power (*P*), travel speed (*V*), and powder feed rate (*F*) as the input variables and bead width (*W*), height (*H*), and depth (*D*) as the output targets. Three machine learning algorithms were used to develop geometric prediction models. The Non-dominated Sorting Genetic Algorithm II (NSGA-II) was then used to find Pareto-optimal combinations of process parameters, maximizing bead width and height while minimizing depth. Ultimately, experimental validation was performed using the optimal parameters predicted by the model.

2. Materials and Methods

2.1. Materials

The experiments employed Inconel 690 powder with a particle size distribution of 53–106 μm, the morphology of which is demonstrated in Figure 1. AISI 1045 steel substrates (200 mm × 150 mm × 10 mm) were used, with the chemical compositions of both the powder and the substrates provided in Table 1. Prior to the process of deposition, the surfaces of the substrate were subjected to a grinding procedure using abrasive paper, which was then followed by a cleaning process involving the use of ethanol.

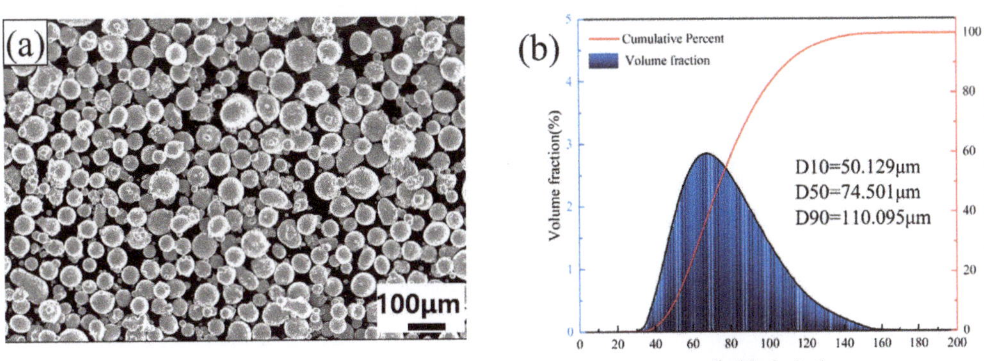

Figure 1. (**a**) SEM morphology of Inconel 690 alloy powder; (**b**) powder particle size distribution diagram.

Table 1. Chemical composition of the powder and substrate (wt.%).

Material	C	Ni	Cr	Fe	Mn	Si
Inconel 690	0.02	Bal.	29.8	9.86	0.053	0.20
1045 steel	0.45	0.15	0.15	Bal.	0.7	0.27

2.2. Laser Direct Energy Deposition System

In this study, all samples were fabricated using a 6-axis L-DED system developed in-house. As illustrated in Figure 2, the motion unit incorporates an industrial robot (RB20, GSK CNC Equipment Co., Ltd., Guangzhou, China). The heat source uses a fiber laser (GW 5M-060-HC, GW (Shanghai) Laser Technology Co., Ltd., Shanghai, China) with a maximum power output of 6000 W and a wavelength of 1070 nm. The forming nozzle uses a coaxial three-point nozzle (CT-1317-01, Nanjing Huirui Photoelectric Technology Co., Ltd., Nanjing, China) with 99.999% pure argon gas to prevent oxidation. Throughout the deposition process, the argon gas maintained a stable flow rate of 20 L/min, and the distance between the substrate and the nozzle was consistently maintained at 17 mm with a laser beam spot diameter of 4 mm. During the forming process, the laser power (P), travel speed (v) and powder feed rate (F) are controlled by the programming language in the robot teach pendant.

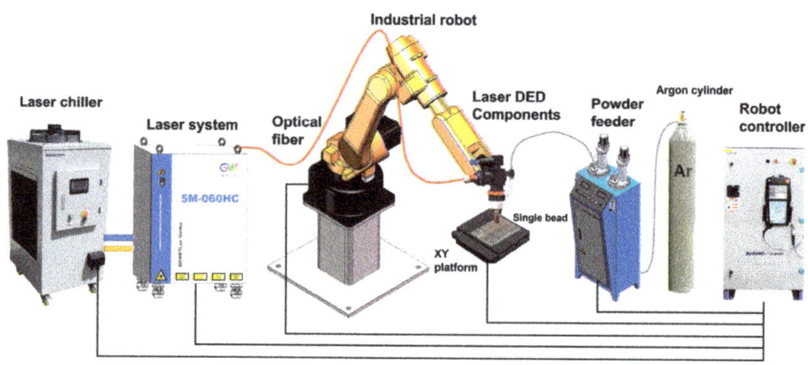

Figure 2. Schematic diagram of a laser direct energy deposition system.

2.3. Experiments and Data Collection

To obtain the training dataset for the machine learning models, we selected process parameters that covered a wide processing window (see Table 2). The laser power range was 1200–3000 W, in steps of 600 W; the travel speed range was 4–12 mm/s, in steps of 4 mm/s; and the powder feed rate range was 1.8–7.2 g/min, in steps of 1.8 g/min. This covered a wide range of linear energy density values, resulting in a single-bead dataset comprising 80 distinct bead geometries. Here, Linear energy density (LED) is defined as follows:

$$LED = \frac{P}{v \times F}, \quad (1)$$

where P is the laser power in watts, v is the travel speed in mm/s, and F is the powder feed rate in g/min.

The length of the single deposition bead is 40 mm. After the cladding process is completed, the prepared specimens are cut into 10 mm × 50 mm × 10 mm specimens using an electric spark wire cutter (each specimen contains 5 deposition beads). After embedding in epoxy resin, the specimens were polished with sandpaper ranging from 200# to 3000#, followed by polishing with polishing paste and a polishing machine until a mirror-like finish was achieved. The specimens were then rinsed with anhydrous ethanol and dried. The polished specimens were etched with 5% nitric acid in ethanol until the coating boundaries were visible. Finally, the polished samples were imaged using an OLYMPUS SZX7 macro-style microscope (Evident Corporation, Tokyo, Japan). Typical laser-deposited single bead morphologies are shown in Figure 3. The bead width W, bead height H, bead depth D, and bead area A were measured using Image-Pro Plus 6.0 software, with each measurement repeated three times and the average value taken.

Table 2. Process parameters for L-DED.

Process Parameters (Symbol, Unit)	Values
Laser power (P, W)	1200, 1800, 2400, 3000
Travel speed (v, mm/s)	4, 6, 8, 10, 12
Powder feed rate (F, g/min)	1.8, 3.6, 5.4, 7.2
Laser spot diameter (d, mm)	4
Argon gas flux (Q, L/min)	20

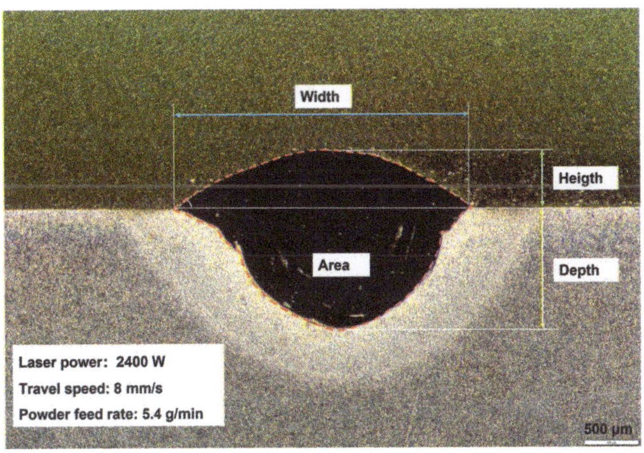

Figure 3. Cross-sectional morphology of laser-deposited single bead.

2.4. Machine Learning Model Development

This study selected three prevalent machine learning (ML) models capable of regression analysis—Support Vector Regression (SVR), Backpropagation Neural Network (BPNN), and eXtreme Gradient Boosting (XGBoost)—to predict bead width (W), height (H), depth (D), and area (A). Prior to model training, data normalization was performed using the MinMaxScaler() function to rescale features within the (0, 1) range. This preprocessing step mitigates model sensitivity to input noise, standardizes feature scales, accelerates convergence, alleviates gradient issues, and enhances generalization capability.

The dataset was partitioned into a training set (70%, 56 samples) and a test set (30%, 24 samples) via the train_test_split() function from Scikit-learn's model_selection module in Python 3.9.7. Data analysis and modeling were conducted using Python 3.9.7 with key libraries (NumPy, Pandas, Scikit-learn, XGBoost, TensorFlow). The hyperparameters of the model were optimized through the implementation of a grid search. In order to avoid the problem of model overfitting, each set of hyperparameters was evaluated using 5-fold cross-validation. Finally, the optimal parameter combination was selected to establish the model. Optimized models were serialized using Python's pickle module for subsequent multi-objective optimization.

Three metrics were adopted to evaluate model accuracy: Root Mean Squared Error (*RMSE*), which quantifies deviations between predicted and actual values, with lower values (closer to 0) indicating superior performance; Mean Absolute Error (*MAE*), which measures expected absolute error loss, with lower values denoting better regression; and Coefficient of Determination (R^2), which assesses model fit, ranging (0, 1), with higher values reflect greater explanatory power of independent variables.

$$RMSE = \sqrt{\frac{1}{n}\sum_{i=1}^{n}(y_i - \hat{y}_i)^2}, \tag{2}$$

$$MAE = \frac{1}{n}\sum_{i=1}^{n}|y_i - \hat{y}_i|, \tag{3}$$

$$R^2 = 1 - \sum_{i=1}^{n}(y_i - \hat{y}_i)^2 \Big/ \sum_{i=1}^{n}(y_i - \overline{y}_i)^2, \tag{4}$$

where y_i is the actual value, $\hat{y}_i$ is the predicted value, $\overline{y}_i$ is the average value, and n is the number of samples.

2.5. Multi-Objective Optimization

In the L-DED process, maximum deposition efficiency is achieved by selecting the maximum bead width and height, while avoiding excessive melt depth, which could lead to a coarse-grain structure. Optimizing L-DED parameters to simultaneously maximize bead width (*W*) and height (*H*) while minimizing depth (*D*) constitutes a classic multi-objective problem. The developed ML models served as fitness functions within the Non-dominated Sorting Genetic Algorithm II (NSGA-II) to identify Pareto-optimal solutions. Since NSGA-II minimizes objectives, the goals for W and H were negated ($-f_W$, $-f_H$). The optimization formulation is the following:

$$\begin{aligned} & X = (P, v, F) \\ & \text{Min } f(X) = (-f_W, -f_H, f_D) \\ & s.t. \begin{cases} P_{min} \le P \le P_{max} \\ v_{min} \le v \le v_{max} \\ F_{min} \le F \le F_{max} \end{cases} \end{aligned} \tag{5}$$

In the equation, X represents the input process parameters, including laser power (P), travel speed (v) and powder feed rate (F). f_W denotes the bead width prediction model, f_H the bead height prediction model, and f_D the bead depth prediction model.

The constraints represent the permissible ranges for each process parameter in laser-directed energy deposition. The parameters of the NSGA-II algorithm were set as follows: Initial sampling was set to FloatRandomSampling, the population size to 100, the crossover probability to 0.9, the mutation probability to 0.15, and the number of generations to 100. The Pareto front was then plotted. This is constructed from the search space, highlighting trade-offs and optimal parameter design schemes. Finally, typical optimization schemes were selected for experimental verification.

3. Results and Discussion

3.1. Effect of Process Parameters on the Geometry of the Deposited Bead

Figure 4 shows the cross-sections of deposited samples prepared using different process parameters. The deposition beads exhibit good metallurgical bonding with the substrate and have regular, well-filled shapes. In additive manufacturing processes, the width of the deposition beads controls the wall thickness of formed thin-walled parts directly, while their height controls deposition efficiency. The depth of the deposition beads primarily influences the size of the remelting zone between layers, which affects grain growth morphology and mechanical properties. Under constant travel speed and powder feed rate, the width, height and area of the deposition beads gradually increase as the laser power increases. Similarly, at a constant laser power and powder feed rate, a decrease in travel speed results in a gradual increase in the width, height, and area of the deposition beads. Similar influence patterns have been observed in other processes and materials [19]. The melting energy of the substrate and powder is determined by the joint effect of laser power and travel speed, thereby influencing the geometric shape and properties of the deposited material. Insufficient power can result in porosity and poor fusion, whereas excessive power can lead to deformation and residual stress. A travel speed that is too slow results in high dilution, whereas a speed that is too fast can cause poor fusion and porosity issues [23]. The bead morphology is further influenced due to the shielding effect of L-DED metal powders on the laser and the convective heat transfer process of the carrier gas flow on the melt pool. Therefore, it is difficult to accurately predict the bead morphology from cross-sectional photographs of the deposition beads alone.

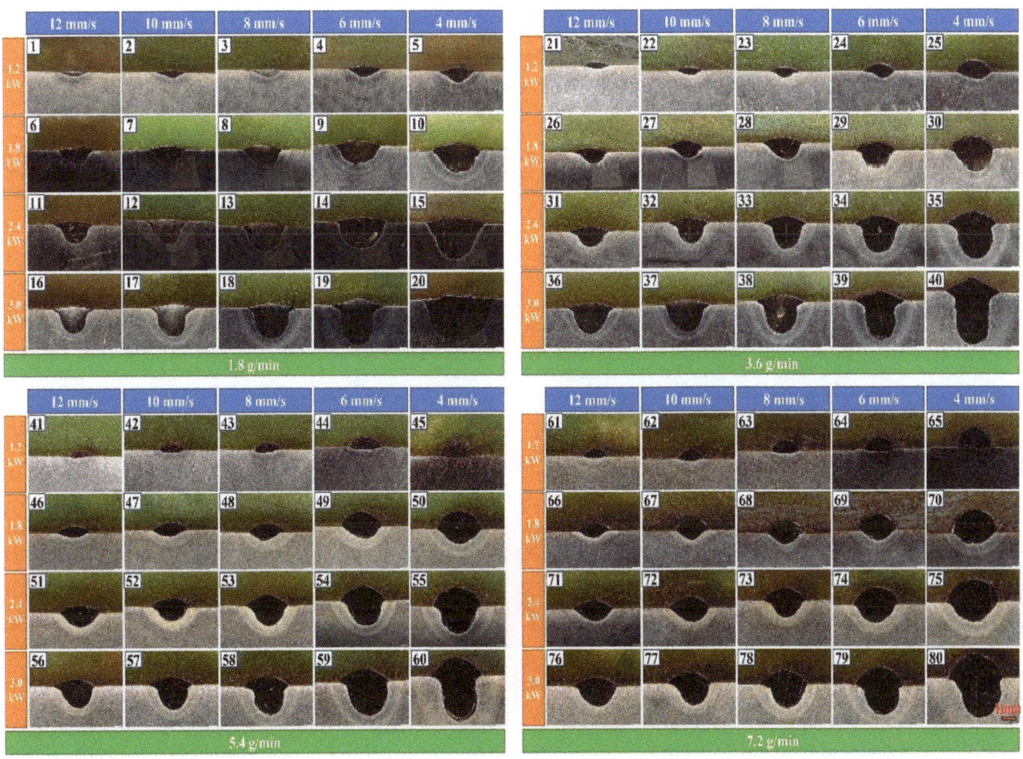

Figure 4. Macrographic morphology of bead cross-sections prepared using different processing parameters.

3.2. Correlation Analysis of Process Parameters

A correlation analysis was conducted on the three process parameters and four bead morphology indicators based on the results of the full factorial experiment. Pearson's correlation coefficient was used to measure the correlation between the process parameters and the bead morphology characteristics. This coefficient can accurately measure the linear correlation between two variables [24]. It can be expressed as follows:

$$\rho_{XY} = \frac{cov(X,Y)}{\sigma_X \sigma_Y} = \frac{E(X-\mu_X)(Y-\mu_Y)}{\sigma_X \sigma_Y} \tag{6}$$

In the formula, cov(*X*, *Y*) denotes the covariance between X and Y, and σ_X and σ_Y denote the sample standard deviations of X and Y, respectively. The Pearson correlation coefficient ranges from −1 to 1. A value of 1 indicates a perfect positive correlation between the two variables, a value of −1 indicates a perfect negative correlation, and a value close to 0 indicates almost no linear relationship.

Figure 5 shows the Pearson correlation coefficient matrix between the three process parameters (laser power, travel speed, and powder feed rate) and the bead height, width, depth, and area. Figure 5 shows that laser power exhibits strong positive correlations with bead width, depth, and area, with respective correlation coefficients of 0.82, 0.85, and 0.73. Laser power shows a weak positive correlation with bead height, with a correlation coefficient of 0.14. Travel speed exhibits weak negative correlations with bead width, depth, height, and area, with respective correlation coefficients of −0.47, −0.35, −0.58, and −0.58. The powder feed rate is strongly positively correlated with bead height (with a correlation coefficient of 0.7) while exhibiting a weak negative correlation with bead width and depth, with respective correlation coefficients of −0.11 and −0.24. These patterns are consistent with those observed in Section 3.1 through analysis of bead cross-sectional photographs. Pearson correlation coefficient analysis demonstrates that laser power and travel speed exert a significant influence on the width, depth, and area of the bead, while powder feed rate exerts a substantial influence on the height of the bead. Laser power, travel speed and powder feed rate can be used as input variables to train a machine learning model to predict the geometric dimensions of the deposited layer.

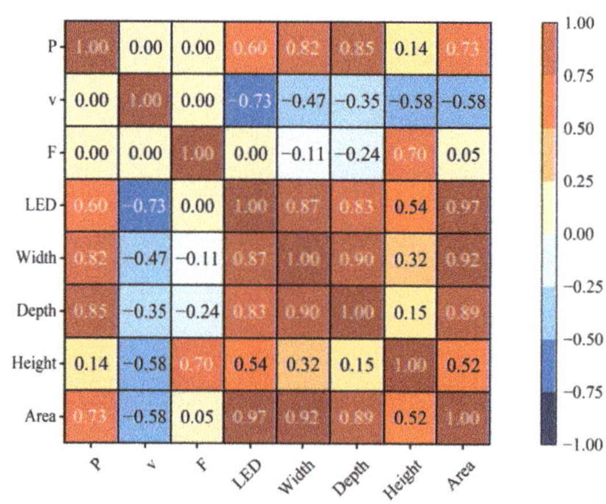

Figure 5. Pearson correlation coefficient matrix between process parameters and bead geometry.

The width and height of the bead directly impact the dimensional accuracy of the LMD process. The establishment of an appropriate depth is a prerequisite for ensuring interlayer metallurgical bond strength. In the context of additive manufacturing processes, the dimensions of width and height can be readily measured non-destructively, with the capability to obtain these measurements online during the forming process. The bead area is defined as the mathematical integral of width, depth, and height, and it has been demonstrated to exhibit a strong positive correlation with these parameters

(Pearson correlation coefficients of 0.92, 0.89, and 0.52, respectively). In light of the practical imperatives inherent to additive manufacturing processes and the necessity to streamline the intricacies of multi-objective optimization algorithms, thereby reducing computation time, this study proposes the adoption of bead width, height, and depth as the multi-objective optimization objectives.

3.3. Machine Learning Modeling

Figure 6 shows a comparison of the prediction performance of the SVR, XGBoost, and BPNN models on the training and test sets for the bead width. The diagonal line represents the data predicted using the ideal model. As can be seen from the figure, the data in the training and test sets are close to the diagonal line. It can be seen from the figure that the three machine learning models generally fit the bead width well, although some individual points show large differences between the actual and predicted values, which may be related to the algorithm's inability to adequately respond to local outliers in the overall trend or an imbalance in the sample distribution of the dataset. To select the optimal bead width prediction model, the three machine learning models were evaluated, and their prediction accuracy on the bead width test set is shown in Table 3. The BPNN model achieved the highest prediction accuracy and fitting degree, with an R^2 of 0.955 and an RMSE of 0.114 mm, while the XGBoost model demonstrated similarly high accuracy and goodness of fit, with an R^2 of 0.949 and an RMSE of 0.122 mm. The SVR model performed significantly worse than the other two models in terms of accuracy and goodness of fit, with an R^2 of 0.926 and an RMSE of 0.147 mm. Taking into account the model's generalization ability, the BPNN model was selected as the bead width prediction model.

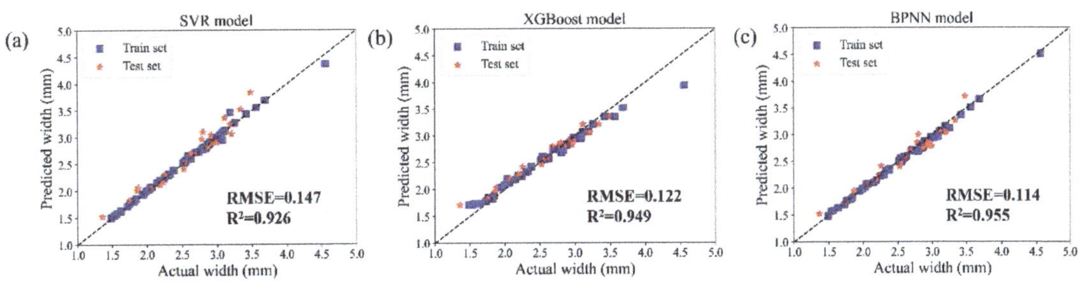

Figure 6. Comparison of prediction accuracy of different machine learning models for bead width: (**a**) SVR model; (**b**) XGBoost model; (**c**) BPNN model.

Table 3. Comparison of the performance metrics of three machine learning algorithms for bead width.

Model	MAE	RMSE	R^2	Hyperparameter
SVR	0.123	0.147	0.926	*kernel* = *rbf*, C = 10, ε = 0.05
XGBoost	0.099	0.122	0.949	*Booster* = *gbtree*, *n_estimators* = 250, *learning_rate* = 0.01, *max_depth* = 5
BPNN	0.086	0.114	0.955	*Neurons* = 64, *learning_rate* = 0.01, *epochs* = 300, *batch_size* = 2

Figure 7 illustrates how the SVR, XGBoost, and BPNN models perform in predicting bead height in the training and test sets. As can be seen from the figure, the data in the training and test sets are close to the diagonal line. It can be seen from the figure that all three machine learning models fit the bead height data well, with an R^2 value exceeding 0.9, indicating excellent overall fitting ability. To select the optimal bead width prediction model, the three machine learning models were evaluated, and their prediction accuracy on the bead width test set is shown in Table 4. The BPNN model achieved the highest prediction accuracy and fit, with R^2 reaching 0.949 and RMSE reaching 0.069 mm, while the XGBoost model exhibited a similarly high level of accuracy and fitting degree, with R^2 reaching 0.946 and RMSE reaching 0.071 mm. The SVR model shows significantly lower accuracy and fitting degree than the previous two models, with R^2 at 0.912 and RMSE at 0.090 mm. Considering the models' generalization ability, the BPNN model was selected as the bead height prediction model.

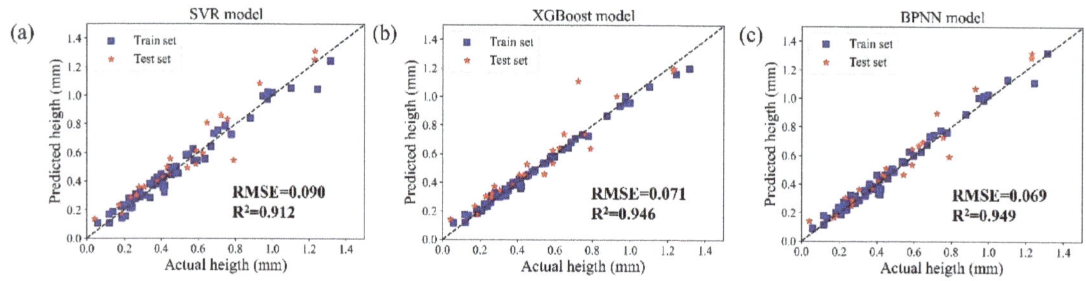

Figure 7. Comparison of prediction accuracy of different machine learning models for bead height: (**a**) SVR model; (**b**) XGBoost model; (**c**) BPNN model.

Table 4. Comparison of the performance metrics of three machine learning algorithms for bead height.

Model	MAE	RMSE	R^2	Hyperparameter
SVR	0.072	0.090	0.912	*kernel* = *rbf*, C = 1, ε = 0.05
XGBoost	0.061	0.071	0.946	*Booster* = *gbtree*, *n_estimators* = 400, *learning_rate* = 0.01, *max_depth* = 5
BPNN	0.052	0.069	0.949	*Neurons* = 132, *learning_rate* = 0.01, *epochs* = 200, *batch_size* = 8

Figure 8 illustrates the predictive capabilities of the SVR, XGBoost, and BPNN models on the training and test sets with respect to bead depth. As can be seen from the figure, the data in the training and test sets are close to the diagonal line, and the three machine learning models generally perform well in fitting the bead depth, with R^2 values exceeding 0.9. Table 5 shows the prediction accuracy of the three models on the test set for bead depth. The BPNN model has the highest prediction accuracy and fitting quality, with R^2 reaching 0.962 and RMSE reaching 0.132 mm, while the SVR model exhibits high prediction accuracy and quality of fit, with R^2 reaching 0.959 and RMSE reaching 0.137 mm. In contrast, the XGBoost model shows significantly lower accuracy and fitting quality, with an R^2 of 0.940 and an RMSE of 0.165 mm. Taking into account the models' generalization ability, the BPNN model was selected as the prediction model for bead depth.

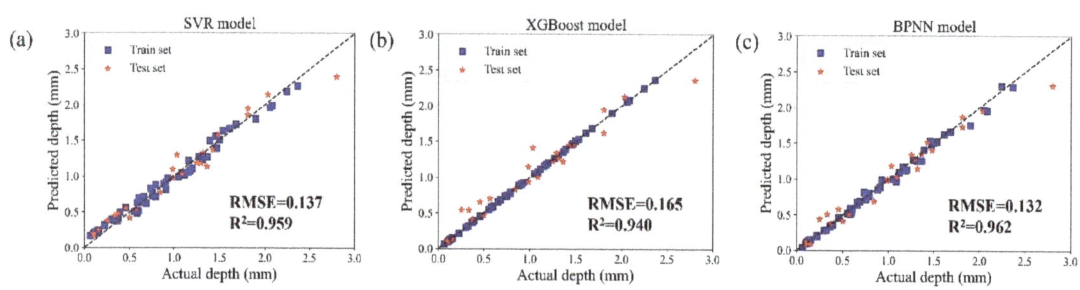

Figure 8. Comparison of prediction accuracy of different machine learning models for bead depth: (**a**) SVR model; (**b**) XGBoost model; (**c**) BPNN model.

Table 5. Comparison of the performance metrics of three machine learning algorithms for bead depth.

Model	MAE	RMSE	R^2	Hyperparameter
SVR	0.108	0.137	0.959	*kernel* = *linear*, C = 10, ε = 0.05
XGBoost	0.117	0.165	0.940	*Booster* = *gblinear*, *n_estimators* = 1000, *learning_rate* = 0.05, *max_depth* = 5
BPNN	0.097	0.132	0.962	*Neurons* = 128, *learning_rate* = 0.01, *epochs* = 100, *batch_size* = 4

In order to further verify the applicability of machine learning algorithms without predicting targets, area predictions were also made, with the prediction model not being used as target optimization input. Figure 9 illustrates the predictive capabilities of the SVR, XGBoost, and BPNN models on the training and test sets for bead area. As can be seen from the figure, the data in the training and test sets are close to the diagonal line, and the three machine learning models generally perform well in fitting the bead area, with R^2 values exceeding 0.95. Table 6 shows the prediction accuracy of the three models on the bead area test set. The XGBoost model has the highest prediction accuracy and R^2 value, with R^2 reaching 0.974 and RMSE reaching 0.283 mm^2. The SVR model follows with an R^2 of 0.967 and RMSE of 0.315 mm^2. The BPNN model lags slightly behind the first two with an R^2 value of 0.964 and an RMSE value of 0.329 mm^2. Taking into account the model's generalization ability, the XGBoost model was selected as the prediction model for bead area.

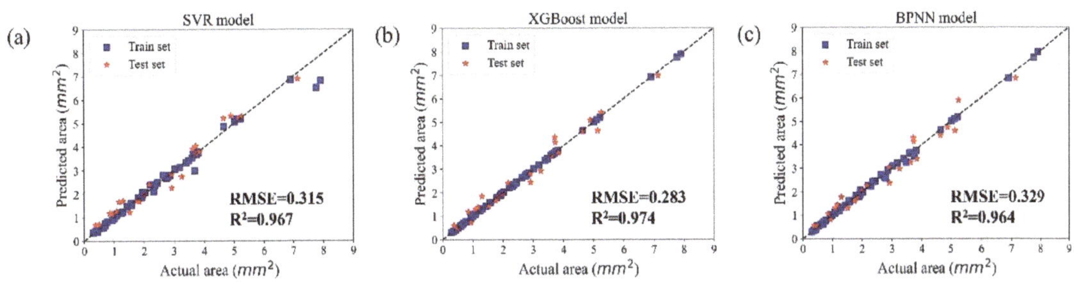

Figure 9. Comparison of prediction accuracy of different machine learning models for bead area: (**a**) SVR model; (**b**) XGBoost model; (**c**) BPNN model.

Table 6. Comparison of the performance metrics of three machine learning algorithms for bead area.

Model	MAE	RMSE	R^2	Hyperparameter
SVR	0.256	0.315	0.967	*kernel* = *rbf*, C = 10, ε = 0.05
XGBoost	0.221	0.283	0.974	*Booster* = *gbtree*, *n_estimators* = 300, *learning_rate* = 0.05, *max_depth* = 5
BPNN	0.279	0.329	0.964	*Neurons* = 32, *learning_rate* = 0.01, *epochs* = 200, *batch_size* = 1

In summary, the R^2 values of the three machine learning algorithms, SVR, XGBoost, and BPNN, are all greater than 90%, which proves that machine learning models are indeed suitable for predicting the geometric shape of additive manufacturing deposition beads. Due to the varying performance of different machine learning models when processing datasets, this study employed the BPNN model to predict bead width, height, and depth, and the XGBoost model to predict bead area.

3.4. Multi-Objective Optimization of Process Parameters

Figure 10 shows the optimal process parameter solutions for maximizing bead width and height while minimizing depth. As can be seen in the figure, there are 100 solution sets uniformly distributed on the Pareto front. In multi-objective optimization problems, each solution set represents an optimization scheme that cannot be dominated by other solution sets. This reflects the trade-offs between different objectives. To verify the accuracy of the model optimization results, experimental verification was conducted in this study. As shown in Figure 10c, based on curvature analysis of the Pareto frontier, the Pareto front has an inflection point, which was selected for verification. Similarly, two additional points were selected to form the verification set. The process parameter values of the verification set were obtained by denormalizing using the inverse_transform() function. Experiments were conducted using these values, and the geometric dimensions of the deposited layer cross-section were measured.

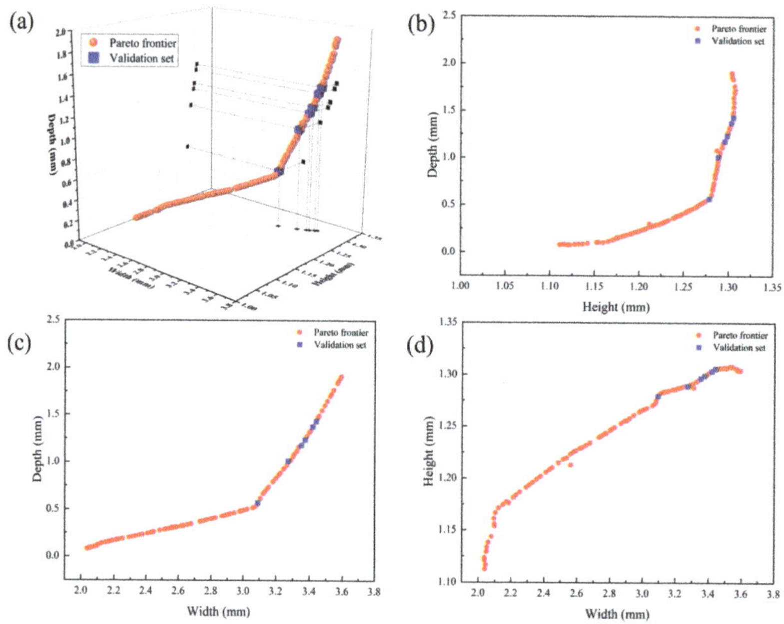

Figure 10. (**a**) Pareto solution set obtained by the NSGA-II algorithm; (**b**–**d**) Projection of the Pareto frontier onto (**b**) the depth/height plane; (**c**) the depth/width plane; (**d**) the height/width plane.

The cross-sectional metallographic images of the verification set deposition beads are displayed in Figure 11, wherein the bead width, bead height, and bead depth were meticulously measured. The experimental results were then compared to the predictions derived from the machine learning model, thus validating the accuracy of the latter. As illustrated in Table 7, a comparison is presented between the output quality characteristics as predicted by the model and the experimental values obtained under the optimal process parameters. The results demonstrate that the relative errors of bead width and bead depth are all controlled within 5% under the six process conditions, with only the height error in the first group reaching 5.34%. This may be attributed to underfitting of the model in specific regions and random errors caused by melt pool fluctuations during laser additive manufacturing, but the model has good overall prediction accuracy. The machine learning and multi-objective optimization algorithm proposed in this study has been shown to be highly effective in predicting and optimizing the geometric morphology of laser additive manufacturing deposition beads, as evidenced by the low error rate.

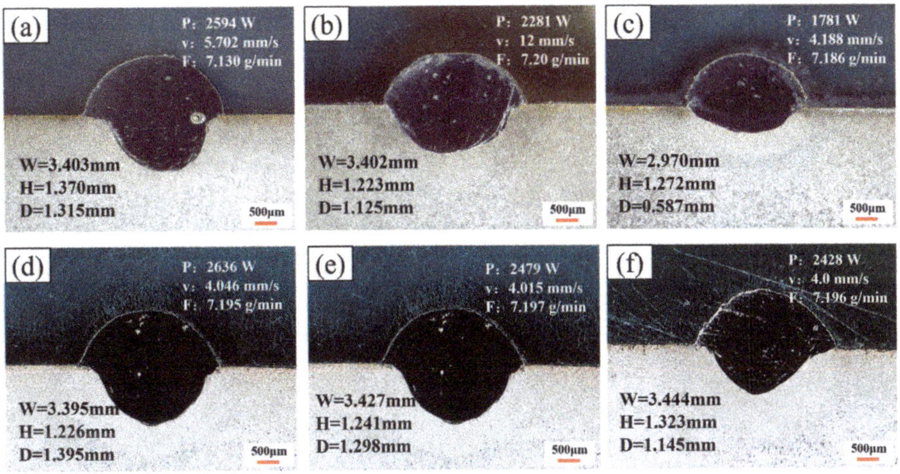

Figure 11. Metallographic microstructure of the deposition path in the verification test: (**a**) No. 1; (**b**) No. 2; (**c**) No. 3; (**d**) No. 4; (**e**) No. 5; (**f**) No. 6.

Table 7. Comparison of predicted and actual values for the optimized process.

Sample	Target	Predicted Value	Actual Value	Error
1	Width	3.422	3.403	0.50%
	Height	1.304	1.370	4.82%
	Depth	1.375	1.315	4.56%
2	Width	3.276	3.402	3.70%
	Height	1.289	1.223	5.34%
	Depth	1.013	1.125	3.55%
3	Width	3.096	2.970	4.24%
	Height	1.280	1.272	0.63%
	Depth	0.569	0.587	3.07%
4	Width	3.444	3.395	1.43%
	Height	1.306	1.266	3.18%
	Depth	1.432	1.395	2.69%
5	Width	3.378	3.427	1.44%
	Height	1.299	1.241	4.69%
	Depth	1.238	1.298	4.64%
6	Width	3.356	3.444	2.56%
	Height	1.297	1.323	1.97%
	Depth	1.180	1.145	3.08%

4. Conclusions

This study pioneered an integrated ML-NSGA-II framework for predicting and optimizing bead geometry in Inconel 690 laser-directed energy deposition (L-DED). Key findings include the following:

(1) Strong process-geometry correlations: Laser power positively correlated with bead width ($r = 0.82$), depth ($r = 0.85$), and area ($r = 0.73$); powder feed rate correlated with height ($r = 0.70$); travel speed showed negative correlations (−0.47~−0.58).

(2) Hybrid ML model superiority: BPNN achieved optimal prediction for width ($R^2 = 0.955$), height ($R^2 = 0.949$), and depth ($R^2 = 0.962$), while XGBoost excelled in area prediction ($R^2 = 0.974$), collectively forming a high-precision prediction system.

(3) Validation of Pareto solutions: NSGA-II generated 100 optimized parameter sets, with experimental verification confirming ≤5% error for width/depth and a maximum error of 5.34% for height.

The framework establishes an effective data-driven optimization pathway for nuclear-grade L-DED applications. Future work will expand the model to incorporate auxiliary process parameters (spot size, gas flow, incidence angle, modulation frequency, etc.) and mechanical properties (hardness, strength, defects, etc.), with transfer learning applications targeting nuclear dissimilar-material welding and turbine blade repair.

Author Contributions: Conceptualization, C.L. (Chen Liu) and C.L. (Changsheng Liu); methodology, C.L. (Chen Liu); software, C.L. (Chen Liu) and J.L.; validation, C.L. (Chen Liu) and J.L.; formal analysis, C.L. (Chen Liu) and S.S.; investigation, C.L. (Chen Liu) and X.Y.; resources, C.L. (Chen Liu) and X.Y.; data curation, C.L. (Chen Liu) and S.S.; writing—original draft preparation, C.L. (Chen Liu) and X.Z.; writing—review and editing, C.L. (Chen Liu) and X.Z.; visualization, X.Z.; supervision, C.L. (Changsheng Liu) and S.S.; project administration, C.L. (Changsheng Liu); funding acquisition, C.L. (Changsheng Liu). All authors have read and agreed to the published version of the manuscript.

Funding: This study was financially supported by the Joint Funds of NSFC-Liaoning (Grant No. U1508213).

Data Availability Statement: The original contributions presented in this study are included in the article. Further inquiries can be directed to the corresponding author.

Conflicts of Interest: The authors declare no conflicts of interest.

References

1. Debroy, T.; Wei, H.L.; Zuback, J.S.; Mukherjee, T.; Elmer, J.W.; Milewski, J.O.; Beese, A.M.; Wilson-Heid, A.; De, A.; Zhang, W. Additive manufacturing of metallic components—Process, structure and properties. *Prog. Mater. Sci.* **2018**, *92*, 112–224. [CrossRef]
2. Srivastava, M.; Rathee, S.; Patel, V.; Kumar, A.; Koppad, P.G. A review of various materials for additive manufacturing: Recent trends and processing issues. *J. Mater. Res. Technol.* **2022**, *21*, 2612–2641. [CrossRef]
3. Li, Z.; Sui, S.; Ma, X.; Tan, H.; Zhong, C.; Bi, G.; Clare, A.T.; Gasser, A.; Chen, J. High deposition rate powder- and wire-based laser directed energy deposition of metallic materials: A review. *Int. J. Mach. Tools Manuf.* **2022**, *181*, 103942. [CrossRef]
4. Saito, S.; Baba, S.; Kinoshita, K.; Takada, N.; Someya, S. Thermal–hydraulic performance of additively manufactured crossflow minichannel heat exchangers. *Appl. Therm. Eng.* **2025**, *274*, 126543. [CrossRef]
5. Srivastava, M.; Jayakumar, V.; Udayan, Y.; Sathishkumar, M.; Muthu, S.M.; Gautam, P.; Nag, A. Additive manufacturing of Titanium alloy for aerospace applications: Insights into the process, microstructure, and mechanical properties. *Appl. Mater. Today* **2024**, *41*, 102481. [CrossRef]
6. Palaniappan, S.; Joshi, S.S.; Sharma, S.; Radhakrishnan, M.; Krishna, K.V.M.; Dahotre, N.B. Additive manufacturing of FeCrAl alloys for nuclear applications—A focused review. *Nucl. Mater. Energy* **2024**, *40*, 101702. [CrossRef]
7. Jiang, D.; Thissen, H.; Hughes, T.C.; Yang, K.; Wilson, R.; Murphy, A.B.; Nguyen, V. Advances in additive manufacturing of auxetic structures for biomedical applications. *Mater. Today Commun.* **2024**, *40*, 110045. [CrossRef]
8. Gao, G.; Wang, Y.; Wang, Y.; Zhang, K.; Xiang, D.; Ma, J. Development on shape and performance control of aeronautical parts in additive manufacturing. *J. Manuf. Process.* **2025**, *141*, 351–374. [CrossRef]
9. Veiga, F.; Suarez, A.; Aldalur, E.; Artaza, T. Wire arc additive manufacturing of invar parts: Bead geometry and melt pool monitoring. *Measurement* **2022**, *189*, 110452. [CrossRef]
10. Afshari, M.; Khandaei, M.; Farzadi, A.; Razavi, R.S.; Barekat, M.; Rezaei, M. A new approach to evaluation of bead geometry and cladded layer in laser additive manufacturing of Inconel 718 using the experimental design, inverse analysis and numerical simulation. *Opt. Laser Technol.* **2025**, *184*, 112464. [CrossRef]
11. Chigilipalli, B.K.; Veeramani, A. An experimental investigation and neuro-fuzzy modeling to ascertain metal deposition parameters for the wire arc additive manufacturing of Incoloy 825. *CIRP J. Manuf. Sci. Technol.* **2022**, *38*, 386–400. [CrossRef]
12. Shen, Q.; Xue, J.; Zheng, Z.; Yu, X.; Ou, N. Machine learning-based prediction of CoCrFeNiMo0.2 high-entropy alloy weld bead dimensions in wire arc additive manufacturing. *Mater. Today Commun.* **2024**, *41*, 110359. [CrossRef]
13. Li, R.; Ma, H.; Zeng, K.; Suo, H.; Li, C.; Fu, Y.; Zhang, M.; Zhang, M.; Fang, X. Differential Evolution-Optimized Multi-Output Support Vector Regression-Based Prediction of Weld Bead Morphology in Wire-Fed Laser-Arc Directed Energy Deposition of 2319 Aluminum Alloy. *Addit. Manuf. Front.* **2025**, *4*, 200203. [CrossRef]

14. Lee, J.A.; Sagong, M.J.; Jung, J.; Kim, E.S.; Kim, H.S. Explainable machine learning for understanding and predicting geometry and defect types in Fe-Ni alloys fabricated by laser metal deposition additive manufacturing. *J. Mater. Res. Technol.* **2023**, *22*, 413–423. [CrossRef]
15. Wang, J.; Feng, L.; Xu, J.; Zha, X.; Cui, Y.; Li, K.; Xie, T. Optimal process parameter combinations search for desired deposited layer geometry in laser-arc hybrid additive manufacturing based on multi-pass overlapping deposited layer contour prediction model and improved NSGA-II algorithm. *Opt. Laser Technol.* **2025**, *187*, 112700. [CrossRef]
16. Chen, K.; Jiang, X.; Liu, W.; Zhang, F.; Liu, A.; Xu, X.; Song, B.; Xi, W.; Bian, H.; Yang, G. A novel multi-objective optimization of high-power laser directed energy deposition green processes: A case study of titanium alloy. *J. Clean. Prod.* **2025**, *494*, 144877. [CrossRef]
17. Aghaei, H.; Larki, H.A.; Taqizadeh, M.; Aoudia, M.; Aich, W.; Kolsi, L.; Dehkordi, M.H.R. Multi- objective modeling and optimization of dissimilar laser welding by integrating an artificial intelligence predictive model with NSGA-II algorithm. *Mater. Today Commun.* **2024**, *40*, 109765. [CrossRef]
18. Panico, A.; Corvi, A.; Collini, L.; Sciancalepore, C. Multi objective optimization of FDM 3D printing parameters set via design of experiments and machine learning algorithms. *Sci. Rep.* **2025**, *15*, 16753. [CrossRef]
19. Yu, A.; Pan, Y.; Wan, F.; Kuang, F.; Lu, X. Multi-objective optimization of laser powder bed fused titanium considering strength and ductility: A new framework based on explainable stacking ensemble learning and NSGA-II. *J. Mater. Sci. Technol.* **2025**, *228*, 241–255. [CrossRef]
20. Kim, J.; Kim, J.; Park, J.; Baek, S.; Ma, N.; Lee, S.; Kim, D. Continuous tempering effect induced PWHT alternative technology using wire arc additive manufacturing for application in replacing nuclear pressurized water reactor system repairing: CALPHAD, FEM simulation, and EBSD investigation. *J. Mater. Res. Technol.* **2023**, *25*, 2961–2988. [CrossRef]
21. Zhong, Y.; Zheng, Z.; Li, J.; Wang, C. Fabrication of 316L nuclear nozzles on the main pipeline with large curvature by CMT wire arc additive manufacturing and self-developed slicing algorithm. *Mater. Sci. Eng. A* **2021**, *820*, 141539. [CrossRef]
22. Jeong, S.G.; Kim, E.S.; Ahn, S.Y.; Chun, J.H.; Ryu, J.Y.; Woo, H.G.; Yoo, S.H.; Kang, S.H.; Kim, H.S. Prototyping of nuclear fuel assembly parts by laser powder bed fusion of Inconel 718. *Prog. Nucl. Energy* **2025**, *184*, 105694. [CrossRef]
23. Wang, J.; Zhao, B.; Liu, Y.; Zhao, J.; Ma, G. Research Progress in Shape-Control Methods for Wire-Arc-Directed Energy Deposition. *Materials* **2024**, *17*, 5704. [CrossRef]
24. Edelmann, D.; Móri, T.F.; Székely, G.J. On relationships between the Pearson and the distance correlation coefficients. *Stat. Probab. Lett.* **2021**, *169*, 108960. [CrossRef]

metals

MDPI

Article

Effect of Heat Treatments and Related Microstructural Modifications on High-Cycle Fatigue Behavior of Powder Bed Fusion–Laser Beam-Fabricated Ti-6Al-2Sn-4Zr-6Mo Alloy

Gianluca Pirro [1,*], Alessandro Morri [1], Alessandra Martucci [2], Mariangela Lombardi [2] and Lorella Ceschini [1]

1 Department of Industrial Engineering (DIN), Alma Mater Studiorum—University of Bologna, Viale del Risorgimento 4, 40136 Bologna, Italy; alessandro.morri4@unibo.it (A.M.); lorella.ceschini@unibo.it (L.C.)
2 Department of Applied Science and Technology, Politecnico di Torino, Corso Duca degli Abruzzi 24, 10129 Torino, Italy; alessandra.martucci@polito.it (A.M.); mariangela.lombardi@polito.it (M.L.)
* Correspondence: gianluca.pirro2@unibo.it

Abstract

The study investigates the influence of microstructures on fatigue behavior and failure mechanisms of the α-β titanium alloy Ti6246, fabricated via Powder Bed Fusion-Laser Beam (PBF-LB). In particular, the investigation assesses the effect of two post-processing heat treatments, namely α-β annealing at 875 °C (AN875) and solution treatment at 825 °C followed by aging at 500 °C (STA825), on the alloy's rotating and bending fatigue behavior. The results indicate that the STA825 condition provides superior fatigue resistance (+25%) compared to AN875, due to the presence of a finer bilamellar microstructure, characterized by thinner primary α lamellae (α_p) and a more homogeneous distribution of secondary α lamellae (α_s) within the β matrix. Additionally, an investigation conducted using the Kitagawa–Takahashi (KT) approach and the El-Haddad model, based on the relationship between the fatigue limit and defect sensitivity, revealed improved crack propagation resistance from pre-existing defects (ΔK_{th}) for the STA825 condition compared to AN875. Notably, the presence of fine α_s after aging for STA825 is effective in delaying crack nucleation and propagation at early stages, while refined α_p contributes to hindering macrocrack growth. The fatigue behavior of the STA825-treated Ti6246 alloy was even superior to that of the PBF-LB-processed Ti64, representing a viable alternative for the production of high-performance components in the automotive and aerospace sectors.

Keywords: Powder Bed Fusion–Laser Beam; titanium alloys; heat treatment; microstructures; high cycle fatigue; defects

1. Introduction

The properties of titanium alloys, including their high strength-to-weight ratio, optimal fracture toughness, and excellent high-temperature and oxidation resistance, have led to their widespread adoption in sectors such as aerospace, motorsport, and automotive [1–4]. The growing demand for high-performance components, combined with the need for complex geometries unachievable through conventional machining technologies [5], has driven the adoption of additive manufacturing (AM) [6,7]. Among AM technologies, Powder Bed Fusion–Laser Beam (PBF-LB) is the most extensively studied and widely used, as it enables the fabrication of strong, lightweight, and geometrically complex titanium components, including those with internal channels, directly from Computer-Aided models.

Currently, the α-β titanium alloy Ti-6Al-4V (Ti64) is the most widely used in AM due to the following: (i) its long-standing application in the aerospace and automotive industries; (ii) its well-documented excellent mechanical performance; (iii) its availability on the market. For these reasons, PBF-LB Ti64 has been the subject of extensive research [3,8], particularly aimed at maximizing fatigue strength, as fatigue represents the primary cause of failure in automotive and aerospace components (e.g., airframe components, connecting rods, etc.).

However, the growing demand for enhanced performance in AM-produced components has sparked significant interest in alternative α-β titanium alloys, such as Ti-6Al-2Sn-4Zr-6Mo (Ti6246), which exhibits a higher β-phase fraction and enhanced mechanical potential compared to Ti64.

For conventionally processed Ti6246, the optimization of tailored heat treatments enables the achievement of high tensile properties [9,10]. In particular, these properties are attained through solution treatment below the β transus temperature, followed by aging, which promotes the formation of a fine bimodal microstructure consisting of a globular primary α phase (α_p) in a fine α-β lamellar matrix [11]. Beyond static properties, Peters et al. [12] reported that Ti6246 exhibits superior fatigue strength compared to Ti64 in both high-cycle fatigue (HCF) and very-high-cycle fatigue (VHCF) regimes. This improvement is primarily attributed to the higher α phase fraction present in Ti64 compared to Ti6246 under similar heat treatment conditions, resulting from its higher content of α-stabilizing elements. In wrought titanium alloys, which are typically free of macroscopic defects, fatigue crack nucleation preferentially occurs within the α_p, with subsequent propagation occurring either transgranularly within the α phase (especially when coarse) or along the α/β interfaces [13,14].

To date, no specific studies have been carried out on AM processed Ti6246. Nevertheless, it is well established that the unique microstructural features induced by PBF-LB in α-β Ti alloys significantly affect both tensile and fatigue behavior [15,16]. Many researchers have observed that high cooling rates typical of these processes can lead to the formation of metastable phases, such as α' [17] and α'' martensite [6], depending on the β-stabilizer content. These metastable microstructures can negatively affect their reliability during service [18,19]. Although optimized heat treatment can increase the tensile properties of the alloys, making them comparable and even superior to those of conventionally processed alloys, similar improvements in fatigue resistance remain challenging. In fact, the high surface roughness, gas porosity, and lack of fusion (LoF) defects of PBF-LB parts generate localized stress concentrations that promote crack nucleation, significantly reducing fatigue resistance [20–22].

Leuders et al. [18] observed that for a Ti64 alloy produced by PBF-LB, α+β annealing treatment improved crack propagation resistance compared to the as-built state, due to the transformation of α' martensite into a lamellar α+β structure. However, the heat treatment had no significant effect on crack nucleation resistance, which primarily occurred near pores. Yu et al. [15] extensively studied the effect of Solution Treating and Aging (STA) heat treatment within the α+β phase field on a Ti64 alloy fabricated via PBF-LB. Their findings revealed that this treatment promotes the formation of a distinct bilamellar microstructure, consisting of α_p lamellae within a fine matrix of secondary α (α_s) lamellae and a β phase, which exhibited enhanced fatigue resistance compared to the lamellar structure resulting from α+β annealing. This highlights the critical role of heat treatment optimization in improving fatigue performance.

Although the influence of microstructures on the fatigue behavior of α-β Ti64 alloys has been widely investigated, to the best of the authors' knowledge, a research gap exists regarding the effect of heat treatment on the fatigue strength of PBF-LB Ti6246 alloys that requires further investigations. In previous works [23,24], the authors optimized the heat treatment parameters for PBF-LB Ti6246 and established correlations between its microstructural features and tensile properties. Building on these findings, the present study aims to investigate the effects of two heat treatments, STA and α+β annealing, on the fatigue behavior of PBF-LB Ti6246. Moreover, the correlation between fatigue failure and defects in the PBF-LB Ti6246 was investigated using the Kitagawa–Takahashi (KT) approach and the El-Haddad model [25,26], aiming to assess the potential for enhancing defect tolerance in this alloy. Indeed, enhancing defect tolerance could enable the elimination of the costly and time-consuming Hot Isostatic Pressing (HIP) process, which is commonly applied to titanium PBF-LB components to reduce internal defects and improve fatigue performance.

2. Materials and Methods

2.1. Sample Production

The PBF-LB samples were produced by gas-atomized Ti6246 powders supplied by Eckart TLS (Hartenstein, Zwickau, Germany). Glow Discharge Optical Emission Spectroscopy (GD-OES, GDA-650, Spectruma Analytik GmbH, Hof, Germany) was employed to verify that the chemical composition of the samples met the required standards. The chemical composition of the supplied powders, the printed samples, and the standard UNS R56260 [27] is reported in Table 1.

Table 1. Ti6246 alloy powders' chemical composition (wt.%), as declared by the supplier and reported in the UNS R56260 standard.

Elements [% wt.]	Ti	Al	Sn	Zr	Mo	Fe	N	O	H	Others
Supplier composition	Bal.	5.96	1.85	3.67	5.89	0.03	0.01	0.12	0.002	Total max < 0.30
Sample GD-OES	Bal.	6.36 ± 0.13	2.06 ± 0.08	4.21 ± 0.09	5.85 ± 0.14	0.14 ± 0.01	-	-	-	Total: 0.20 ± 0.06
Standard composition	Bal.	5.5 ÷ 6.5	1.75 ÷ 2.25	3.5 ÷ 4.5	5.5 ÷ 6.5	max 0.15	max. 0.04	max. 0.15	max. 0.0125	Each max 0.10 Total max 0.40

An analysis of the powder's particle size distribution was conducted, showing D10, D50, and D90 values of 24.1, 38.0, and 54.7 μm, respectively. Cylindrical samples (120 mm in height and 12 mm in diameter) were made using an EOS M270 Dual Mode (EOS GmbH, Krailling, Germany) 3D metal laser sintering system under a high-purity argon atmosphere. The system is equipped with a continuous Nd:YAG fiber laser capable of delivering up to 200 W of power, with a laser spot diameter of 100 μm. The PBF-LB process parameters, summarized in Table 2, were defined based on a previous study [6].

Table 2. PBF-LB scanning parameters for Ti6246 samples production.

Power [W]	Scanning Speed [mm/s]	Hatch Distance [μm]	Layer Thickness [μm]	Scanning Strategy	Platform Temperature [°C]
190	1250	100	30	Standard 67° EOS strategy	100

The axisymmetric samples were fabricated with their axes oriented perpendicular to the pre-heated build platform, an industrial practice that minimizes the required support structure. To prevent distortions associated with mechanical removal, cylindrical specimens were detached from the build plate using electrical discharge machining. Based on previous studies, the specimens underwent two distinct heat treatments. The first treatment, optimized by Carrozza et al. [24], was an annealing process conducted in a Pro.Ba VF8000/S high-vacuum furnace (TAV Vacuum Furnaces S.p.A, Caravaggio, Bergamo, Italy) within the α+β phase field at 875 °C for 2 h, followed by slow furnace cooling at a controlled rate of 5 °C/min. This treatment aimed to enhance the alloy's ductility and toughness. The second treatment, optimized by the authors [23], was an STA process performed in a Nabertherm LT 9/14 muffle furnace (Nabertherm GmbH, Lilienthal, Germany). It involved solution treatment at 825 °C for 2 h, rapid water quenching, and subsequent aging at 500 °C for 3 h to achieve high strength. For clarity, these heat treatments will hereafter be referred to as AN875 and STA825, respectively (Figure 1). Table 3 summarizes the tensile test data, in terms of Yield Strength (YS), ultimate tensile strength (UTS), and elongation to failure (EL), carried out on the alloy and reported in [23]. These data were averaged over three samples for each condition.

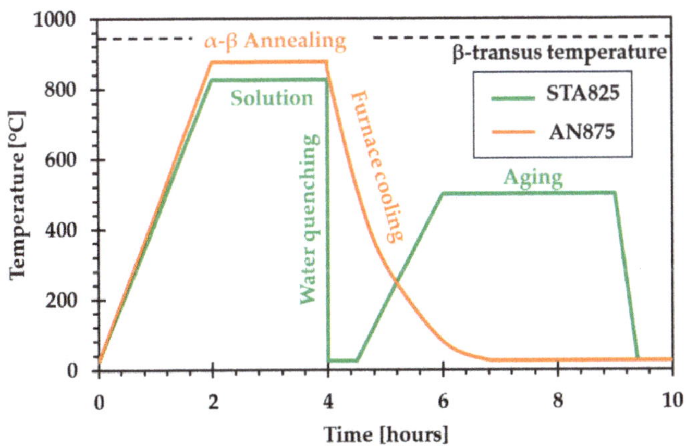

Figure 1. Representation of STA825 (green) and AN875 (orange) thermal cycles.

Table 3. A summary of tensile properties for machined tensile specimens subjected to AN875 and STA825 treatments (Adapted from ref. [23]).

	YS [MPa]	UTS [MPa]	EL [%]
STA825	1118 ± 12	1327 ± 1	10.1 ± 0.4
AN875	936 ± 9	1085 ± 1	22.3 ± 0.5

The cylindrical specimens were machined and polished to produce 4-point bend fatigue samples according to BS ISO 1143:2021 [28], with a Φ10 mm grip diameter and a Φ5 mm gauge section. Surface roughness measurements were carried out on the gauge section of the fatigue samples using a JR25 Nanovea (Irvine, CA, USA) optical profilometer to verify compliance with the specifications. The samples showed a smooth and regular surface with average Ra values of 0.32 ± 0.02 μm. Figure 2 illustrates the geometry and dimensions of the samples used in this study.

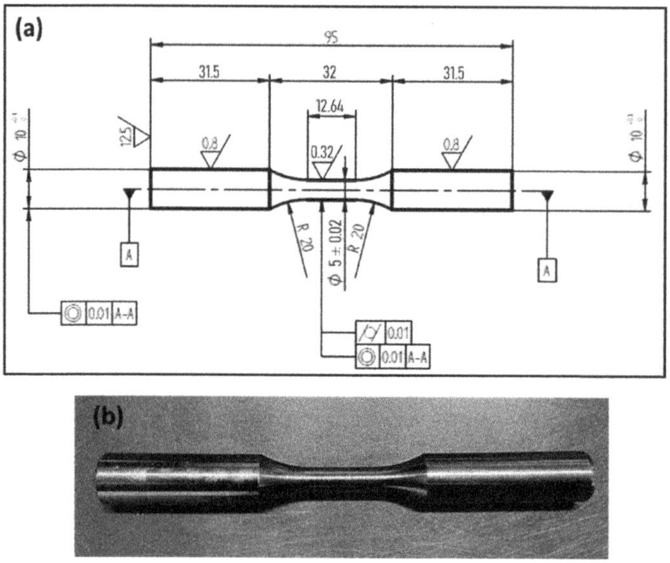

Figure 2. The geometry and dimensions [mm] (**a**) and an image of the fatigue samples (**b**) according to BS ISO 1143:2021.

2.2. Microstructural Characterization

Microstructural characterization was performed on the longitudinal sections of the cylindrical specimens. Three samples for each heat treatment condition were analyzed. One cylinder with a diameter of 10 mm and a length of 10 mm was extracted from each grip end, hot-mounted in conductive resin, ground with abrasive papers up to P1200 grit, and polished with diamond suspensions (9 µm and 3 µm), followed by final polishing with a water-based colloidal silica and hydrogen peroxide solution to achieve a mirror finish [29]. Polished cross-sections were examined using a Zeiss Axio Imager A.1M optical microscope (OM, Oberkochen, Germany) to identify process-related defects present in the samples. For detailed microstructural analysis, samples were etched with Kroll's reagent (93% H_2O, 5% HNO_3, 2% HF) and imaged at low magnification with OM and at high magnification using a Tescan Mira 3 Field Emission Gun–Scanning Electron Microscope (FEG-SEM, Brno, Czech Republic). Image analysis was performed using ImageJ v1.52a [30] to assess sample density, defect content and size, as well as key microstructural features, including the α_p lamellae content and thickness. The α_p content was defined as the α_p (dark area in Figure 3a) area fraction relative to the total analyzed area. α_p lamellae were isolated in the microstructure using the "Outlines" function of the ImageJ software v1.52a (Figure 3b), which then aggregates the contributions from the delineated areas. The "Ellipse" ImageJ tool was instead applied to approximate the morphology of the α_p lamellae via elliptical fitting (Figure 3c). The arithmetic mean of the minor axes of the fitted ellipses was defined as the characteristic lamellar thickness, as described in [22]. Microstructural analyses were conducted on two samples for each heat treatment condition, with five images for each sample. Images were taken with varying magnifications for each condition (5kx for AN875 and 30kx for STA825), due to their differing microstructural fineness.

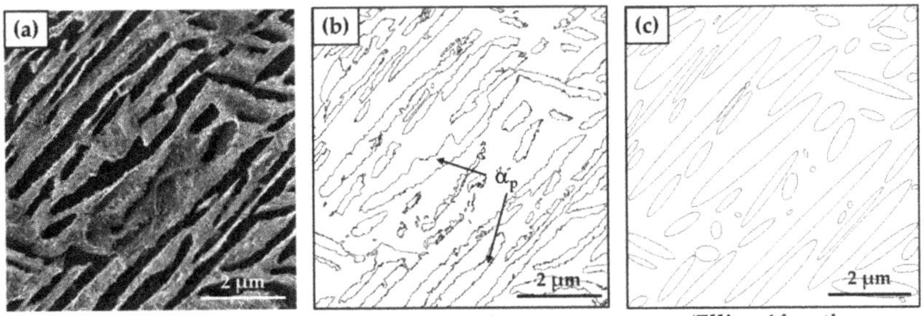

FEG-SEM Image **'Outlines' function** **'Ellipse' function**

Figure 3. FEG-SEM microstructure (**a**) for image analysis made by using the 'Outlines' Function (**b**) to determine the α_p content and the 'Ellipse' Function (**c**) to quantify the α_p lamella thickness.

2.3. Fatigue, Hardness Tests, and Fractographic Analyses

High-cycle fatigue strength was evaluated through fully reversed (R = −1) rotating–bending tests conducted in a four-point configuration machine (RB35, Italsigma, Forlì, Italy), according to the ISO 1143:2021 standard. The tests were performed at a frequency of 50 Hz, with a run-out limit of 2×10^6 cycles. Fatigue strength at a 50% probability of failure ($\sigma_{d,50\%}$) and the associated standard deviation (σ_γ) were determined using the staircase method (ISO 12107:2012 [31]). Based on preliminary tests, a stress increment of 20 MPa (approximately 5% of the estimated fatigue strength) was selected, and 15 specimens were tested for each heat treatment condition. Vickers hardness measurements were performed on the grip sections of the fatigue specimens using a 30 kg load and a 15 s dwell time to ensure consistency in hardness across all the samples. For both treatment conditions, the values were obtained by averaging 5 measurements performed on two different samples.

All the run-out specimens were subsequently retested until failure. Fractographic analyses were conducted using the FEG-SEM to identify crack nucleation sites, dominant failure mechanisms, and the size and nature of the defect responsible for initiation, referred to as the "killer defect". For each specimen tested, the stress range ($\Delta\sigma$)—a function of the maximum (σ_{max}) and minimum (σ_{min}) stress—and the stress intensity factor range (ΔK) at the crack tip were determined, calculated using Equations (1) and (2), respectively. Since the tests were conducted under symmetric alternating load conditions (R = −1), σ_{max} and σ_{min} are equal in magnitude and opposite in sign, and $\Delta\sigma$ is twice the alternating stress (σ_a). The stress intensity factor is a function of the stress range, the critical defect size (in terms of $\sqrt{}$area, also known as the Murakami parameter), and the defect's location (where coefficient C is 0.5 for internal defects or 0.65 for surface defects) [26,32,33].

$$\Delta\sigma = \sigma_{max} - \sigma_{min} = 2\sigma_a, \quad (1)$$

$$\Delta K = C \cdot \Delta\sigma \cdot \sqrt{\pi \cdot \sqrt{area}}, \quad (2)$$

KT diagrams were developed based on the fracture surface analysis data, in accordance with established methodologies for additively manufactured alloys [26]. These diagrams plot $\Delta\sigma$ and ΔK as functions of the $\sqrt{}$area, establishing a correlation between high-cycle fatigue strength and the critical defect size for the Ti6246 alloy. The diagrams include both run-out conditions, marked with "O", and failure conditions, indicated by "X". The data points on the KT diagram were then fitted using the El-Haddad model [34]. This model generates threshold curves that identify regions of failure (above the curve) from non-failure (below the curve). Specifically, the obtained curves are described by Equations (3)–(5), which represent the relationship between the fatigue threshold stress range ($\Delta\sigma_{th}$), the threshold crack propagation resistance range (ΔK_{th}), and the critical defect size ($\sqrt{area}$):

$$\Delta\sigma_{th}(\sqrt{area}) = \Delta\sigma_{th,0}\sqrt{\frac{\sqrt{area_0}}{\sqrt{area}+\sqrt{area_0}}}, \quad (3)$$

$$\Delta K_{th}(\sqrt{area}) = \Delta K_{th,lc}\sqrt{\frac{\sqrt{area}}{\sqrt{area}+\sqrt{area_0}}}, \quad (4)$$

$$\sqrt{area_0} = \frac{1}{\pi}\cdot\left(\frac{\Delta K_{th,lc}}{Y\cdot\Delta\sigma_{th,0}}\right)^2 \quad (5)$$

$\Delta\sigma_{th,0}$ represents the limiting stress range for defect-free specimens; $\Delta K_{th,lc}$ represents the resistance limit to long crack (macrocrack) propagation; and $\sqrt{area_0}$ is a modeling parameter proportional to the critical crack length, distinguishing short from long cracks, and it is dependent on $\Delta\sigma_{th,0}$ and $\Delta K_{th,lc}$, which best fit the experimental data.

3. Results and Discussion

3.1. Microstructural Characterization

Figure 4a presents a representative OM image showing small, circular defects indicative of gas porosity (marked by arrows) and a larger, irregular defect characteristic of a LoF, further detailed in the FEG-SEM image in Figure 4b. Porosity analyses for the AN875 and STA825 samples revealed no significant deviations from the previously reported values in [23]. The average density measurements are consistent with prior results, with densities of 99.81 ± 0.05% and 99.87 ± 0.09%, respectively.

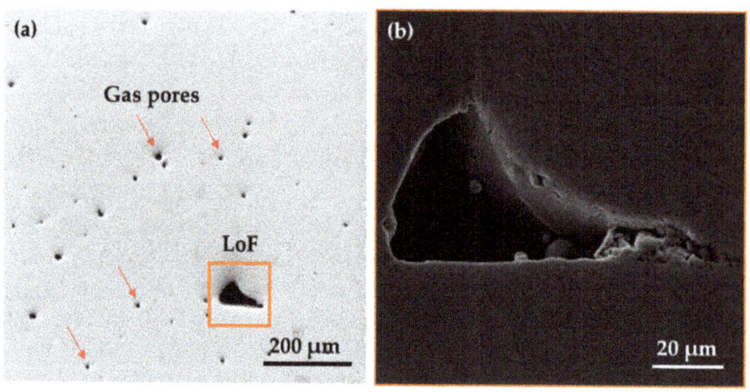

Figure 4. An OM image of a Ti6246 sample showing gas porosities (indicated by red arrows) and a larger LoF defect (**a**), with a corresponding high-magnification FEG-SEM image (**b**).

The defect size distribution, derived from image analysis of OM micrographs, is illustrated in Figure 5a. The data represents the number of defects per unit area as a function of the square root of the defect area. The accompanying table reports the 50th (D_{50}) and 90th (D_{90}) percentiles of defect size and the maximum defect size (D_{max}). In the AN875 heat-treated alloy, a lower relative frequency of defects is observed, accompanied by a higher proportion of larger defects compared to the STA825 condition. This behavior is attributed to the higher solution treatment temperature used in the AN875 process, which may promote slight growth of gas-induced porosities, as evidenced in previous works on AM-processed alloys [35–37]. Nonetheless, the variation in defect size is limited to just a few microns and is considered negligible in terms of its impact on the mechanical performance of the material, particularly its fatigue behavior.

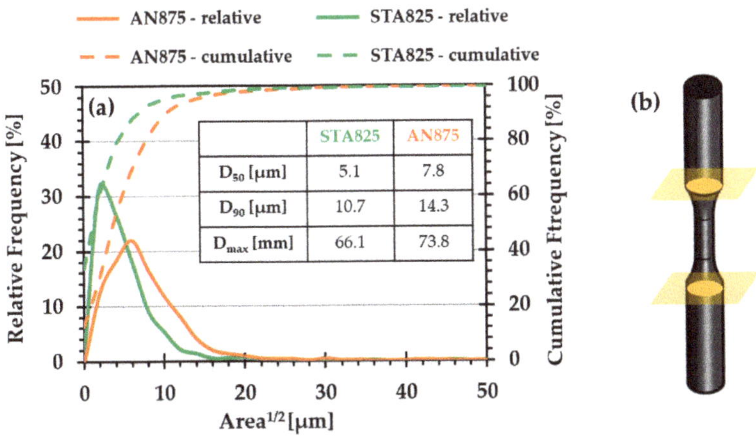

Figure 5. The relative and cumulative frequency distributions of defect sizes (in terms of the square root of the area) (**a**) obtained from image analysis on OM micrographs of polished cross sections (**b**).

Figure 5b shows a schematic representation of the cross-sections analyzed for each sample.

Figure 6 presents FEG-SEM micrographs of chemically etched samples following heat treatment. Both conditions exhibit columnar prior β grains, generated from PBF-LB printing, and a bilamellar microstructure characterized by the formation of α_s within the β phase, located between α_p lamellae. However, the STA825 condition exhibits a significantly finer arrangement of both α_p and α_s lamellae compared to the AN875 condition. These microstructural differences arise from the distinct solution treatment parameters and cooling rates applied in each heat treatment, which influence phase transformation kinetics and lamellar refinement. As reported by Galarraga et al. [38], the growth of the α phase is primarily governed by two factors: solution temperature and cooling rate. Higher solution temperatures, approaching the β-transus, combined with slower cooling rates, promote coarsening of α_p lamellae. In the AN875 condition, the microstructure consists of large α_p lamellae with rounded tips, along with finer α_s lamellae that nucleate within the β phase (α_s+β) separating the α_p lamellae (Figure 6d). However, the α_s fraction is relatively low and inhomogeneously distributed. During isothermal holding at 875 °C, accelerated diffusion processes drive the coarsening of α_p. This effect is further amplified by slow furnace cooling, which promotes additional coarsening of α_p and the formation and consequent coarsening of α_s lamellae [39]. Conversely, the STA825 condition exhibits a finer bilamellar microstructure with a significantly reduced size and volume fraction of α_p lamellae, primarily influenced by the solution temperature (825 °C) and holding time. Additionally, α_s lamellae are extremely fine and uniformly distributed within the β phase (α_s + β). Rapid water quenching, in fact, inhibits diffusive transformations, preserving the α_p amount achieved at 825 °C and leading to the formation of metastable β phase, as discussed by the authors in [23]. The subsequent aging process induces transformation of the metastable β phase, leading to the formation of a finely distributed α_s lamellar structure within the β phase, as indicated in Figure 6b, and the precipitation of very fine strengthening α_2 (Ti_3Al) within α_p, in agreement with the results of the XRD analysis reported in our previous study [23]. Similar findings for α-β alloys containing a minimum of 6% Al after STA heat treatment are also described in [38,40–42].

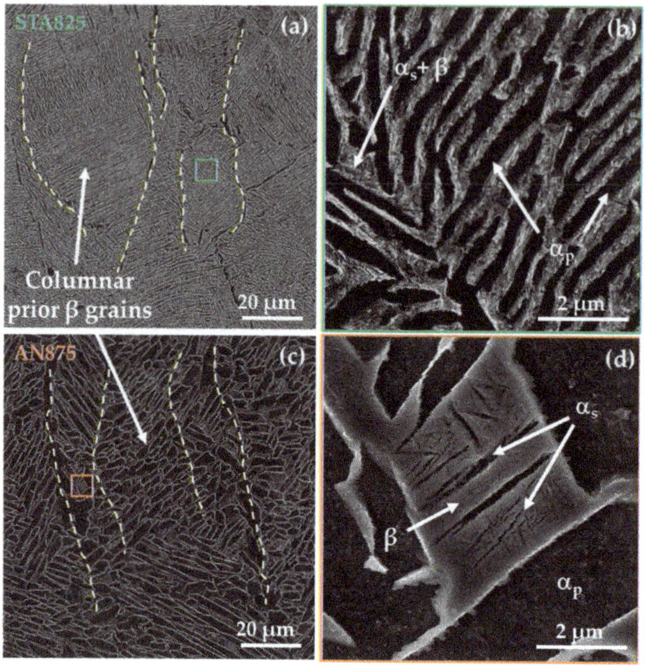

Figure 6. Low- and high-magnification FEG-SEM images of STA825 (**a**,**b**) and AN875 microstructures (**c**,**d**). The yellow dotted lines and the white arrows highlight, respectively, the columnar prior β-grain boundaries and the principal microstructural features.

Table 4 summarizes the key microstructural characteristics, including the α_p lamellae content and thickness, for each heat treatment condition. These values agree with those obtained during the heat treatment optimization phase [23].

Table 4. Summary of microstructural features (α_p lamellae content and thickness) for STA825 (measured after aging) and AN875 specimens.

	α_p Content [%]	α_p Thickness [μm]
STA825	39.32 ± 0.69	0.29 ± 0.02
AN875	69.99 ± 1.28	1.46 ± 0.10

3.2. Fatigue Behavior

The results of the fatigue tests are reported in Table 5. The STA825 samples ($HV_{30} = 425 \pm 5$) exhibited higher fatigue strength compared to the AN875 ones ($HV_{30} = 351 \pm 9$), with the 50% probability of failure ($\sigma_{d,50\%}$) being approximately 25% higher (474 ± 15 MPa vs. 382 ± 15 MPa). For both heat treatment conditions, $\sigma_{d,50\%}$ was found to be approximately 35–36% of the ultimate tensile strength (UTS) (Table 5), highlighting a correlation between the static and fatigue properties of the alloy, regardless of the heat treatment conditions.

Table 5. A summary of fatigue strength at the 50% failure probability ($\sigma_{d,50\%}$) and standard deviation (σ_γ), and the $\sigma_{d,50\%}$/UTS ratio of the investigated heat treatment conditions.

	$\sigma_{d,50\%}$ [MPa]	σ_γ [MPa]	$\frac{\sigma_{d,50\%}}{UTS}$ [%]
STA825	474	15	35.7
AN875	382	15	35.2

This observation confirms the soundness of the design strategy employed for the STA825 heat treatment. Specifically, the STA825 treatment offers enhanced fatigue resistance over the conventional AN875 condition due to the following: (i) enhanced static strength, resulting from the synergistic strengthening effect of a refined and homogeneously distributed bilamellar architecture (comprising α_s lamellae and a retained β phase within the interlamellar regions) and the precipitation of Ti_3Al intermetallics during the aging stage; (ii) improved resistance to early-stage fatigue crack initiation and propagation, particularly along $\alpha_p/(\alpha_s + \beta)$ interfaces, due to the presence of fine α_s lamellae, as confirmed by previous studies [14,43]; and (iii) an increased capability to deflect or arrest macrocrack propagation, attributed to the presence of fine and hardened primary α lamellae, which outperform the coarser counterparts typically formed under the AN875 regime.

Figure 7a,b shows low-magnification images of fracture surfaces of the different fatigue samples, highlighting regions characterized by stable fatigue crack propagation and final overload fractures. For both treatment conditions, nucleation was observed to occur superficially or sub-superficially, mainly from Lof defects, as visible in high-magnification images (Figure 7c,d).

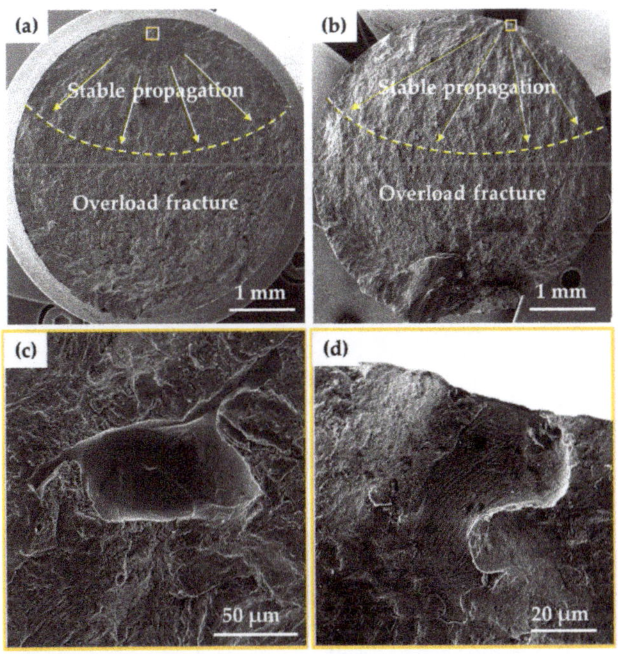

Figure 7. Low-magnification (**a**,**b**) FEG-SEM images of different fracture surfaces with a focus on crack nucleation sites originating from sub-superficial (**c**) or superficial LoFs (**d**). The yellow arrows and dotted lines highlight, respectively, the stable propagation direction and the end of the stable propagation zone.

When nucleation occurs from subsurface killer defects, a low-speed crack propagation zone (LSP) and a high-speed crack propagation zone (HSP) can be identified (Figure 8a). The LSP zone (Figure 8b) can be associated with fish-eye-like crack propagation, as it exhibits an alternating distribution of parallel bright strips and dark grooves, a phenomenon termed the "mackerel pattern". This pattern was identified by Gao et al. for Ti64 [44] and Ti6.5Al-3.5Mo.1.5Zr-0.3Si (TC11) [45] alloys. The mackerel pattern can be attributed to the difficulty encountered by the crack during its early growth stages in propagating along the fine $\alpha_p/(\alpha_s + \beta)$ interface [45]. This effect is much more readily identifiable in the STA825 sample because the crack propagates with greater difficulty due to the presence of uniformly distributed α_s lamellae within the β phase. According to Lütjering et al. [40], α_s lamellae are difficult to deform plastically and can be considered undeformable. Therefore, their presence induces strong resistance to crack propagation, resulting in the formation of grooves and strips within the LSP [44]. In contrast, the HSP in Figure 8c zone exhibits a different morphology, characterized by the presence of tearing ridges and quasi-cleavage facets. This phenomenon is not observed when the crack nucleates and propagates from surface defects, as shown in Figure 8d. In such cases, only HSP zones can be detected, exhibiting quasi-cleavage propagation, characterized by tearing ridges and the presence of microcracks both at the crack tip (Figure 8e) and further along the propagation path (Figure 8f), in which the fatigue striations typical of a fracture under cyclic loading are detectable also. These observations agree with the findings of Li et al. [46], who attributed these peculiar fracture surface morphologies to differences in the ΔK values associated with fracture initiation at the surface versus the subsurface.

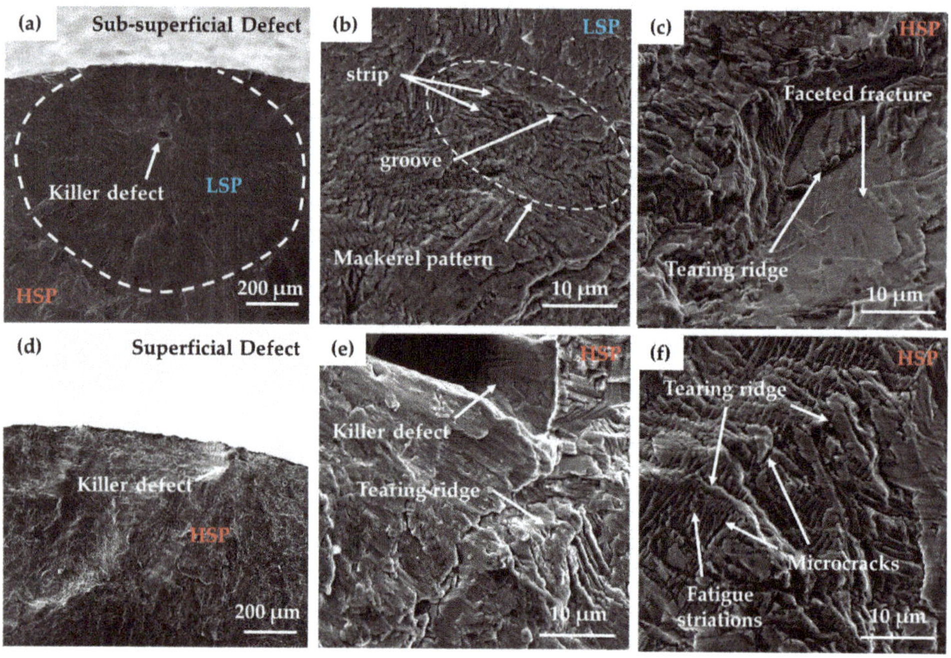

Figure 8. Crack propagation from subsurface defects (**a**), characterized by LSP (**b**) and HSP (**c**) zones, and from surface defects (**d**) with quasi-cleavage zones both at the crack tip (**e**) and in the propagation zone (**f**).

Micrographs of the overload fracture zones (Figure 9) reveal a ductile morphology, characterized by the presence of large dimples for the AN875 condition and shallow dimples for the STA825, consistent with the higher elongation to failure of the AN875 alloy than the STA825 alloy and the fracture surfaces observed in the tensile specimens documented in [23].

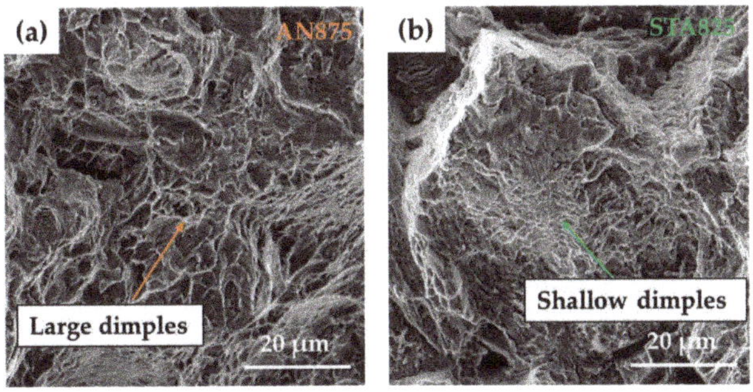

Figure 9. High-magnification images of the overload fracture zone of the specimens AN875 (**a**) and STA825 (**b**).

Figure 10a,b present the KT diagrams. These diagrams illustrate, respectively, the stress range as a function of the defect size ($\sqrt{area}$; $\Delta\sigma$) and the stress intensity factor range versus the defect size ($\sqrt{area}$; ΔK), complete with their respective El-Haddad best-fit curves. The best-fit parameters derived from the El-Haddad model for the two tested conditions are listed in Table 6.

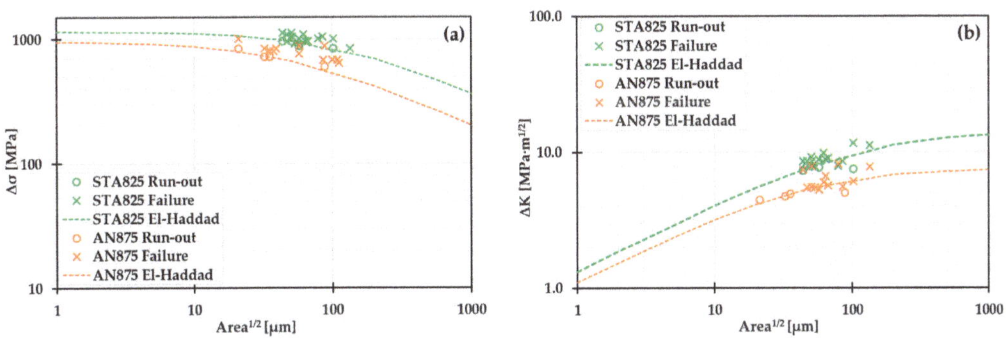

Figure 10. KT diagrams of the stress range ($\Delta\sigma$) (**a**) and the stress intensity factor (ΔK) (**b**) as a function of the defect size ($\sqrt{area}$). "X" indicates failed specimens, while "O" refers to run-out specimens.

Table 6. El-Haddad best-fit parameters for the analyzed conditions.

	$\Delta\sigma_{th,0}$ [MPa]	$\Delta K_{th,lc}$ [MPa $\sqrt{m}$]	$\sqrt{area_0}$ [µm]
STA825	1170	14.5	116
AN875	980	8.2	53

The results obtained from the KT diagrams revealed that the $\Delta\sigma_{th,0}$ and $\Delta K_{th,lc}$ resistances were 19 and 77% higher, respectively, for the STA825 sample compared to the AN875. For samples produced with additive manufacturing technologies, the latter value is particularly significant since crack nucleation almost always originates from pre-existing defects generated during the process. This confirms the positive effect of the fine bilamellar microstructure on the fatigue performance of the alloy, resulting from the synergistic combination of high tensile strength, good ductility, and enhanced resistance to crack propagation.

Although the heat-treated PBF-LB Ti6246 alloy can achieve a higher UTS compared to its wrought counterpart (1327 MPa [23] vs. 1279 MPa [10]), its fatigue performance is significantly lower due to the presence of defects, such as LoF and pores [14,21], which act as preferential sites for crack initiation. In contrast, in conventionally processed alloys, cracks generally initiate within large α grains or α grain clusters, where localized microstructural features promote irreversible slip accumulation [43], rather than from manufacturing-induced defects. Peters et al. [12] reported that the STA-treated alloy with a fine bimodal microstructure exhibited HCF strength at stress levels up to 625 MPa, approximately 37 and 21% higher than the AN875 and STA825 conditions, respectively, due to the different crack initiation mechanisms.

Figure 11 provides a comparative analysis of the fatigue behavior of the Ti6246 alloy produced via PBF-LB in this study, the same alloy manufactured through conventional methods, and the Ti64 alloy fabricated with PBF-LB technology and subjected to different heat treatments. The reported data for the Ti64 alloy refer to tests conducted under a stress ratio of R = −1, using surface-polished specimens, consistent with the testing conditions applied to the Ti6246 alloy in this work. When comparing the PBF-LB STA825 Ti6246 with the PBF-LB Ti64 alloy, the STA825 Ti6246 shows superior fatigue performance regardless of the heat treatment conditions, as highlighted in Figure 10.

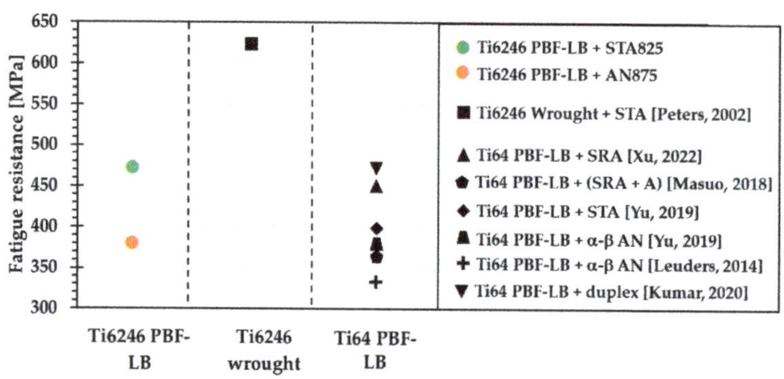

Figure 11. A comparison of the fatigue behavior of the following: (i) the Ti6246 alloy produced via PBF-LB in this study; (ii) the same alloy manufactured through conventional methods (Adapted from [12]); (iii) the Ti64 alloy fabricated with PBF-LB technology and subjected to various heat treatments, including Stress-Relieving Annealing (SRA) (Adapted from [16]), SRA and aging (SRA+A) (Adapted from [47]), α+β annealing (α-β AN) (Adapted from [15,48]), STA (Adapted from [15]), and double step annealing (duplex) (Adapted from [49]).

4. Conclusions

This study focused on assessing the effect of heat treatments and the resulting microstructures on the fatigue resistance of a Ti6246 alloy produced via PBF-LB. Two heat treatment conditions were investigated: α+β annealing at 875 °C for 2 h (AN875) and α+β solution treatment at 825 °C, followed by water quenching and aging at 500 °C for 3 h (STA825). Based on microstructural characterization, fatigue testing, and fractographic analysis, the following conclusions can be drawn:

(i) The AN875-treated samples exhibited the lowest fatigue strength ($\sigma_{d,50\%}$ = 382 ± 15 MPa), primarily due to the low mechanical properties associated with significant microstructural coarsening caused by the high annealing temperature, slow furnace cooling rate, and the absence of strengthening precipitates. The low strength primarily affected crack initiation, while the coarse primary α (α_p) phase facilitated crack propagation.

(ii) The STA825-treated samples demonstrated the highest fatigue strength ($\sigma_{d,50\%}$ = 474 ± 15 MPa), corresponding to an improvement of nearly 25% over the AN875 condition. This enhancement is attributed to the presence of strengthening precipitates, the refinement of α_p lamellae, and the formation of ultrafine, homogeneously distributed secondary α lamellae within the β phase ($\alpha_s + \beta$), which hinder fatigue crack nucleation and early-stage propagation.

(iii) Both heat treatment conditions exhibited a similar $\sigma_{D,50\%}/UTS$ ratio, ranging between 35% and 36%, demonstrating a clear correlation between static and fatigue properties.

(iv) In both heat treatment conditions, fatigue crack nucleation consistently occurred at surface or subsurface lack of fusion defects. These process-induced defects represent the primary limiting factor in the fatigue performance of PBF-BF Ti6246 compared to its conventionally produced counterpart.

(v) Crack propagation behavior was strongly influenced by defect location. Subsurface defects triggered an initial low-speed propagation (LSP) region characterized by a "mackerel pattern", likely due to the high microcrack growth resistance in the early propagation stage at $\alpha_p/(\alpha_s + \beta)$ interfaces. This was followed by a high-speed propagation (HSP) region exhibiting quasi-cleavage fracture. In contrast, surface-initiated cracks propagated directly through quasi-cleavage mechanisms from the crack tip. Fatigue striations were observed in the HSP zone.

(vi) El-Haddad's best-fit parameters derived from the Kitagawa–Takahashi diagrams confirmed the beneficial role of the fine bilamellar microstructure of the STA825 condition in enhancing fatigue performance, with an increase of 19% in the intrinsic fatigue limit ($\Delta\sigma_{th,0}$) and 77% in the threshold stress intensity factor ($\Delta K_{th,lc}$).

(vii) The STA825-treated Ti6246 alloy exhibited superior fatigue strength compared to PBF-LB Ti-6Al-4V subjected to various heat treatments.

This study addresses a critical gap in the current literature concerning the influence of heat treatments on the microstructure and fatigue performance of Ti-6Al-2Sn-4Zr-6Mo (Ti6246) alloy processed via Powder Bed Fusion—Laser Beam (PBF-LB). Future investigations will focus on the role of surface finishing and Hot Isostatic Pressing (HIP) as potential post-processing strategies to mitigate the detrimental effects of process-induced defects and further improve fatigue resistance. Expanding the knowledge base on this alloy and optimizing its post-processing treatments are essential steps toward enabling its broader adoption in high-performance structural applications, such as fan blades, engine shafts, or connecting rods.

Author Contributions: Conceptualization, G.P. and A.M. (Alessandro Morri); software G.P.; methodology, G.P. and A.M. (Alessandro Morri); validation, A.M. (Alessandro Morri) and G.P.; formal analysis, G.P. and A.M. (Alessandro Morri); investigation, G.P.; resources, A.M. (Alessandra Martucci) and M.L.; data curation, G.P.; writing—original draft preparation, G.P.; writing—review and editing, A.M. (Alessandro Morri), L.C., A.M. (Alessandra Martucci), and M.L.; visualization, L.C. and A.M. (Alessandro Morri); supervision, L.C.; project administration, A.M. (Alessandro Morri) and L.C.; funding acquisition, L.C. All authors have read and agreed to the published version of the manuscript.

Funding: This study was financed by the European Union-NextGenerationEU (National Sustainable Mobility Center CN00000023, Italian Ministry of University and Research Decree n. 1033-17/06/2022, Spoke 11-Innovative Materials & Lightweighting). The opinions expressed are those of the authors only and should not be considered as representative of the European Union or the European Commission's official position. Neither the European Union nor the European Commission can be held responsible for them.

Data Availability Statement: The data presented in this study are available on request from the corresponding author. (Due to industrial secrecy, the data are not publicly released. However, they can be shared upon specific request).

Conflicts of Interest: The authors declare no conflicts of interest.

Abbreviations

The following abbreviations are used in this manuscript:

AM	additive manufacturing
PBF-LB	Powder Bed Fusion–Laser Beam
HCF	high cycle fatigue
STA	solution treating and aging
LoF	lack of fusion
KT	Kitagawa–Takahashi
GD-OES	Glow Discharge Optical Emission Spectroscopy
OM	optical microscope
FEG-SEM	Field Emission Gun–Scanning Electron Microscope
LSP	low-speed propagation
HSP	high-speed propagation

References

1. Zhao, Q.; Sun, Q.; Xin, S.; Chen, Y.; Wu, C.; Wang, H.; Xu, J.; Wan, M.; Zeng, W.; Zhao, Y. High-strength titanium alloys for aerospace engineering applications: A review on melting-forging process. *Mater. Sci. Eng. A* **2022**, *845*, 143260. [CrossRef]
2. Nyamekye, P.; Rahimpour Golroudbary, S.; Piili, H.; Luukka, P.; Kraslawski, A. Impact of additive manufacturing on titanium supply chain: Case of titanium alloys in automotive and aerospace industries. *Adv. Ind. Manuf. Eng.* **2023**, *6*, 100112. [CrossRef]
3. Liu, S.; Shin, Y.C. Additive manufacturing of Ti6Al4V alloy: A review. *Mater. Des.* **2019**, *164*, 107552. [CrossRef]
4. Yang, N.; Shan, M.; Shi, G.; Guo, H.; Xiong, S.; Luo, M.; Zhang, D.; Cui, M. Physicochemical State Classification of Heat-treated TC4 Samples Based on Laser-induced Breakdown Spectroscopy (LIBS). *At. Spectrosc.* **2024**, *45*, 324–335. [CrossRef]
5. Li, G.; Chandra, S.; Rahman Rashid, A.; Palanisamy, S.; Ding, S. Machinability of additively manufactured titanium alloys: A comprehensive review. *J. Manuf. Process.* **2022**, *75*, 72–99. [CrossRef]
6. Carrozza, A.; Aversa, A.; Fino, P.; Lombardi, M. A study on the microstructure and mechanical properties of the Ti-6Al-2Sn-4Zr-6Mo alloy produced via Laser Powder Bed Fusion. *J. Alloys Compd.* **2021**, *870*, 159329. [CrossRef]
7. Herzog, D.; Seyda, V.; Wycisk, E.; Emmelmann, C. Additive manufacturing of metals. *Acta Mater.* **2016**, *117*, 371–392. [CrossRef]
8. Rani, S.U.; Kesavan, D.; Kamaraj, M. Evaluation of influence of microstructural features of LPBF Ti-6Al-4 V on mechanical properties for an optimal strength and ductility. *J. Alloys Compd.* **2023**, *960*, 170575. [CrossRef]
9. Yumak, N.; Aslantaş, K. A review on heat treatment efficiency in metastable β titanium alloys: The role of treatment process and parameters. *J. Mater. Res. Technol.* **2020**, *9*, 15360–16280. [CrossRef]
10. Alluaibi, M.H.I.; Cojocaru, E.M.; Rusea, A.; Şerban, N.; Coman, G.; Cojocaru, V.D. Microstructure and mechanical properties evolution during solution and ageing treatment for a hot deformed, above β-transus, Ti-6246 alloy. *Metals* **2020**, *10*, 1114. [CrossRef]
11. Sauer, C.; Lütjering, G. Processing, Microstructure and Properties of Ti-6246. 1999. Available online: https://cdn.ymaws.com/titanium.org/resource/resmgr/ZZ-WTCP1999-VOL1/1999_Vol.1-4-Processing,_Mic.pdf (accessed on 1 April 2024).
12. Peters, J.O.; Lütjering, G.; Nalla, R.K.; Altenberger, I.; Ritchie, R.O. High-Cycle Fatigue of Beta Titanium Alloys. 2002. Available online: https://www2.lbl.gov/ritchie/Library/poster/1_Fatigue2002-HCFofBetaTitaniumAlloys.pdf (accessed on 6 February 2025).
13. Jeong, D.; Kwon, Y.; Goto, M.; Kim, S. High cycle fatigue and fatigue crack propagation behaviors of β-annealed Ti-6Al-4V alloy. *Int. J. Mech. Mater. Eng.* **2017**, *12*, 1. [CrossRef]
14. Liu, Z.; Dash, S.S.; Zhang, J.; Lyu, T.; Lang, L.; Chen, D.; Zou, Y. Fatigue crack growth behavior of an additively manufactured titanium alloy: Effects of spatial and crystallographic orientations of α lamellae. *Int. J. Plast.* **2024**, *172*, 103819. [CrossRef]
15. Yu, H.; Li, F.; Wang, Z.; Zeng, X. Fatigue performances of selective laser melted Ti-6Al-4V alloy: Influence of surface finishing, hot isostatic pressing and heat treatments. *Int. J. Fatigue* **2019**, *120*, 175–183. [CrossRef]
16. Xu, Z.; Liu, A.; Wang, X. Fatigue performance differences between rolled and selective laser melted Ti6Al4V alloys. *Mater. Charact.* **2022**, *189*, 111963. [CrossRef]
17. Sing, S.L.; Huang, S.; Yeong, W.Y. Additive Manufacturing of Titanium and Titanium Alloy Biomedical De-vices. In *Additive Manufacturing in Biomedical Applications*; ASM International: West Conshohocken, PA, USA, 2022; pp. 192–200. [CrossRef]
18. Leuders, S.; Thöne, M.; Riemer, A.; Niendorf, T.; Tröster, T.; Richard, H.; Maier, H. On the mechanical behaviour of titanium alloy TiAl6V4 manufactured by selective laser melting: Fatigue resistance and crack growth performance. *Int. J. Fatigue* **2013**, *48*, 300–307. [CrossRef]

19. Zhang, X.; Xue, K.; Liu, L.; Xiao, J.; Liu, J.; Jin, F.; Luo, M.; Zhan, M.; Li, H. Effect of electrical pulse treatment on the microstructural and mechanical responses of heterogeneous linear friction welded TC17/TC4 dissimilar joint. *J. Mater. Res. Technol.* **2025**, *35*, 1–12. [CrossRef]
20. Li, P.; Warner, D.; Fatemi, A.; Phan, N. Critical assessment of the fatigue performance of additively manufactured Ti-6Al-4V and perspective for future research. *Int. J. Fatigue* **2016**, *85*, 130–143. [CrossRef]
21. Butt, M.M.; Laieghi, H.; Kvvssn, V.; Uddin, Z.; Shah, M.; Ansari, P.; Salamci, M.U.; Patterson, A.E.; Kizil, H. Fatigue performance in additively manufactured metal alloys. *Prog. Addit. Manuf.* **2024**, *10*, 1809–1841. [CrossRef]
22. Wycisk, E.; Solbach, A.; Siddique, S.; Herzog, D.; Walther, F.; Emmelmann, C. Effects of defects in laser additive manufactured Ti-6Al-4V on fatigue properties. *Phys. Procedia* **2014**, *56*, 371–378. [CrossRef]
23. Pirro, G.; Martucci, A.; Morri, A.; Lombardi, M.; Ceschini, L. A novel solution treatment and aging for powder bed fusion-laser beam Ti-6Al-2Sn-4Zr-6Mo alloy: Microstructural and mechanical characterization. *Int. J. Miner. Met. Mater.* **2025**, *32*, 414–424. [CrossRef]
24. Carrozza, A.; Aversa, A.; Fino, P.; Lombardi, M. Towards customized heat treatments and mechanical properties in the LPBF-processed Ti-6Al-2Sn-4Zr-6Mo alloy. *Mater. Des.* **2022**, *215*, 110512. [CrossRef]
25. Cao, F.; Zhang, T.; Ryder, M.A.; Lados, D.A. A Review of the Fatigue Properties of Additively Manufactured Ti-6Al-4V. *JOM* **2018**, *70*, 349–357. [CrossRef]
26. Zanni, M.; Di Egidio, G.; Tonelli, L.; Morri, A.; Ceschini, L. An analysis of the influence of heat treatments and surface finishing conditions on the high cycle fatigue behaviour of W360 hot work tool steel produced by Powder Bed Fusion–Laser Beam. *Mater. Sci. Eng. A* **2025**, *932*, 148237. [CrossRef]
27. Lampman, S. Wrought Titanium and Titanium Alloys. In *Properties and Selection: Nonferrous Alloys and Special-Purpose Materials*; ASM International: West Conshohocken, PA, USA, 1990; pp. 592–633. [CrossRef]
28. *BS ISO 1143:2021*; Rotating Bar Bending Fatigue Testing. European Standards: Bruxelles, Belgium, 2021.
29. Gammon, L.M.; Briggs, R.D.; Packard, J.M.; Batson, K.W.; Boyer, R.; Domby, C.W. Metallography and Microstructures of Titanium and Its Alloys. In *Metallography and Microstructures*; ASM International: West Conshohocken, PA, USA, 2004; pp. 899–917. [CrossRef]
30. Schneider, C.A.; Rasband, W.S.; Eliceiri, K.W. NIH Image to ImageJ: 25 years of image analysis. *Nat. Methods* **2012**, *9*, 671–675. [CrossRef]
31. *ISO 12107:2012*; Metallic Materials-Fatigue Testing-Statistical Planning and Analysis of Data. European Standards: Bruxelles, Belgium, 2012.
32. Murakami, Y.; Masuo, H.; Tanaka, Y.; Nakatani, M. Defect Analysis for Additively Manufactured Materials in Fatigue from the Viewpoint of Quality Control and Statistics of Extremes. *Procedia Struct. Integr.* **2019**, *19*, 113–122. [CrossRef]
33. Murakami, Y.; Beretta, S. Small Defects and Inhomogeneities in Fatigue Strength: Experiments, Models and Statistical Implications. *Extremes* **1999**, *2*, 123–147. [CrossRef]
34. El Haddad, M.H.; Smith, K.N.; Topper, T.H. Fatigue crack propagation of short cracks. *J. Eng. Mater. Technol.* **1979**, *101*, 42–46. [CrossRef]
35. Bu, H.; Chen, L.; Duan, Y. Effect of Solution Heat Treatment on the Porosity Growth of Nickel-Based P/M Superalloys. *Metals* **2022**, *12*, 1973. [CrossRef]
36. Di Egidio, G.; Ceschini, L.; Morri, A.; Martini, C.; Merlin, M. A Novel T6 Rapid Heat Treatment for AlSi10Mg Alloy Produced by Laser-Based Powder Bed Fusion: Comparison with T5 and Conventional T6 Heat Treatments. *Metall. Mater. Trans. B Process Metall. Mater. Process. Sci.* **2022**, *53*, 284–303. [CrossRef]
37. Tammas-Williams, S.; Withers, P.; Todd, I.; Prangnell, P. Porosity regrowth during heat treatment of hot isostatically pressed additively manufactured titanium components. *Scr. Mater.* **2016**, *122*, 72–76. [CrossRef]
38. Galarraga, H.; Warren, R.J.; Lados, D.A.; Dehoff, R.R.; Kirka, M.M.; Nandwana, P. Effects of heat treatments on microstructure and properties of Ti-6Al-4V ELI alloy fabricated by electron beam melting (EBM). *Mater. Sci. Eng. A* **2017**, *685*, 417–428. [CrossRef]
39. Ter Haar, G.M.; Becker, T.H. Selective laser melting produced Ti-6Al-4V: Post-process heat treatments to achieve superior tensile properties. *Materials* **2018**, *11*, 146. [CrossRef]
40. Lutjering, G.; Williams, J.C. *Titanium*, 2nd ed.; Springer: Berlin/Heidelberg, Germany, 2007; Chapter 7; pp. 302–314.
41. Vrancken, B.; Thijs, L.; Kruth, J.-P.; Van Humbeeck, J. Heat treatment of Ti6Al4V produced by Selective Laser Melting: Microstructure and mechanical properties. *J. Alloys Compd.* **2012**, *541*, 177–185. [CrossRef]
42. Yu, J.; Yin, Z.; Huang, Z.; Zhao, S.; Huang, H.; Yu, K.; Zhou, R.; Xiao, H. Effect of Aging Treatment on Microstructural Evolution and Mechanical Properties of the Electron Beam Cold Hearth Melting Ti-6Al-4V Alloy. *Materials* **2022**, *15*, 7122. [CrossRef]
43. Szczepanski, C.; Jha, S.; Larsen, J.; Jones, J. Microstructural influences on very-high-cycle fatigue-crack initiation in Ti-6246. *Metall. Mater. Trans. A Phys. Metall Mater Sci.* **2008**, *39*, 2841–2851. [CrossRef]
44. Gao, T.; Zhao, X.; Xue, H.; Sun, Z. Characteristics and micromechanisms of fish-eye crack initiation of a Ti-6Al-4V alloy in very high cycle fatigue regime. *J. Mater. Res. Technol.* **2022**, *21*, 3140–3153. [CrossRef]

45. Gao, T.; Xue, H.; Sun, Z.; Retraint, D. Investigation of crack initiation mechanism of a precipitation hardened TC11 titanium alloy under very high cycle fatigue loading. *Mater. Sci. Eng. A* **2020**, *776*, 138989. [CrossRef]
46. Li, W.; Zhao, H.; Nehila, A.; Zhang, Z.; Sakai, T. Very high cycle fatigue of TC4 titanium alloy under variable stress ratio: Failure mechanism and life prediction. *Int. J. Fatigue* **2017**, *104*, 342–354. [CrossRef]
47. Masuo, H.; Tanaka, Y.; Morokoshi, S.; Yagura, H.; Uchida, T.; Yamamoto, Y.; Murakami, Y. Influence of defects, surface roughness and HIP on the fatigue strength of Ti-6Al-4V manufactured by additive manufacturing. *Int. J. Fatigue* **2018**, *117*, 163–179. [CrossRef]
48. Leuders, S.; Lieneke, T.; Lammers, S.; Tröster, T.; Niendorf, T. On the fatigue properties of metals manufactured by selective laser melting—The role of ductility. *J. Mater. Res.* **2014**, *29*, 1911–1919. [CrossRef]
49. Kumar, P.; Ramamurty, U. High cycle fatigue in selective laser melted Ti-6Al-4V. *Acta Mater.* **2020**, *194*, 305–320. [CrossRef]

metals

MDPI

Article

Toward an Efficient and Robust Process–Structure Prediction Framework for Filigree L-PBF 316L Stainless Steel Structures

Yu Qiao *,†, Marius Grad † and Aida Nonn †

Computational Mechanics and Materials Lab, Faculty of Mechanical Engineering, OTH Regensburg, Galgenbergstraße 30, 93053 Regensburg, Germany; marius.grad@oth-regensburg.de (M.G.); aida.nonn@oth-regensburg.de (A.N.)

* Correspondence: yu.qiao@oth-regensburg.de

† These authors contributed equally to this work.

Abstract

Additive manufacturing (AM), particularly laser powder bed fusion (L-PBF), provides unmatched design flexibility for creating intricate steel structures with minimal post-processing. However, adopting L-PBF for high-performance applications is difficult due to the challenge of predicting microstructure evolution. This is because the process is sensitive to many parameters and has a complex thermal history. Thin-walled geometries present an added challenge because their dimensions often approach the scale of individual grains. Thus, microstructure becomes a critical factor in the overall integrity of the component. This study focuses on applying cellular automata (CA) modeling to establish robust and efficient process–structure relationships in L-PBF of 316L stainless steel. The CA framework simulates solidification-driven grain evolution and texture development across various processing conditions. Model predictions are evaluated against experimental electron backscatter diffraction (EBSD) data, with additional quantitative comparisons based on texture and morphology metrics. The results demonstrate that CA simulations calibrated with relevant process parameters can effectively reproduce key microstructural features, including grain size distributions, aspect ratios, and texture components, observed in thin-walled L-PBF structures. This work highlights the strengths and limitations of CA-based modeling and supports its role in reliably designing and optimizing complex L-PBF components.

Keywords: L-PBF; texture; microstructure simulation; cellular automata; EBSD validation

1. Introduction

Additive manufacturing (AM) has established itself as a promising, resource-efficient technology that is opening up new possibilities for the production of metal components in various industries. Laser powder bed fusion (L-PBF) is a widely used AM process in which the component is built up layer by layer by locally melting raw material powder with a laser. With virtually unlimited degrees of geometric freedom, the L-PBF process enables the fabrication of complex, customized, filigree components with significantly reduced machining requirements compared to conventional manufacturing (CM) [1–8]. Despite its potential, AM, particularly L-PBF, faces significant obstacles to becoming a reliable and sustainable method for producing high-strength, safety-critical components. A key technological barrier is the complexity of the process, which is characterized by its dependence on a large number of parameters and a complex thermal history during solidification, making it difficult to predict microstructures and mechanical properties. For L-PBF filigree structures, the challenge is compounded by geometric and mechanical property deviations from desired specifications, exacerbated by their thin-walled design. In addition, the dimensions of these structures are often on the scale of grain sizes, placing the microstructure at the forefront of mechanical behavior assessment. Addressing these challenges requires not only a comprehensive understanding but also accurate and efficient predictive tools to assess the effects of process parameters and microstructures.

Investigating how different process parameters affect grain morphology is not only costly but also time-consuming. With increasing computing power, progress has been made in developing predictive physics-based multiscale modeling approaches to link process, microstructure, and properties by deriving so-called process-structure-property (PSP) relationships. Depending on L-PBF process parameters such as laser power, scan speed, and scan strategy, the resulting microstructures are highly complex and can include long, curved columnar grains with varying orientations and combinations of planar and cellular features [1]. There are several multiscale modeling approaches for simulating L-PBF solidification processes and establishing process-structure (PS) relationships: (1) cellular automata (CA) [2,3] integrated with finite element [4–6], finite difference [7,8], finite volume [9–12], or lattice Boltzmann methods [13] to model energy transfer; (2) kinetic Potts-type Monte Carlo (KMC) [14]; and (3) phase-field models (PFM) [15–19]. Each model offers unique perspectives and challenges in capturing the complexities of microstructural changes during AM processes. CA and KMC models operate at the mesoscale, focusing primarily on simulating a large number of grains with attention to their overall dimensions and shape. On the other hand, PFMs operate at the microscale, capable of detailed simulations down to the subgrain level, including aspects such as solute concentrations and the shapes of precipitates and dendrites. The CA approach provides a simplified, coarser view of cellular or dendritic microstructure by considering only the convex hull of a dendrite and ignoring internal segregation within a grain. The advantage of a CA model over a PF or KMC model is that it combines computational simplicity and scalability with physical rigor. The growth kinetics of dendrite tips in CA models are determined by local undercooling, but these models lack detailed information about the internal structure of grains. Due to the steep thermal gradients in AM melt pools that result in a narrow mushy zone, some CA models for additive solidification neglect heterogeneous nucleation to improve computational efficiency [8,20]. In [21], a detailed 3D numerical model was developed to investigate grain growth mechanisms in L-PBF 316L stainless steel using the CA method, with inputs from powder-scale thermo-fluid dynamics, including the temperature field and molten powder bed configuration. KMC models are used to simulate curvature-driven phase boundary growth and can address large-scale problems such as modeling entire devices. Although KMC models provide some insight into grain structures similar to those observed experimentally, they struggle to accurately simulate grain selection and texture formation in additive manufacturing (AM) processes. Nevertheless, KMC models have the potential to effectively model aspects such as grain coarsening in the heat-affected zone (HAZ) during post-solidification heat treatments. The PF model, based on irreversible thermodynamics, analyzes microstructural changes, taking into account the effects of solute concentration and temperature gradients, which are particularly relevant for AM. For certain alloys in AM, complexities like dendritic growth and solute diffusion may be omitted, allowing PFMs to focus more on polycrystalline microstructure evolution, similar to CA and KMC models [22,23]. Recent advances in 3D grain-scale PF modeling have enabled detailed simulations of the AM solidification process, including nucleation, orientation, grain growth kinetics, and grain coarsening [24,25].

Xue et al. [26] recently introduced the Physics-Embedded Graph Network (PEGN), a novel method that represents grain structure by graphs and incorporates classical PF theory into these networks, providing a scalable and time-efficient way to accurately predict microstructure evolution. However, despite the strengths of phase-field modeling in simulating additively manufactured (AM) microstructures, many studies still neglect texture evolution—either by limiting simulations to a small set of crystallographic orientations [27] or by omitting crystallographic orientation entirely. Chen et al. [28] used coupled FEM-PFM simulations to study how inter-track overlap ratio affects texture evolution in L-PBF SS316L. To reduce computational cost, they modeled 45 crystallographic orientations derived from <100>, <101>, and <111>. They identified hatch distance—linked to overlap ratio—as a key factor in controlling texture.

Xie et al. [29] extended a cellular automata model to better capture the microstructure evolution in additive manufacturing of Inconel 625, particularly across samples with varying cross-sectional areas. By incorporating crystallographically preferential nucleation based on local thermal gradients and allowing grain growth in mushy zones via solid-solid interface migration, the model more accurately represented texture development. Their simulations and EBSD measurements revealed that narrow-section samples exhibit nearly equiaxed grains with strong <100>//BD texture, while broad-section samples showed coarser, more heterogeneous microstructures. To enhance the computational efficiency of cellular automata (CA) models in additive manufacturing, Liang et al. [30] introduced a multi-level capture algorithm that prunes redundant grain growth paths through level-by-level pre- and post-pruning, paired with time-resolved growth tracking. This approach allows for larger time steps without sacrificing accuracy in grain competition modeling. In contrast, ExaCA, a CA framework recently updated by Rolchigo et al. [31], incorporates remelting, cumulative multi-pass thermal history, and scan path rotation using a sparse temperature-time format. With GPU acceleration and scalable architecture, ExaCA enables high-throughput, large-scale simulations, though it relies on approximate grain capture logic.

Despite recent advances in computational efficiency and modeling accuracy, robust and predictive simulation workflows for complex use cases—such as thin-walled filigree structures in biomedical and aerospace applications—remain limited. Determining suitable L-PBF parameters and scan strategies to achieve targeted microstructures and properties is still largely reliant on trial and error, with no assurance of achieving long-term performance. While morphological features of microstructures can often be qualitatively reproduced in current simulations, texture prediction is typically neglected or oversimplified, especially for intricate geometries and realistic scanning strategies, and rarely validated against EBSD data. Yet texture plays a critical role in determining mechanical properties and failure modes, particularly when structure dimensions approach grain size.

The questions of which modeling techniques to use, how process parameters influence both morphological and crystallographic features, and what limitations still hinder predictive capability remain open. Addressing these questions is vital for enabling efficient, simulation-driven process optimization for L-PBF of filigree components.

This work aims to address these gaps by developing a robust and time-efficient approach for predicting the microstructure of L-PBF 316L stainless steel manufactured with parameters optimized for thin-walled, load-bearing applications. The specific objectives are (1) to adapt and apply a state-of-the-art cellular automata (CA) modeling framework for L-PBF process simulation; (2) to evaluate the resulting microstructure predictions in terms of both morphology and crystallographic texture; and (3) to quantify the influence of key process variables—specifically laser absorption rate and nucleation density—on texture evolution. Process simulation is carried out using the latest ExaCA framework, driven by temperature fields generated from AdditiveFoam thermal simulations. The predicted textures are validated against experimental electron backscatter diffraction (EBSD) data for 316L stainless steel, enabling a detailed assessment of the model's predictive capability.

2. Methodology

This section details the methodology for predicting grain morphology and texture of additive manufactured 316L through the combined use of ExaCA [32,33] and AdditiveFoam [34]. ExaCA is a cellular automaton developed to efficiently simulate grain growth in additive manufacturing (AM) processes. It is grounded on the pioneering work of Gandin and Rappaz [35] and decouples the CA from the temperature field simulation, significantly speeding up the overall simulation process by reusing temperature profiles. Efficiency is further improved by utilizing Kokkos and MPI, which enable single-node CPU parallel computation, GPU acceleration, and cross-node parallel computation. Meanwhile, AdditiveFoam, a customer computational fluid dynamic (CFD) solver for temperature field simulation, has been adopted. AdditiveFoam is based on OpenFOAM and is specifically tailored for additive manufacturing applications. The two models are loosely coupled via a CSV-like sparse temperature history file whose format is defined by ExaCA. In the following sections, we provide a summary of the theoretical and algorithmic framework of these two models.

2.1. Celluar Automata

2.1.1. Algorithm for Decentered Octahedrons

The Decentered Octahedron algorithm is a computational technique employed in cellular automaton models to simulate grain growth during solidification processes. This algorithm was initially proposed and implemented by Gandin and Rappaz [35]. The advantage of this algorithm lies in its ability to effectively track the envelope of arbitrarily oriented growing grains while minimizing the artificial anisotropy introduced by a regular grid. This is accomplished by utilizing octahedrons to represent and track the envelope of growing grains, with the six half-diagonals corresponding to the preferred <100> growth directions of dendrites. In the context of the Decentered Octahedron algorithm, the cells within the simulation domain are classified into three categories: liquid, solid, and active cells. Active cells separate the liquid and solid cells, with each active cell being assigned an independently tracked octahedron. During the simulation process, the octahedrons of active cells expand at a rate determined by the local temperature and undercooling conditions, which are modeled by the interface reaction function (see the following section). The centers of neighboring liquid cells of an active cell can be impinged upon by the octahedron of the active cell, leading to a capture event. This event is accompanied by the transition of the captured cell state from liquid to active, during which a new octahedron is assigned to the cell. The new octahedron is positioned to overlap the corner of the capturing octahedron closest to the center of the captured cell while maintaining the same orientation as the capturing octahedron. A cell becomes inactive when there are no more neighboring liquid cells. Consequently, the octahedron associated with this inactive cell is no longer tracked. If the temperature of a liquid cell falls to the critical undercooling threshold prior to when it is captured, a nucleation event occurs. The subsequent section elaborates on the specific algorithmic procedure for nucleation.

2.1.2. Interface Reaction Function

At the grain growth front, where the solid phase region grows into the undercooling area, a range of complex multi-physical phenomena occur, typically encompassing solute segregation, interfacial energy, and local non-equilibrium effects. The use of computational models to simulate these phenomena on a large scale is prohibitively expensive, if not practically infeasible. Consequently, the Interfacial Response Function (IRF) is employed. The IRF serves as a mathematical representation and approximation of the interface behavior between the solid and liquid phases during solidification, relating the growth velocity of the grain envelope to the local undercooling.

In ExaCA, the analytical approach of the Kurz, Giovanola, and Trivedi (KGT) model [36] is adopted. The interface response function has the form of

$$V = A(\Delta T)^3 + B(\Delta T)^2 + C(\Delta T) + D$$

where ΔT is the local undercooling within the active cell. $A, B, C,$ and D are fitting parameters; in this work, we use the values of the parameters that come with ExaCA's source code for 316L. The half-diagonals are then the growth at each time step by the amount of $V\Delta t/\Delta x$, where Δt is the time step size and Δx is the cell size.

2.2. Nucleation of Heterogeneous Grains

In laser powder bed fusion (L-PBF) additive manufacturing, the molten metal liquid exhibits inhomogeneity due to the presence of gas bubbles, unmelted or partially melted metal powders, or fragmented dendrite tips from growing grains. This inhomogeneity results in a heterogeneous nucleation. ExaCA has adapted the original nucleation approach of Gandin and Rappaz by coupling the temperature profile with the predetermined knowledge of the location and duration of a cell in the liquid state prior to simulation initialization. In ExaCA, a fraction n_0, representing the nucleation density (i.e., the number of potential nuclei per unit volume), is randomly selected from the set of cells destined to become liquid. Each selected cell is assigned a critical undercooling ΔT_{crit}, sampled from a Gaussian distribution.

$$p(\Delta T_{crit}) = \frac{1}{\Delta T_\sigma \sqrt{2\pi}} e^{-\frac{(\Delta T_{crit} - \Delta T_N)^2}{2\Delta T_\sigma^2}}$$

where ΔT_N is the mean nucleation undercooling and ΔT_σ the standard deviation. If a liquid cell is chosen as a potential nucleus and is not captured by a neighboring grain before reaching its critical undercooling ΔT_{crit}, a nucleation event will occur, resulting in the formation of a new nucleus with a random orientation.

2.3. Temperature Profile with CFD Simulation

The cellular automaton (CA) model simulates the growth of dendritic and cellular structures by measuring local undercooling conditions through an interfacial response function (IRF) that correlates the grain boundary velocity with the thermodynamic driving force. The CFD simulation provides the thermal field that drives the microstructural evolution by solving a set of conservation equations that include the conservation of mass, momentum, and energy. The temperature profile is subsequently then coupled to the CA model through a sparse representation as specified by ExaCA. This section details the formulation of the coupled CFD simulation model.

The conservation of mass ensures that the mass of the fluid is conserved as it moves through space and time.

$$\frac{\partial \rho}{\partial t} + \nabla \cdot (\rho \mathbf{u}) = 0 \quad (1)$$

where ρ denotes the mixture density of solid and liquid phases, t represents time, and $\mathbf{u}$ is the velocity vector field.

The momentum conservation equation incorporates viscous stresses, pressure gradients, and body forces:

$$\rho\frac{\partial \mathbf{u}}{\partial t} + \nabla \cdot \rho\mathbf{u}\mathbf{u} = \nabla \cdot (\mu\nabla\mathbf{u}) - \nabla p + \mathbf{S_u} \tag{2}$$

where p is the pressure field, μ the dynamic viscosity, and $\mathbf{S_u}$ the body force term that includes both buoyancy effects (modeled through the Boussinesq approximation) and mushy zone resistance.

The conservation of energy equation accounts for the change in the thermal energy of the system, considering both heat conduction and other sources of energy.

$$\frac{\partial \rho C_p T}{\partial t} + \mathbf{u} \cdot \nabla\rho C_p T = \nabla \cdot (\kappa\nabla T) + S_T + Q \tag{3}$$

where c_p represents the specific heat capacity, T denotes the temperature field, and κ is the thermal conductivity.

In the case of additive manufacturing, the body force $\mathbf{S_u}$ in the Equation (2) can be formulated as

$$\mathbf{S_u} = \rho_{ref}\left(1 - \beta\left(T - T_{ref}\right)\right)\mathbf{g} - \frac{\mu}{K}\mathbf{u} \tag{4}$$

where ρ_{ref} is the reference mixture density at reference temperature T_{ref}. β is the coefficient of thermal expansion. $\mathbf{g}$ is the gravitational acceleration vector. μ and $\mathbf{u}$ are the viscosity and velocity, respectively. K is the permeability of the mushy zone, given by

$$K = \frac{\lambda^2}{180}\frac{f_l^3}{(1 - f_l)^2} \tag{5}$$

where λ is the dendrite arm spacing, and f_l is the mass fraction of liquid in the solid liquid mixture in the mushy zone.

S_T in Equation (3) accounts for the latent heat duration phase transformation

$$S_T = -L_f\frac{\partial}{\partial t}(\rho f_l) \tag{6}$$

In the above equations, f_l is modeled by

$$f_l = 0.5 \cdot \text{erf}\left(\frac{4(T - T_m)}{(T_l - T_e)}\right) \tag{7}$$

where T_m is the arithmetic mean of liquidus temperature T_l and eutectic temperature T_e.

2.3.1. Heat Source

A volumetric Gaussian heat source with a circular beam in the x-y plane and a finite penetration depth, as proposed by Goldak et al. [37], is used:

$$Q(x, y, z) = \frac{2\eta P}{r^2 d(\pi/3)^{3/2}}\exp\left(\frac{-3(x^2 + y^2)}{r^2} + \frac{-3z^2}{d^2}\right) \tag{8}$$

where x, y, z are the coordinates of a location using the center of the laser beam as the origin. η is a temperature-independent absorption rate of the surface of the powder bed. P is the laser power. r is the beam size in terms of radius. d is the penetration depth.

2.3.2. Boundary Conditions

The thermal boundary conditions at the top surface include conductive and radiative heat transfer.

$$h_{cond} = -\lambda \nabla T \tag{9}$$

$$h_{radi} = \epsilon\sigma\left(T^4 - T_\infty^4\right) \tag{10}$$

where λ is the thermal conductivity, ϵ is the emissivity, T_∞ is the far-field temperature, and $\sigma = 5.67 \times 10^{-8}$ W/m^2/K^4 is the Stefan-Boltzmann constant. For the other sides of the simulation domain, Neumann BC is applied with a value of zero.

The main driving force of fluid flow in the melt pool is the Marangoni effect, which is caused by differences in surface tension due to the large temperature gradient (for an illustration, see Figure 1). The velocity boundary condition is equal to

$$\mu \frac{\partial \mathbf{u}}{\partial \mathbf{n}} = \frac{\partial \gamma}{\partial T} \frac{\partial T}{\partial \boldsymbol{\tau}} \tag{11}$$

where $\frac{\partial \gamma}{\partial T}$ is the surface tension gradient and $\boldsymbol{\tau}$ is the tangential vector. The Marangoni effect plays a critical role in laser-based additive manufacturing due to the steep temperature gradients induced by localized heating. As surface tension decreases with increasing temperature, fluid is driven from the hotter center of the melt pool toward the cooler periphery. This thermocapillary-driven surface flow alters the shape of the melt pool, enhances convective mixing, and changes the direction and rate of heat extraction—all of which influence the resulting grain morphology. Depending on the local solidification conditions, Marangoni convection may either suppress or promote grain refinement [38].

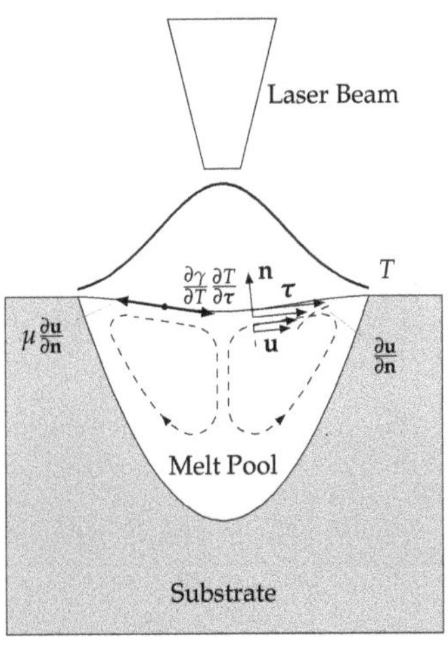

Figure 1. Schematic illustration of the Marangoni effect observed during laser-based additive manufacturing.

2.4. Parameters of Numerical Experiments

The simulation domain and mesh sizes are crucial to computational modeling in additive manufacturing. In this study, both CFD and cellular automata are performed in a size domain of 1 mm × 1 mm × 1 mm. In the CFD simulations, adjustable time step sizes were used, with maximum Courant numbers, maximum interface Courant numbers, and maximum diffusion numbers set to 1, 1, and 10, respectively, which is consistent with the default settings in AdditiveFoam. The mesh size for the CFD domain was set to 5 μm. For the CA simulations, a fixed time step of 0.1 μs was applied, which is one-third of the 0.3 μs used in [32]. The CA mesh size was set to 2 μm, slightly coarser than the 1.667 μm used in the same study, to reduce computational cost while maintaining sufficient accuracy. Tables 1 and 2 contain the parameters for both simulations. To lower the computation cost, fluid flow is turned off in AdditiveFoam. The simulations used 448 Intel(R) Xeon(R) Platinum 8480+ (Sapphire Rapids) CPUs and were run on the Linux Cluster at Leibniz–Rechenzentrum (https://www.lrz.de/ (accessed on 24 May 2025)).

Table 1. Thermophysical properties of SS316L, values without citation: from [39]. *: as is recommended by AdditiveFoam document.

Properties	SS316L	Units
Reference Density, ρ_{pref} [40]	7569.92	$kg\,m^{-1}$
Thermal conductivity of solid metal, κ_s	$9.248 + 0.01571T$	$kg\,m\,s^{-3}\,K^{-1}$
Thermal conductivity of liquid metal, κ_l	$12.41 + 0.003279T$	$kg\,m\,s^{-3}\,K^{-1}$
Thermal conductivity of powder metal, κ_p [41]	0.07467	$kg\,m\,s^{-3}\,K^{-1}$
Specific heat of solid metal, C_{ps}	$462 + 0.134T$	$m^2\,s^{-2}\,K^{-1}$
Specific heat of liquid metal, C_{pl}	775	$m^2\,s^{-2}\,K^{-1}$
Specific heat of powder metal, C_{pp}	$462 + 0.134T$	$m^2\,s^{-2}\,K^{-1}$
Solidus temperature, T_s	1648	K
Liquidus temperature, T_l	1673	K
Surface tension, σ	1.6	$kg\,s^{-2}$
Dendrite arm spacing, λ *	10×10^{-6}	m
Latent heat of fusion, L_f [42]	2.6×10^5	$m^2\,s^{-2}$
Viscosity of liquid metal, μ	0.00345	$kg\,m^{-1}\,s^{-1}$
Temperature of surface tension, $\frac{d\sigma}{dT}$	0.8×10^{-3}	$kg\,m^{-2}\,K^{-1}$
Emmisivity, ϵ [43]	0.28	-
Convective heat transfer coefcient, h [43]	10	$W\,m^{-2}\,K^{-1}$
Absorption Rate, η	$[0.3, 0.35, 0.4, 0.45]$	-
SuperGassian k	2	-
SuperGassian dim	$(0.04, 0.04, 0.03) \times 10^{-3}$	m

Table 2. CA related simulation parameters.

Parameter	Value	Units
Nucleation Density, n_0 [4]	$[300, 600, 1200] \times 10^{12}$	m^{-3}
Nucleation Undercooling (mean)	5	K
Nucleation Undercooling (std)	0.5	K
Substrate grain mean size	30×10^{-6}	m
Substrate powder density	100×10^{12}	m^{-3}
Interfacial function coefficient		
A	0.000007325	-
B	3.12	-
C	0	-

3. Experimental Details

3.1. Specimen Fabrication

Three rectangular blocks, each with six equally sized compartments arranged in two columns and three rows, were fabricated using an AMCM M290 FDR L-PBF system (AMCM GmbH, Starnberg, Germany) under an argon atmosphere. Oxygen levels were maintained below 0.1% to minimize oxidation. The process parameters used during fabrication are summarized in Table 3 and were selected based on prior experimental studies [44], which optimized L-PBF settings for high-resolution 316L stainless steel structures. A standard meander scanning strategy with a 67° layer rotation was applied to reduce texture anisotropy. The platform temperature was controlled at 353 K. The blocks were printed with three wall thicknesses: 0.5 mm, 0.75 mm, and 1.0 mm. The parallel-sheet geometry facilitates the precise extraction of thin metal specimens for microstructural characterization using wire electrical discharge machining (EDM). A schematic of the scan strategy is provided in Figure 2.

For the EBSD analysis in this study, the 1.0 mm samples were selected for two primary reasons. First, the current process-structure simulations are the first step in a process-structure-property-performance (PSPP) modeling framework that aims to predict the mechanical behavior of 316L L-PBF aortic stents, which have a nominal wall thickness of 1 mm. Second, Wiesent et al. [44] demonstrated that 1 mm-thick samples more accurately reflect bulk-like thermal gradients during solidification. Thinner geometries tend to exaggerate edge effects. Future studies are planned to include thinner geometries to capture size effects both experimentally and computationally.

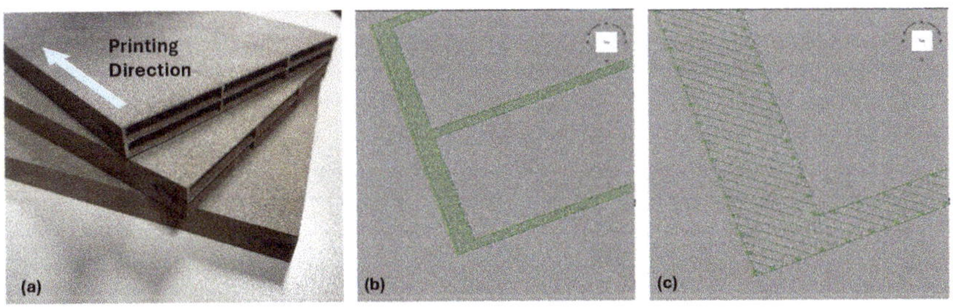

Figure 2. (**a**) Printed rectangular blocks with a height of 155 mm, width of 124 mm, and wall thicknesses of 0.5 mm, 0.75 mm, and 1.0 mm; (**b**) schematic illustration of the scanning strategy; (**c**) enlarged view showing the laser scan path.

Table 3. Process parameters of experiment.

Process Parameters	Values	Units
Layer Thickness	0.02	mm
Laser Power	195	W
Beam Size	0.04	mm
Scanning Speed	1083	mm/s
Chamber Temperature	353	K
Layer Rotation	67	deg

3.2. *Electron Backscatter Diffraction Analysis*

Samples for electron backscatter diffraction (EBSD) analysis were prepared using a conventional metallographic technique involving sequential grinding and polishing steps, followed by 5 min polishing with SiO_2 nanoparticles (OPS). Characterization was conducted on planes normal and parallel to the build direction (BD). Two experimental EBSD datasets, referred to as Experimental 1 and Experimental 2, were obtained from two adjacent 1 mm $\times$ 1 mm $\times$ 1 mm specimens extracted from the same 1.0 mm-thick wall block, as described in Section 3.1. The use of adjacent volumes enables the assessment of local microstructural variability under identical processing conditions. Experimental 1 was acquired using a TESCAN Mira field emission scanning electron microscope (SEM). Grain orientation and texture were determined via the EBSD technique using a Velocity Pro (EDAX AMETEK) detector. The SEM operated at U = 20 kV and I = 1 nA current during EBSD analysis, with a sample tilt angle of 70° and a working distance of WD = 15 mm. The detection configuration set to 1 $\times$ 1 binning with a region of interest (ROI) of 500 $\times$ 500 μm^2 was set. Experimental 2 was obtained using a Zeiss EVO MA15 SEM (Carl Zeiss Microscopy

GmbH, Jena, Germany) coupled with an EDAX EBSD system. The parameters were set as follows: U = 20 kV, I = 4 nA, and WD ≈ 20 mm. The raw EBSD data was processed using commercially available ATEX software (Version 4.13) [45]. Grain boundaries were classified into low angle (2–5°), medium angle (5–15°), and high angle (15–63.5°) based on their misorientation range. For grain size distribution and grain aspect ratio analysis, a grain tolerance angle of 5° was set, and grains with a size of five pixels were excluded.

4. Results

4.1. Comparison of Experimental and Simulated 3D Crystallographic Textures

As described in Section 3.2, two experimental datasets (Experimental 1 and 2) were obtained from two adjacent 1 mm × 1 mm × 1 mm cubes extracted from the same 1.0 mm-thick wall block. This setup enables a direct comparison of local microstructural variation under identical L-PBF processing conditions. A total of 12 simulations were performed, systematically varying laser absorption rate η and nucleation density n_0. Figure 3 presents a three-dimensional crystallographic textures derived from electron backscatter diffraction (EBSD) and compares the crystallographic textures of Experimental 1 and 2 with two simulation cases (Simulation 9: η = 0.40; Simulation 6: η = 0.35), all evaluated at a constant nucleation density n_0 of 1200×10^{12} m^{-3}. For each simulation, microstructural samples were extracted from the center region of the computational domain to avoid boundary effects. The images show representative volumes of 500 × 500 × 500 μm^3, highlighting the three principal sample planes—top (XY), front (XZ), and side (YZ)—of L-PBF-fabricated 316L stainless steel specimens. All textures are displayed using an inverse pole figure (IPF) color scheme with respect to the build direction (BD), where red, green, and blue correspond to <001>, <101>, and <111> orientations, respectively.

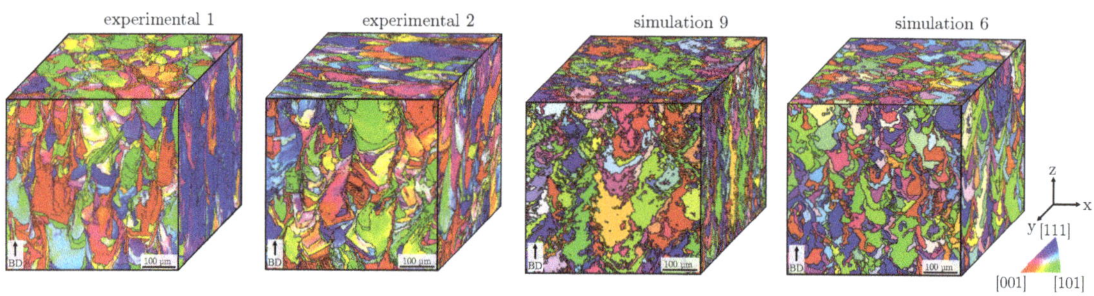

Figure 3. 3D crystallographic texture of two L-PBF-fabricated specimens compared to two simulated crystallographic textures.

The experimental textures reveal characteristic features of L-PBF grain structures, including elongated grains aligned with the build direction and strong crystallographic orientation preferences. Experimental Dataset 1 shows more pronounced texture alignment and fewer misorientations, whereas Experimental Dataset 2 displays slightly greater variability and evidence of grain competition, likely due to localized differences in thermal history during fabrication. Additionally, subtle variations in EBSD sample preparation or surface quality may influence the detectability of fine grains and grain boundaries, contributing to the observed differences. These findings underscore the inherent stochastic nature of the L-PBF process and highlight the value of analyzing multiple regions to robustly validate simulation predictions against experimental microstructures.

These trends are qualitatively reflected in the simulated textures. Simulation 9, modeled with a higher laser absorption rate (η = 0.40), results in stronger texture alignment along BD with visibly columnar grains and distinct orientation bands. In contrast, Simulation 6 (η = 0.35) produces a more equiaxed and diffuse texture, characterized by increased orientation scatter and less pronounced directionality. Although some differences in grain morphology and sharpness remain, the simulations effectively capture the influence of process parameters—especially absorption rate—on texture evolution. This alignment between simulated and measured textures reinforces the predictive capability of the model for capturing grain orientation development in L-PBF-processed parts.

Figure 4 presents comparative visualization of inverse pole figures (IPFs) obtained from experiments and simulations. Each row corresponds to a specific sample surface orientation (XZ, YZ, and XY planes), with the IPFs showing the distribution of crystallographic orientations relative to the X and Z (build) directions using standard stereographic triangles. Texture intensity is expressed in multiples of a random distribution (MRD), with higher values indicating stronger alignment.

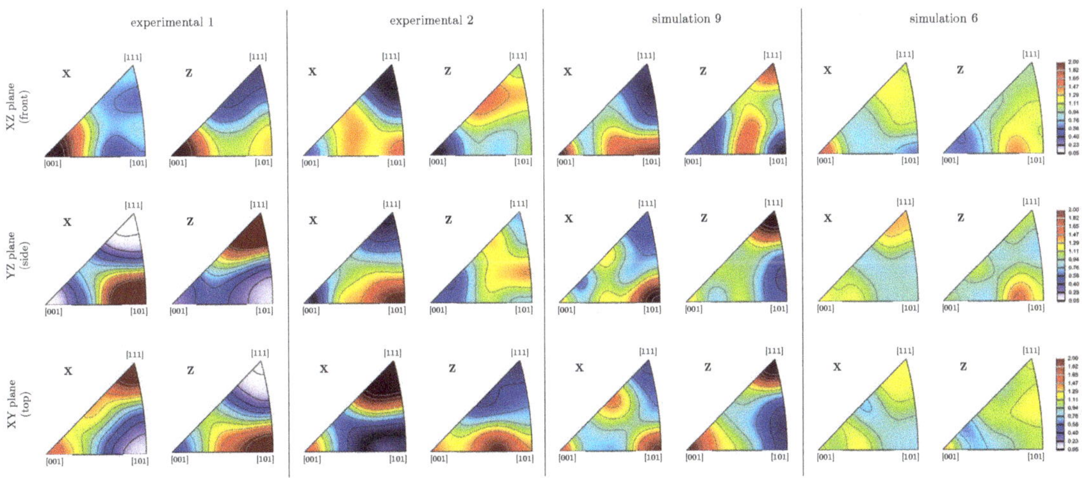

Figure 4. IPF's measured using electron backscatter diffraction. Comparison between two experimental datasets and two simulated data sets.

In the XZ-plane (front face), Experimental Dataset 1 exhibits a pronounced <001> fiber texture along both the X and Z directions, with a secondary peak near <101> in Z. This dual-axis alignment indicates strong directional solidification along the laser scan and build axes, driven by steep thermal gradients. In contrast, Experimental Dataset 2 shows a shift in the X-direction texture toward <101> and a broader distribution in Z, centered closer to <111>, reflecting deviations in heat flow and increased grain competition.

Simulation 9 closely matches Experimental 2, showing dominant <101> alignment in X and <111> in Z, suggesting realistic grain competition and orientation selection under moderate thermal gradients. Simulation 6, modeled with reduced absorption, results in diffuse and less intense textures. While a weak <001> component is still visible in X, the Z texture becomes broader and less defined, centered closer to <101>, consistent with diminished directional solidification.

In the YZ-plane (side face), Experimental Dataset 1 reveals a strong <101> orientation in X and a sharp <111> texture in Z, confirming highly directional solidification. Experimental Dataset 2 follows a similar trend, but with broader and less intense distributions, indicating increased variability in local solidification conditions.

Simulation 9 reproduces these trends well, capturing <101> alignment in X and a strong <111> fiber in Z. Simulation 6, again, exhibits broader and weaker textures. A modest <001> component appears in X, while the Z orientation is more spread out between <101> and <111>, reflecting reduced grain selection due to lower thermal gradients.

In the XY-plane (top surface), both experimental datasets exhibit a dominant <111> texture along the X direction, accompanied by secondary intensity near <001>. shows a pronounced <101> component, whereas Experimental 2 shifts toward a more prominent <001> orientation. These differences likely reflect local variations in thermal gradients and melt pool dynamics, which could be due to effects of the scan strategy, hatch spacing, or subtle differences in melt pool overlap across the analyzed regions.

Simulation 9 predicts a primary <101> orientation in the X direction, accompanied by secondary intensity near <001> and a broad spread between <001> and <111>. In the Z direction, it captures a dual-peak texture, with a dominant <111> component and a weaker <001> contribution. In contrast, Simulation 6 exhibits overall weaker and more diffuse textures. A <111> peak is present in the X direction, along with a secondary maximum spanning between <001> and <101>. The Z direction reveals a modest $\langle 001 \rangle$ component and a broader distribution between <101> and <111>.

Overall, both simulations qualitatively replicate the main trends observed experimentally across all planes, particularly the influence of laser absorption rate on texture sharpness and directional alignment. Simulation 9 shows better agreement with the experimental data, especially in reproducing the <111> texture in the Z direction and capturing <101> alignment in the X direction under competitive grain growth conditions. Simulation 6, while reflecting general orientation tendencies, exhibits weaker and more diffuse textures due to reduced thermal gradients and less directional solidification. These findings highlight the sensitivity of crystallographic texture to processing parameters and underscore the importance of accurate thermal field modeling and orientation selection mechanisms in predictive microstructure simulations.

4.2. Normalized Texture Index

We used the Normalized Texture Index to quantitatively evaluate the similarity between the grain textures obtained from simulations and those obtained from experiments. The Normalized Texture Index (NTI) is defined as

$$TI_{norm} = \frac{\int_{SO(3)} \left[f_{ref}(g) - f(g)\right]^2 dg}{\int_{SO(3)} \left[f_{ref}(g)\right]^2 dg}$$

Here, f_{ref} represents the orientation distribution function (ODF) of the grain textures obtained from experimental data, and f corresponds to those obtained from simulation results. A smaller value of the Normalized Texture Index (NTI) indicates closer alignment between the simulated grain texture and the experimentally measured one. To evaluate overall texture agreement, the NTI is averaged across the three principal planes—XY, XZ, and YZ—for each combination of absorption rate η and nucleation density n_0.

$$TI_{norm,avg}^{\eta,n_0} = \frac{1}{3} \sum_{s \in \{XY,XZ,YZ\}} TI_{norm}^{s,\eta,n_0}$$

The parameter $s \in \{XY, XZ, YZ\}$ specifies the plane of analysis. The parameter ranges are defined as $\eta \in \{0.30, 0.35, 0.40, 0.45\}$ for the absorption rate and $n_0 \in \{300, 600, 1200\}$ (in units of $\times 10^{12}$ m^{-3}) for the nucleation density. The term $TI_{norm}^{s,\eta,n}$ is the Normalized Texture Index (NTI) as defined above and is calculated by comparing the simulated orientation distribution function (ODF) for the specified parameters with the corresponding experimental ODF on the same plane s. The resulting scores are presented in Table 4.

Table 4. Macro average of Normalized Texture Index (NTI) and Earth Mover's Distance (EMD) of the chord length distribution (CLD) for each combination of absorption rate η and nucleation density n_0. Lower values of NTI or EMD indicate closer alignment between the simulated and experimental grain texture or morphology, respectively. To mitigate the effect of skewness in the CLD during EMD computation, the fourth root of the chord length is used, and the resulting distribution is denoted as $\sqrt[4]{CLD}$. We use $EMD(\cdot)$ to denote the Earth Mover's Distance between the corresponding distributions. Highlighted values indicate the best match in each column. The simulation with $\eta = 0.35$, $n_0 = 1200$ yields the closest texture to both Exp 1 and Exp 2. Morphologically, the simulation with $\eta = 0.45$, $n_0 = 1200$ is closest to Exp 1, while $\eta = 0.45$, $n_0 = 300$ shows the closest match to Exp 2. An overview of the parameter combinations is provided in Table 5.

sim. Nr.	η	n_0	$TI^{\eta,n_0}_{norm,avg}$		$EMD(\sqrt[4]{CLD})$	
			Exp 1	Exp 2	Exp 1	Exp 2
6	0.35	1200	**0.437**	**0.545**	0.138	0.087
5	0.35	600	0.450	0.595	0.118	0.052
3	0.30	1200	0.457	0.547	0.127	0.081
2	0.30	600	0.501	0.570	0.501	0.057
4	0.35	300	0.508	0.639	0.110	0.041
1	0.30	300	0.521	0.615	0.116	0.047
8	0.40	600	0.533	0.632	0.115	0.048
9	0.40	1200	0.553	0.637	0.137	0.079
7	0.40	300	0.563	0.664	0.120	0.039
12	0.45	1200	0.594	0.740	**0.101**	0.030
11	0.45	600	0.696	0.853	0.103	0.031
10	0.45	300	0.702	0.854	0.107	**0.027**

The NTI values show that Simulation 6 most closely matches the experimental orientation distribution functions overall. However, Simulation 9 more accurately captures the dominant texture features observed in the IPFs, such as the <111> orientation in the Z direction and the <101> orientation in the X direction. These distinctions highlight the value of combining NTI-based analysis with visual texture evaluation to improve understanding of model accuracy.

4.3. Quantitative Grain Morphology Analysis: Comparison Between Experiments and Simulations

Figure 5 shows the grain size distributions for three orthogonal orientations—YZ, XZ, and XY planes—comparing experimental results (top row) with simulated data (bottom row). In each subplot, Experimental 1 is shown in grey, and Experimental 2 in red; in the simulations, Simulation 9 is grey, and Simulation 6 is red. The histograms display the probability density of grain sizes within the range of 0 to 100 μm.

All distributions are right-skewed, indicating a predominance of fine grains and a decreasing frequency of larger grains. In the experimental data, a higher fraction of very fine grains (<10 μm) is observed across all planes, particularly in the XY-plane (top surface), which may reflect rapid cooling and nucleation at the melt pool boundary. Experimental 2 consistently shows a broader distribution and more frequent occurrence of larger grains compared to Experimental 1, suggesting more pronounced grain growth, potentially due to localized heat accumulation or reduced thermal gradients.

In the simulated datasets, Simulation 9 (higher absorption rate) produces grain size distributions that more closely match the experimental curves, particularly in the XZ and YZ planes. Simulation 6, with a lower absorption rate, exhibits slightly broader and more right-shifted distributions, reflecting weaker thermal gradients and reduced directional solidification. Across all planes, the simulations slightly underpredict the fraction of the smallest grains (<10 μm), which may be attributed to limitations in the nucleation model or mesh resolution effects.

Overall, the simulations reproduce the general statistical characteristics of grain size evolution in L-PBF, capturing the dominant trend of fine-grained structures with trailing populations of larger grains.

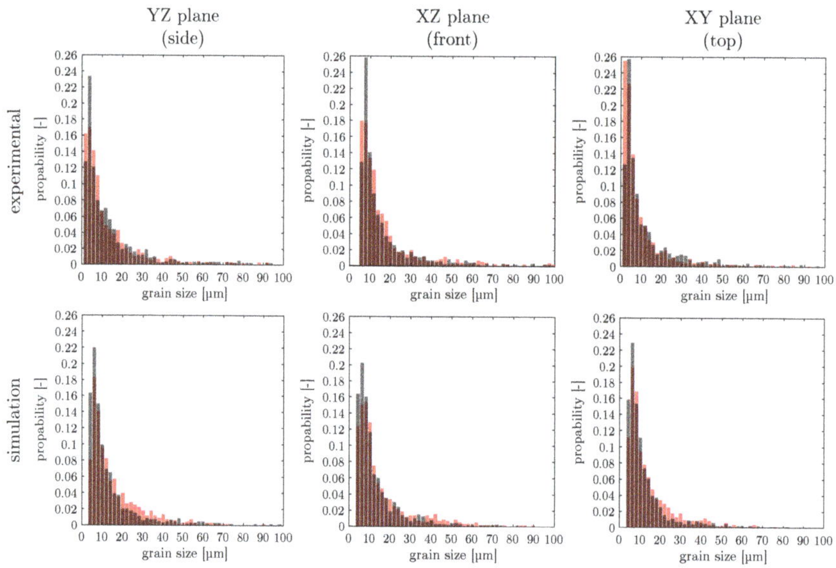

Figure 5. Grain size distribution for the different orientations (XZ, YZ, and XY plane). Comparison between experimental data and simulated data for the two different datasets (grey: Experimental 1 and simulation 9; red: Experimental 2 and simulation 6).

Figure 6 presents the grain aspect ratio distributions as observed from three orthogonal orientations—YZ, XZ, and XY planes—comparing experimental datasets (top row) with simulation results (bottom row). Aspect ratio serves as an indicator of grain elongation and morphological anisotropy, where values closer to 1 represent more equiaxed grains and higher values indicate increasingly elongated structures.

In the experimental datasets, aspect ratios exhibit broad distributions, particularly in the YZ and XZ planes, where values extend up to 9–10. This indicates a high degree of grain elongation along the build direction and suggests significant morphological variability within the fabricated parts. The XY plane, representing the top surface, shows narrower distributions and a higher frequency of low aspect ratios, which is consistent with the expected transition to more equiaxed grains near the top of the melt pool due to reduced thermal gradients.

The simulated aspect ratio distributions are more compact than those observed experimentally, with most grains exhibiting values below 4 across all planes. Peak frequencies in both experimental and simulated datasets occur within the 1.5–2.5 range, indicating that the model accurately captures the dominant grain shape characteristics. However, the simulations slightly underrepresent grains with higher aspect ratios (>5), particularly in the YZ and XZ planes, where more elongated grains are experimentally observed. This modest discrepancy likely arises from simplifications in the simulation framework, such as the treatment of thermal gradients, melt pool dynamics, or curvature-driven growth mechanisms.

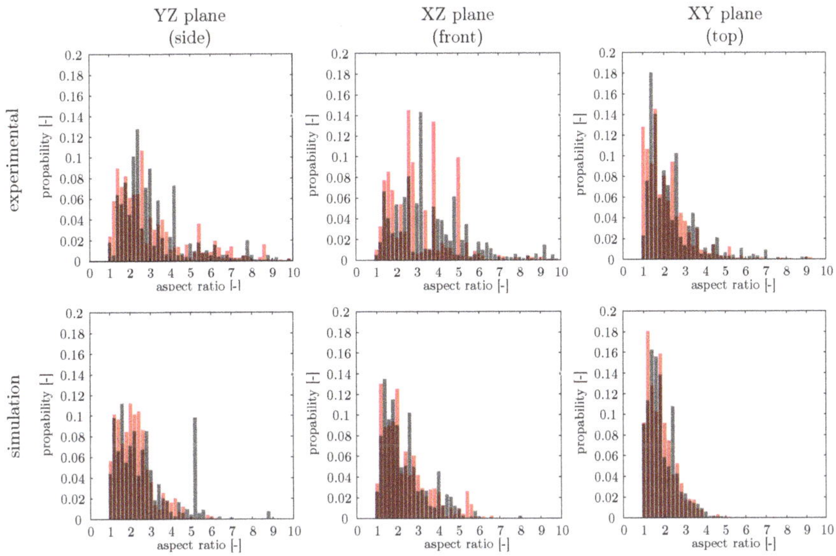

Figure 6. Aspect ratio distribution for the different orientations (XZ, YZ, and XY planes). Comparison between experimental data and simulated data for the two different datasets (grey: experimental Dataset 1 and simulation 9; red: experimental Dataset 2 and simulation 6).

A comparison between simulation cases reveals that Simulation 9, with a higher laser absorption rate ($\eta = 0.40$), produces a slightly broader aspect ratio distribution than Simulation 6 ($\eta = 0.35$), suggesting enhanced directional grain growth under higher energy input. Despite the slight underestimation of extreme elongation, the simulations reproduce the overall trends well, supporting the model's capacity to reflect key morphological features observed in L-PBF-fabricated materials. These results highlight the effectiveness of the current modeling approach while also pointing to opportunities for further refinement to better capture grain growth anisotropy.

Table 5 provides a statistical summary of grain size and aspect ratio distributions for both simulated and experimental datasets across the three orthogonal planes (XY, XZ, and YZ). The simulated results are reported for various laser absorption rates (η = 0.30–0.45) and nucleation densities (n_0 = 300–1200 $\times 10^{12}$ m^{-3}), while the experimental data correspond to Experimental 1 and 2.

In the simulations, the minimum grain size consistently remains around 6 µm, reflecting the mesh resolution limit and post-processing threshold. The mean grain size (μ) exhibits clear dependence on both absorption rate η and nucleation density n_0. For example, in the XY plane at $\eta = 0.35$, increasing η from 300 to 1200 $\times 10^{12}$ m^{-3} leads to a reduction in μ from 16.91 µm to 15.01 µm, consistent with expectations that higher nucleation promotes finer grains.

However, the influence of absorption rate on grain size is not strictly monotonic. These nonlinear variations suggest a complex interplay between thermal gradients, melt pool dynamics, and grain competition during solidification. Nevertheless, higher absorption rates tend to be associated with coarser grains, particularly in the XZ and YZ planes, where melt pool depth and thermal exposure are more influential. The largest mean grain size among the simulated data occurs at $\eta = 0.45$ and $n_0 = 300 \times 10^{12}$ m^{-3} in the YZ plane (μ = 20.18 µm), while finer microstructures are observed at lower energy inputs and higher nucleation densities.

Table 5. Statistics about grain morphology of XY, XZ, and YZ planes at center of the simulation domain under absorption rate $\eta = \{0.30, 0.35, 0.40, 0.45\}$ and nucleation density at $n_0 = \{300, 600, 1200\} \times 10^{12}\ \text{m}^{-3}$. The highlighted row represents the simulation setups in which we conducted in-depth texture analysis.

plane	Sim. Nr.	η	n_0 $\times 10^{12}$ m^{-3}	Grain Size (μm)				Aspect Ratio	
				Min	Max	μ	σ	Mean	Std
XY	1	0.30	300	6.04	92.28	17.82	13.82	0.54	0.17
	2	0.30	600	6.01	73.77	17.57	12.72	0.54	0.17
	3	0.30	1200	6.01	83.61	15.83	11.59	0.55	0.17
	4	0.35	300	6.01	85.30	16.91	13.87	0.55	0.17
	5	0.35	600	6.00	97.93	16.48	12.55	0.55	0.16
	6	**0.35**	**1200**	**6.01**	**82.43**	**15.01**	**10.79**	**0.57**	**0.17**
	7	0.40	300	6.00	106.73	17.47	15.66	0.57	0.17
	8	0.40	600	6.02	84.14	16.94	13.68	0.56	0.17
	9	**0.40**	**1200**	**6.01**	**134.02**	**15.00**	**12.20**	**0.56**	**0.17**
	10	0.45	300	6.02	113.45	17.53	16.29	0.55	0.16
	11	0.45	600	6.01	134.31	17.16	16.06	0.57	0.18
	12	0.45	1200	6.00	112.45	16.15	14.24	0.57	0.16
	-	Exp 1	Exp 1	6.01	97.63	16.41	13.75	0.44	0.18
	-	Exp 2	Exp 2	6.04	134.14	19.84	17.90	0.40	0.18
XZ	1	0.30	300	6.00	96.89	19.43	16.07	0.51	0.19
	2	0.30	600	6.00	93.14	18.70	14.78	0.51	0.19
	3	0.30	1200	6.01	75.05	17.15	13.07	0.49	0.18
	4	0.35	300	6.04	120.59	18.99	18.12	0.50	0.17
	5	0.35	600	6.04	95.46	19.00	15.53	0.48	0.18
	6	**0.35**	**1200**	**6.01**	**97.16**	**17.00**	**13.81**	**0.49**	**0.17**
	7	0.40	300	6.03	142.12	19.05	18.60	0.48	0.18
	8	0.40	600	6.01	118.26	18.00	17.61	0.49	0.18
	9	**0.40**	**1200**	**6.01**	**141.09**	**17.38**	**15.51**	**0.48**	**0.18**
	10	0.45	300	6.00	231.44	19.66	24.13	0.48	0.18
	11	0.45	600	6.00	193.78	18.98	22.09	0.47	0.18
	12	0.45	1200	6.02	168.51	17.88	18.62	0.46	0.18
	-	Exp 1	Exp 1	6.01	138.62	18.97	17.26	0.36	0.18
	-	Exp 2	Exp 2	6.08	127.06	20.20	17.92	0.37	0.18
YZ	1	0.30	300	6.01	97.33	19.33	15.25	0.51	0.18
	2	0.30	600	6.00	85.05	18.24	13.68	0.51	0.18
	3	0.30	1200	6.04	87.91	16.82	12.36	0.51	0.17
	4	0.35	300	6.00	132.78	19.20	16.58	0.49	0.18
	5	0.35	600	6.01	105.00	18.58	15.03	0.50	0.17
	6	**0.35**	**1200**	**6.01**	**88.95**	**17.03**	**12.23**	**0.49**	**0.17**
	7	0.40	300	6.03	95.45	19.55	17.33	0.49	0.18
	8	0.40	600	6.00	115.42	18.18	15.71	0.49	0.18
	9	**0.40**	**1200**	**6.02**	**129.86**	**16.80**	**13.75**	**0.49**	**0.18**
	10	0.45	300	6.00	159.16	20.18	18.95	0.47	0.18
	11	0.45	600	6.01	178.48	19.60	21.16	0.48	0.17
	12	0.45	1200	6.04	137.21	18.19	17.88	0.48	0.18
	-	Exp 1	Exp 1	6.01	197.78	19.61	20.96	0.36	0.19
	-	Exp 2	Exp 2	6.04	160.62	18.92	17.74	0.39	0.18

In terms of aspect ratio, the simulated values range from 0.46 to 0.57, with standard deviations between 0.16 and 0.19, indicating more equiaxed grain shapes under the definition used here (aspect ratio = width/length). In contrast, the experimental aspect ratios are lower, between 0.36 and 0.44, suggesting more elongated, columnar grains, particularly along the build direction. These findings are consistent with the 3D EBSD reconstructions, which show vertically extended grains in the experimental datasets, especially in Experimental 2. The discrepancy may stem from differences in local melt pool conditions, remelting, or epitaxial growth behavior in the physical process that are not fully represented in the current CA simulation framework. Additionally, the simulation may underestimate grain elongation due to simplified assumptions about thermal gradients and grain boundary mobility.

Overall, the statistical analysis confirms that both nucleation density and absorption rate are key parameters in controlling grain size and morphology. The simulation results align well with experimental trends in terms of size distributions and anisotropy, supporting the model's validity for predicting grain-level features in L-PBF-processed components.

4.4. Earth Mover's Distance of Chord Length Distribution

The morphological similarities between the grain simulation results and the experimental observations are quantified using the Earth Mover's Distance (EMD) of the Chord Length Distribution (CLD).

Quantifying morphology based on microstructure metrics such as the equivalent diameter or aspect ratio overlooks important mesostructure details [46]. The chord length, defined as the length of line segments extending throughout the grain from one grain boundary to the next, has the ability to address this limitation, as does its distribution in a sample. Following [46], we compute the angularly resolved CLDs by rotating the sample by an angle θ before calculating the CLDs using n scan lines along a fixed global x-axis. In this work, $\theta \in \{0°, -30°, -60°, -90°\}$ and $n = 100$ are used.

We quantitatively compare the CLDs of the simulation results with the experimental observations by employing the EMD. EMD can be mathematically formulated as

$$EMD(P,Q) = \inf_{\gamma \in \Pi(P,Q)} \mathbb{E}_{(x,y)\sim\gamma}(d(x,y))$$

where P and Q are two probability distributions whose similarity is being investigated. $\Pi(P,Q)$ is the set of all joint probability distributions with marginal distributions of P and Q. The distance metric is denoted by $d(x,y)$. In this work, the Euclidean distance is used.

To compute an overall similarity using EMD for each set of simulation results or experimental data, a CLD is calculated based on statistics of chord length from all θ and sides. To reduce the sensitivity of EMD calculations to the significant skewness of the chord length distributions (CLDs), we compute EMDs using the fourth root of the chord length, a monotonic increasing transformation. This approach reduces the need for a high-resolution binning grid when comparing CLDs. The resulting EMD scores are presented in Table 4.

The lowest EMD values in Table 4 indicate that the simulation with $\eta = 0.45$, $n_0 = 1200$ most closely matches the chord length distribution of Experimental Dataset 1, while $\eta = 0.45$, $n_0 = 300$ best aligns with Experimental 2.

5. Discussion

The analysis of crystallographic texture across the three principal planes reveals that the cellular automata (CA) model is capable of qualitatively reproducing key directional trends observed in the experimental data. In particular, Simulation 9 captures dominant features such as the <111> peak in the build direction (Z) and <101> alignment in the scan direction (X), especially in the second experimental dataset. These textures reflect directional solidification driven by thermal gradients and are characteristic of grain growth along preferential crystallographic directions under strong anisotropic conditions in L-PBF. Conversely, Simulation 6 yields more diffuse textures, with broader orientation distributions and reduced intensity, indicating a less pronounced directional growth regime. These observations suggest that the choice of process parameters—especially laser absorption and nucleation density—directly influences the development and sharpness of texture, which is critical for mechanical performance in load-bearing components.

To complement visual inspection of inverse pole figures (IPFs), the normalized texture index (NTI) was employed to quantify the similarity between simulated and experimental orientation distribution functions (ODFs). Interestingly, while Simulation 6 produced the lowest NTI scores (0.437 for Experiment 1 and 0.545 for Experiment 2), visual analysis shows that Simulation 9 more accurately reflects the dominant texture components. This apparent contradiction can be attributed to the NTI's sensitivity to the full texture distribution, including low-intensity regions that may not be perceptible in IPFs. Simulation 6 matches the broader ODF, while Simulation 9 better reproduces the sharp directional textures that dominate material behavior. These findings underline the importance of using both quantitative metrics and visual interpretation to assess texture fidelity comprehensively.

Analysis of grain size and aspect ratio distributions further supports the texture-based conclusions. Simulated grain sizes fall within the experimental range. Aspect ratio distributions in all cases show good agreement around the dominant range of 1.5–2.5. However, the simulations slightly underrepresent the long tail of highly elongated grains (aspect ratios > 5), particularly in cross-sections perpendicular to the build direction. This modest discrepancy likely arises from the simplified thermal gradient representation in the CA model, which does not fully capture the spatial and temporal variability of heat flow and its influence on grain anisotropy.

To evaluate microstructural similarity from a morphological perspective, the Earth Mover's Distance (EMD) was used to compare chord length distributions (CLDs). The simulation with $\eta = 0.45$ and nucleation density of $n_0 = 1200$ yielded the lowest EMD relative to Experiment 1, suggesting a good match for fine-grained, densely packed morphologies. For Experiment 2, the lowest EMD was obtained using the same absorption rate but a lower nucleation density ($n_0 = 300$), consistent with coarser, less uniform grain structures. These results reinforce the known relationships between nucleation density, grain growth, and morphology: higher nucleation densities promote finer structures through increased grain competition, while lower nucleation densities allow for grain coarsening and elongation. EMD thus complements grain size analysis by quantifying morphological similarity in a spatially resolved manner.

Despite the strong overall performance of the CA model, several limitations and opportunities for improvement remain. The current framework does not account for vaporization-induced recoil pressure, and simulation of Marangoni effects is turned off to save computation cost, both of which significantly influence melt pool shape, flow dynamics, and solidification behavior in L-PBF. These fluid flow effects, along with keyhole formation, pore evolution, and gas entrapment, can alter local grain orientation and morphology, especially near melt pool boundaries—features not captured in the present model. Similarly, the model lacks treatment of post-process phenomena, such as grain growth, recrystallization, and coarsening during heat treatment, which can significantly modify texture and morphology in service-ready components.

Furthermore, while the current CA approach assumes simplified thermal gradients and uniform material properties, incorporating anisotropic growth kinetics, orientation-dependent solidification velocities, and curvature-driven grain boundary motion could greatly enhance predictive accuracy. Multiscale coupling with phase-field models or CFD-based thermal simulations could enable more realistic grain evolution pathways, especially in geometrically complex or functionally graded builds.

In summary, the results demonstrate that the cellular automata (CA) model, when calibrated with appropriate process parameters, is capable of capturing the dominant features of L-PBF microstructure—both in texture and morphology. By combining statistical measures such as the normalized texture index (NTI) and Earth Mover's Distance (EMD) with visual and directional analyses, a holistic evaluation of simulation fidelity is achieved. These findings underscore the potential of simulation-guided design for critical thin-walled applications, such as aortic stents, where precise control over grain morphology and texture is essential for ensuring mechanical reliability. Ongoing enhancements that integrate thermal-fluid dynamics, anisotropic grain growth behavior, and post-processing effects will further expand the model's utility for predictive design and optimization in metal additive manufacturing.

6. Conclusions

This study presents a detailed evaluation of grain morphology and texture in L-PBF-processed 316L stainless steel using a cellular automata (CA) simulation framework, validated against experimental EBSD data across three orthogonal planes. By systematically varying laser absorption rate and nucleation density, the model was able to reproduce a wide range of experimentally observed microstructural features.

Key findings include:

- Simulation 9 ($\eta = 0.40$) best reproduced the dominant <111> and <101> texture components seen in the experiments, particularly in the Z and X directions. Although Simulation 6 ($\eta = 0.35$) yielded lower normalized texture index (NTI) scores, visual inspection of inverse pole figures (IPFs) confirmed that Simulation 9 more accurately captured the primary texture features critical for mechanical performance.
- Grain size and aspect ratio distributions from simulations showed good agreement with experiments. A high nucleation density combined with a high absorption rate yielded the closest match to the slightly finer-grained microstructure observed in Experimental 1, whereas Experimental 2 was better matched by simulations with a lower nucleation density at the same absorption rate.
- Earth Mover's Distance (EMD) analysis of chord length distributions confirmed the morphological similarity between simulations and experiments.
- The CA model demonstrates strong capability in predicting both texture and morphology when calibrated with experimentally relevant parameters. However, limitations remain, particularly regarding the modeling of anisotropic grain growth, melt pool dynamics, and post-process effects such as heat treatment.

Future work should focus on enhancing the physical fidelity of the model by incorporating melt pool fluid flow, Marangoni effects, and orientation-dependent growth kinetics. Coupling with thermal-fluid simulations or extending the CA framework to include recrystallization behavior could further improve predictive accuracy and applicability to complex L-PBF geometries, including thin-walled biomedical components such as aortic stents.

Author Contributions: Conceptualization, Y.Q. and A.N.; methodology, Y.Q., M.G. and A.N.; software, Y.Q.; validation, Y.Q. and M.G.; formal analysis, Y.Q. and M.G.; investigation, Y.Q. and M.G.; resources, A.N.; data curation, Y.Q.; writing—original draft preparation, Y.Q.; writing—review and editing, M.G. and A.N.; visualization, Y.Q. and M.G.; supervision, A.N.; project administration, A.N.; funding acquisition, A.N. All authors have read and agreed to the published version of the manuscript.

Funding: This research was funded by the Bavarian State Ministry of Science and the Arts under the project grant "PRESTIGE" as part of the 6th funding round (2021–2024) for applied research and development at universities of applied sciences

Data Availability Statement: Data are contained within the article.

Acknowledgments: We gratefully acknowledge Haneen Daoud and Agata Kulig, from Neue Materialien Bayreuth GmbH for conducting the EBSD analysis, as well as Philipp Lulla, from the FIT Additive Manufacturing Group for providing the material used in this study. We also thank them, along with Philipp Marx, from OTH Regensburg, for their valuable discussions and support throughout the project. Thanks to the library of the University of Applied Sciences Regensburg for taking the publication fee.

Conflicts of Interest: The authors declare no conflicts of interest.

Abbreviations

The following abbreviations are used in this manuscript:

L-PBF	Laser powder bed fusion
AM	Additive manufacturing
PSP	Process-structure-property
CA	Cellular automata
PFM	Phase field models
KMC	Kinetic monte carlo
HAZ	Heat-affected zone

References

1. Ahmed, N.; Barsoum, I.; Haidemenopoulos, G.; Al-Rub, R.K. Process parameter selection and optimization of laser powder bed fusion for 316L stainless steel: A review. *J. Manuf. Process.* **2022**, *75*, 415–434. [CrossRef]
2. Gandin, C.A.; Desbiolles, J.L.; Rappaz, M.; Thévoz, P. A three-dimensional cellular automaton-finite element model for the prediction of solidification grain structures. *Metall. Mater. Trans. A Phys. Metall. Mater. Sci.* **1999**, *30*, 3153–3165. [CrossRef]
3. Rappaz, M.; Gandin, C.A. Probabilistic modelling of microstructure formation in solidification processes. *Acta Metall. Mater.* **1993**, *41*, 345–360. [CrossRef]
4. Teferra, K.; Rowenhorst, D.J. Optimizing the cellular automata finite element model for additive manufacturing to simulate large microstructures. *Acta Mater.* **2021**, *213*, 116930. [CrossRef]
5. Baumard, A.; Ayrault, D.; Fandeur, O.; Bordreuil, C.; Deschaux-Beaume, F. Numerical prediction of grain structure formation during laser powder bed fusion of 316 L stainless steel. *Mater. Des.* **2021**, *199*, 109434. [CrossRef]
6. Panwisawas, C.; Qiu, C.; Anderson, M.J.; Sovani, Y.; Turner, R.P.; Attallah, M.M.; Brooks, J.W.; Basoalto, H.C. Mesoscale modelling of selective laser melting: Thermal fluid dynamics and microstructural evolution. *Comput. Mater. Sci.* **2017**, *126*, 479–490. [CrossRef]
7. Baere, D.D.; Mohanty, S.; Hattel, J.H. Microstructural modelling of above β-transus heat treatment of additively manufactured Ti-6Al-4V using cellular automata. *Mater. Today Commun.* **2020**, *24*, 101031. [CrossRef]
8. Zinovieva, O.; Zinoviev, A.; Ploshikhin, V. Three-dimensional modeling of the microstructure evolution during metal additive manufacturing. *Comput. Mater. Sci.* **2018**, *141*, 207–220. [CrossRef]
9. Herriott, C.; Li, X.; Kouraytem, N.; Tari, V.; Tan, W.; Anglin, B.; Rollett, A.D.; Spear, A.D. A multi-scale, multi-physics modeling framework to predict spatial variation of properties in additive-manufactured metals. *Model. Simul. Mater. Sci. Eng.* **2019**, *27*, 025009. [CrossRef]
10. Lian, Y.; Lin, S.; Yan, W.; Liu, W.K.; Wagner, G.J. A parallelized three-dimensional cellular automaton model for grain growth during additive manufacturing. *Comput. Mech.* **2018**, *61*, 543–558. [CrossRef]
11. Lian, Y.; Gan, Z.; Yu, C.; Kats, D.; Kam, W.; Wagner, G.J. A cellular automaton finite volume method for microstructure evolution during additive manufacturing. *Mater. Des.* **2019**, *169*, 107672. [CrossRef]
12. Li, X.; Tan, W. Numerical investigation of effects of nucleation mechanisms on grain structure in metal additive manufacturing. *Comput. Mater. Sci.* **2018**, *153*, 159–169. [CrossRef]

13. Rai, A.; Markl, M.; Körner, C. A coupled Cellular Automaton–Lattice Boltzmann model for grain structure simulation during additive manufacturing. *Comput. Mater. Sci.* **2016**, *124*, 37–48. [CrossRef]
14. Rodgers, T.M.; Madison, J.D.; Tikare, V. Simulation of metal additive manufacturing microstructures using kinetic Monte Carlo. *Comput. Mater. Sci.* **2017**, *135*, 78–89. [CrossRef]
15. Eiken, J.; Böttger, B.; Steinbach, I. Multiphase-field approach for multicomponent alloys with extrapolation scheme for numerical application. *Phys. Rev. E Stat. Nonlinear Soft Matter Phys.* **2006**, *73*, 066122. [CrossRef]
16. Lu, L.X.; Sridhar, N.; Zhang, Y.W. Phase field simulation of powder bed-based additive manufacturing. *Acta Mater.* **2018**, *144*, 801–809. [CrossRef]
17. Keller, T.; Lindwall, G.; Ghosh, S.; Ma, L.; Lane, B.M.; Zhang, F. Methods to Additive Manufacturing of Ni-based Superalloys. *Acta Mater.* **2017**, *139*, 244–253. [CrossRef] [PubMed]
18. Gong, X.; Chou, K. Phase-field modeling of microstructure evolution in electron beam additive manufacturing. *Jom* **2015**, *67*, 1176–1182. [CrossRef]
19. Tourret, D.; Liu, H.; LLorca, J. Phase-field modeling of microstructure evolution: Recent applications, perspectives and challenges. *Prog. Mater. Sci.* **2022**, *123*, 100810. [CrossRef]
20. Kergaßner, A.; Koepf, J.A.; Markl, M.; Körner, C.; Mergheim, J.; Steinmann, P. A Novel Approach to Predict the Process-Induced Mechanical Behavior of Additively Manufactured Materials. *J. Mater. Eng. Perform.* **2021**, *30*, 5235–5246. [CrossRef]
21. Xu, F.; Xiong, F.; Li, M.J.; Lian, Y. Three-Dimensional Numerical Simulation of Grain Growth during Selective Laser Melting of 316L Stainless Steel. *Materials* **2022**, *15*, 6800. [CrossRef]
22. Pinomaa, T.; Lindroos, M.; Walbrühl, M.; Provatas, N.; Laukkanen, A. The significance of spatial length scales and solute segregation in strengthening rapid solidification microstructures of 316L stainless steel. *Acta Mater.* **2020**, *184*, 1–16. [CrossRef]
23. Karayagiz, K.; Johnson, L.; Seede, R.; Attari, V.; Zhang, B.; Huang, X.; Ghosh, S.; Duong, T.; Karaman, I.; Elwany, A.; et al. Finite interface dissipation phase field modeling of Ni–Nb under additive manufacturing conditions. *Acta Mater.* **2020**, *185*, 320–339. [CrossRef]
24. Chadwick, A.F.; Voorhees, P.W. The development of grain structure during additive manufacturing. *Acta Mater.* **2021**, *211*, 116862. [CrossRef]
25. Yang, M.; Wang, L.; Yan, W. Phase-field modeling of grain evolutions in additive manufacturing from nucleation, growth, to coarsening. *npj Comput. Mater.* **2021**, *7*, 56. [CrossRef]
26. Xue, T.; Gan, Z.; Liao, S.; Cao, J. Physics-embedded graph network for accelerating phase-field simulation of microstructure evolution in additive manufacturing. *npj Comput. Mater.* **2022**, *8*, 201. [CrossRef]
27. Liu, P.; Ji, Y.; Wang, Z.; Qiu, C.; Antonysamy, A.; Chen, L.Q.; Cui, X.; Chen, L. Investigation on evolution mechanisms of site-specific grain structures during metal additive manufacturing. *J. Mater. Process. Technol.* **2018**, *257*, 191–202. [CrossRef]
28. Chen, G.; Wang, X.; Yang, X.; Yang, X.; Zhang, Z.; Dai, R.; Gu, J.; Zhang, T.; Wu, G.; Gong, J. An integrated simulation approach for directing the texture control of austenitic stainless steel through laser beam powder bed fusion. *J. Mater. Process. Technol.* **2025**, *336*, 118707. [CrossRef]
29. Xie, Q.; Yan, X.; Yin, S.; Li, X.; Liu, M.; Wang, Y.; Zhuang, L. An extended version of cellular automata model for powder bed fusion to unravel the dependence of microstructure on printing areas for Inconel 625. *Addit. Manuf.* **2023**, *73*, 103676. [CrossRef]
30. Liang, X.; Zhu, J.; Popovich, V.; Hermans, M.; Richardson, I.; Bos, C. A multi-level capture algorithm for accelerating cellular automata predictions of grain structure and texture in additive manufacturing. *Addit. Manuf.* **2025**, *98*, 104622. [CrossRef]
31. Rolchigo, M.; Coleman, J.; Knapp, G.L.; Plotkowski, A. Grain structure and texture selection regimes in metal powder bed fusion. *Addit. Manuf.* **2024**, *81*, 104024. [CrossRef]
32. Rolchigo, M.; Reeve, S.T.; Stump, B.C.; Knapp, G.L.; Coleman, J.; Plotkowski, A.; Belak, J.F. ExaCA: A performance portable exascale cellular automata application for alloy solidification modeling. *Comput. Mater. Sci.* **2022**, *214*, 111692. [CrossRef]
33. Rolchigo, M.; Reeve, S.T.; Stump, B.C.; Coleman, J.; Plotkowski, A. ExaCA v2.0: A versatile, scalable, and performance portable cellular automata application for additive manufacturing solidification. *Comput. Mater. Sci.* **2025**, *251*, 113734. [CrossRef]
34. Coleman, J.; Kincaid, K.; Knapp, G.L.; Stump, B.; Plotkowski, A.J. ORNL/AdditiveFOAM: Release 1.0. *Zenodo* **2023**. [CrossRef]
35. Gandin, C.A.; Rappaz, M. A 3D Cellular Automaton algorithm for the prediction of dendritic grain growth. *Acta Mater.* **1997**, *45*, 2187–2195. [CrossRef]
36. Trivedi, R.; Kurz, W. Theory of Microstructural Development during Rapid Solidification. *Acta Metall.* **1986**, *34*, 823–830.
37. Goldak, J.A.; Chakravarti, A.P.; Bibby, M. A new finite element model for welding heat sources. *Metall. Trans. B* **1984**, *15*, 299–305. [CrossRef]
38. Lu, N.; Lei, Z.; Yu, X.; Bi, J.; Chen, Y. Effects of melt convection on stray grain formation in single crystal superalloys during directed energy deposition. *Addit. Manuf.* **2021**, *48*, 102429. [CrossRef]
39. Cao, L. Numerical simulation of the impact of laying powder on selective laser melting single-pass formation. *Int. J. Heat Mass Transf.* **2019**. [CrossRef]

40. Coleman, J.; Plotkowski, A.; Stump, B.C.; Raghavan, N.; Sabau, A.S.; Krane, M.J.M.; Heigel, J.C.; Ricker, R.E.; Levine, L.E.; Babu, S.S. Sensitivity of Thermal Predictions to Uncertain Surface Tension Data in Laser Additive Manufacturing. *J. Heat Transf.-Trans. Asme* **2020**, *142*, 122201. [CrossRef]
41. Antony, K.; Arivazhagan, N.; Senthilkumaran, K. Numerical and experimental investigations on laser melting of stainless steel 316L metal powders. *J. Manuf. Process.* **2014**, *16*, 345–355. [CrossRef]
42. Zhang, Y.; Zhang, J. Modeling of solidification microstructure evolution in laser powder bed fusion fabricated 316L stainless steel using combined computational fluid dynamics and cellular automata. *Addit. Manuf.* **2019**, *28*, 750–765. [CrossRef]
43. Cox, B.; Ghayoor, M.; Doyle, R.P.; Pasebani, S.; Gess, J.L. Numerical model of heat transfer during laser powder bed fusion of 316L stainless steel. *Int. J. Adv. Manuf. Technol.* **2022**, *119*, 5715–5725. [CrossRef]
44. Wiesent, L.; Schultheiß, U.; Lulla, P.; Nonn, A.; Noster, U. Mechanical properties of small structures built by selective laser melting 316 L stainless steel–A phenomenological approach to improve component design. *Mater. Werkst.* **2020**, *51*, 1615–1629. [CrossRef]
45. Beausir, B.; Fundenberger, J.-J. *Analysis Tools for Electron and X-Ray Diffraction, ATEX-Software*; Université de Lorraine: Metz, France, 2017. Available online: www.atex-software.eu (accessed on 30 May 2025).
46. Latypov, M.; Kühbach, M.; Beyerlein, I.J.; Stinville, J.C.; Toth, L.S.; Pollock, T.M.; Kalidindi, S.R. Application of chord length distributions and principal component analysis for quantification and representation of diverse polycrystalline microstructures. *Mater. Charact.* **2018**, *145*, 671–685. [CrossRef]

metals MDPI

Article

Microstructure, Porosity, and Bending Fatigue Behaviour of PBF-LB/M SS316L for Biomedical Applications

Conall Kirk [1,†], Weijie Xie [1,2,†], Shubhangi Das [1], Ben Ferguson [1], Chenliang Wu [3], Hau-Chung Man [2] and Chi-Wai Chan [1,*]

1 School of Mechanical and Aerospace Engineering, Queen's University Belfast, Belfast BT9 5AH, UK; ckirk12@qub.ac.uk (C.K.); weijie.xie@connect.polyu.hk (W.X.); sdas@qub.ac.uk (S.D.); bferguson08@qub.ac.uk (B.F.)
2 Department of Industrial and Systems Engineering, Hong Kong Polytechnic University, Hong Kong, China; hc.man@polyu.edu.hk
3 School of Materials Science and Engineering, Shenyang University of Technology, Shenyang 110870, China; mfclwu@sut.edu.cn
* Correspondence: c.w.chan@qub.ac.uk; Tel.: +44-(0)28-9097-4905
† These authors contributed equally to this work.

Abstract: Bending fatigue significantly affects the mechanical stability and lifespan of biomedical implants, such as bone plates and orthopaedic fixation devices, which undergo cyclic loading in the human body. This study examines the microstructure, porosity, and bending fatigue properties of PBF-LB/M SS316L. Samples were analysed across three faces (top, front, and side) using optical microscopy (OM) and scanning electron microscopy (SEM) to observe microstructural features and porosity. Elemental composition was measured by energy-dispersive X-ray spectroscopy (EDX). Phase structures and grain orientations were characterised via X-ray diffraction (XRD) and electron backscatter diffraction (EBSD). Four-point bending fatigue tests, conducted under two loading conditions, below and slightly above the yield point, demonstrated that defects inherent to the PBF-LB/M process, particularly micropores and unmelted powder particles, strongly influence fatigue crack initiation. Real-time monitoring of crack initiation and propagation on the external sample surface was performed using a high-speed digital microscope. These findings indicate the influence of microstructural defects on fatigue performance in PBF-LB/M SS316L, supporting the design and development of more reliable patient-specific biomedical implants.

Keywords: laser powder bed fusion of metals (PBF-LB/M); additive manufacturing; biomedical implants; bending fatigue

1. Introduction

The growing demand for patient-specific medical implants has increased the use of additive manufacturing (AM), or 3D printing, in the biomedical field. Conventionally manufactured (CM) implants usually come in standardised sizes and shapes, often requiring surgical adjustments [1–3]. In contrast, AM manufacturing enables implants to be custom-made according to individual patient anatomies, potentially improving implant integration and clinical outcomes [4–6]. Laser powder bed fusion of metals (PBF-LB/M) is a common AM method for producing metallic implants. In PBF-LB/M, a laser selectively melts layers of metal powder in an inert atmosphere (argon or nitrogen) to minimise oxidation and contamination [7]. PBF-LB/M is well suited for producing complex shapes with desirable mechanical properties, making it ideal for load-bearing implants like bone plates [8,9].

Nevertheless, PBF-LB/M introduces defects such as internal voids or pores, which can compromise mechanical integrity [10]. This is particularly important for implants experiencing repetitive physiological loads, such as bone plates used in fracture stabilisation.

Fatigue failures occur in both AM and CM implants. Fantin et al. [11] reported fatigue failure in a stainless-steel femoral bone plate due to insufficient bone healing, causing increased implant stress. Guerra-Fuentes et al. [12] found fatigue failure in a CM SS316L supracondylar blade plate resulting from screw failures and subsequent stress concentrations. Sudhakar [13] further demonstrated that material flaws, including inclusions in CM implants and porosity in AM implants, significantly affect fatigue life, with porosity particularly acting as a stress concentrator that initiates fatigue cracks.

AM implants generally demonstrate comparable static strengths to CM implants. However, the fatigue performance is less predictable due to manufacturing defects such as porosity, incomplete fusion defects, and variations in phase structure and grain orientation. These factors significantly influence fatigue life by affecting crack initiation and propagation [14–16]. Liang et al. [6] reported reduced fatigue performance in PBF-LB/M SS316L samples under rotational bending due to near-surface fusion defects. Rivolta et al. [17] performed rotating bending tests on as-built and machined PBF-LB/M SS316L samples. They found that subsurface defects had a more significant impact on fatigue life compared to surface defects, owing to the ductility of the material. Concli et al. [18] assessed both high- and low-cycle fatigue behaviours and observed crack nucleation primarily near the surface, attributed to surface characteristics or defects.

Variations in the chemical composition of powders used in PBF-LB/M significantly influence mechanical properties because chemical composition directly affects material microstructure and mechanical response. Blinn et al. [19] demonstrated that even slight differences in chemical composition can result in notable variations in both static and fatigue performance of PBF-LB/M SS316L samples. PBF-LB/M processing conditions involve thermal gradients and complex thermal histories. These conditions create anisotropic grain structures that significantly affect mechanical properties due to directional dependency in grain alignment affecting load-bearing capabilities. Kong et al. [20] conducted uniaxial compressive tests on bulk and lattice PBF-LB/M SS316L samples. They reported reduced compressive strength in the build direction due to the orientation of elongated grains. Marattukalam et al. [21] investigated different laser scanning strategies and found that crystallographic texture significantly affects hardness, Young's modulus, and ultimate tensile strength. Dixit et al. [22] examined the influence of sample orientation during printing (X, Y, XY45°, Z, ZX45°). They observed distinct crystallographic textures across orientations, with the XY45° orientation exhibiting the highest tensile strength due to the predominance of <111> grain orientation.

Fatigue behaviour of 316L stainless steel is influenced by phase transformations occurring under cyclic loading. Although classified as fully austenitic with generally favourable mechanical properties, 316L can undergo martensitic transformations during cyclic loading. These transformations, particularly at grain boundaries, lead to increased brittleness and potential sites for fatigue initiation, thus influencing implant reliability. Ganesh et al. [23] observed strain-induced martensite on fracture surfaces in laser-deposited 316L tested via single edge notch bending. Their findings confirmed that martensitic transformation significantly contributes to fracture behaviour. Previous studies on fatigue in PBF-LB/M SS316L typically focused on high-cycle (elastic regime) or low-cycle (plastic regime) fatigue conditions, or occasionally both. Evaluating both regimes is important to fully understand implant performance under various loading conditions. For example, Jaskari et al. [24] studied high-cycle bending fatigue in samples with varying porosity levels to evaluate fatigue performance under elastic loading conditions. Chen et al. [25]

evaluated low-cycle fatigue under combined mechanical and thermal loads, providing valuable insights into fatigue behaviour within plastic deformation regimes.

However, clinical conditions frequently subject orthopaedic implants to complex mechanical loads that induce deformation spanning both elastic and early plastic regimes. Realistic scenarios such as sudden impacts, accidental falls, or vigorous activities routinely cause implants to experience slight plastic deformation beyond the elastic limit. Winwood et al. [26] highlighted the importance of understanding fatigue behaviour within this elastic-to-early plastic transition, where fatigue damage commonly initiates. Thus, investigating fatigue across elastic and early plastic regimes is critical for realistic implant evaluation. This study employs four-point bending tests, supported by microstructural and porosity analyses, to examine fatigue initiation and crack propagation in high-aspect-ratio PBF-LB/M SS316L samples under clinically relevant loading conditions.

2. Materials and Methods

2.1. Raw Materials and PBF-LB/M Printing

Spherical stainless steel (SS316L) powder (SLM Solutions, Luebeck, Germany) was used as the base material. Particle size analysis was conducted using a Malvern Morphologi 4 particle analyser (Version 10.32) (Malvern Panalytical, Malvern, UK), resulting in a volume-based median particle size (D50) of 30.1 μm, with D10 and D90 values of 19.6 μm and 46.1 μm, respectively. The particle size distribution is presented in Figure 1. Before fabrication, the powder was dried at 80 °C for 24 h to reduce moisture content to approximately 5.4%. High-aspect-ratio samples (100 mm × 20 mm × 3.5 mm) were printed using an SLM 125HL system (SLM Solutions, Germany), equipped with a continuous fibre laser (wavelength: 1070 nm). Optimised printing parameters and scanning strategies previously developed by the authors [27] were employed to produce samples with densities exceeding 99%. The rectangular sample geometry was selected to simulate the shape and dimensions typical of bone plates.

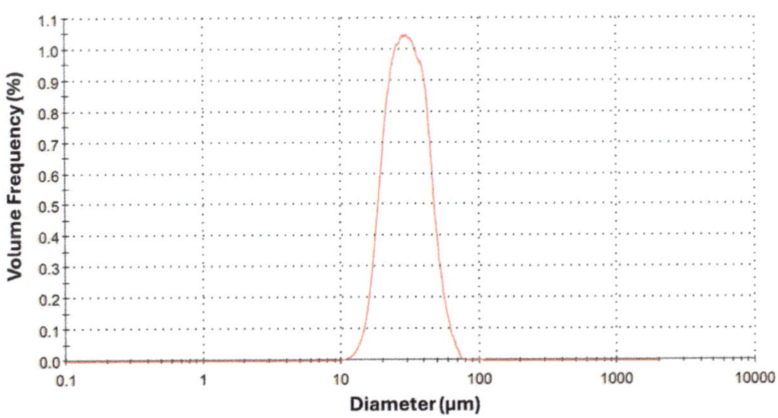

Figure 1. Volume-based particle size distribution (%) of the SS316L powder used for PBF-LB/M.

2.2. Sample Preparation

A representative portion of the printed samples was sectioned to expose three distinct faces: top, front, and side (Figure 2). Each face was individually analysed to evaluate orientation-dependent features. After sectioning, samples were hot-mounted in acrylic resin and polished using a Buehler Double Head Polisher (Buehler, Bluff, IL, USA). Polishing involved sequential grinding with silicon carbide abrasive papers (P320, P400, P600, P800, and P1200 grit), followed by diamond polishing suspensions of progressively finer particle sizes (9 μm, 3 μm, and 1 μm) to achieve a mirror-like finish suitable for microscopy. The polished samples were then chemically etched with aqua regia (12% hydrochloric acid, 28% nitric acid, 60% distilled water) for 5 min to clearly reveal grain boundaries. Finally, samples were gently rinsed with distilled water and carefully cleaned using cotton swabs to avoid surface damage.

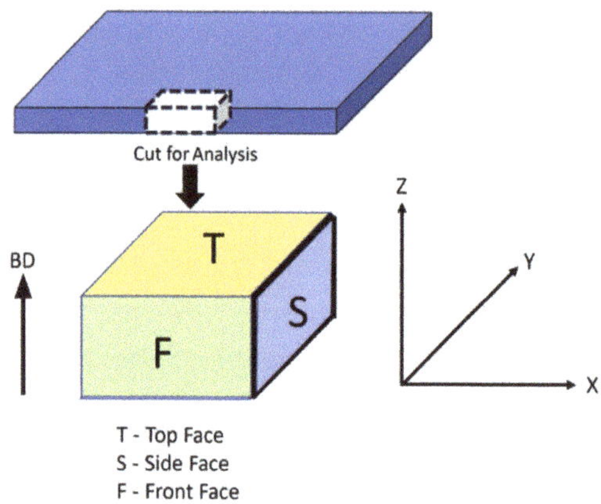

Figure 2. Schematic illustration showing the orientation of the top (T), side (S), and front (F) sample faces relative to the build direction (BD) in the PBF-LB/M process. The coordinate axes (X, Y, Z) provide spatial references for subsequent microstructural characterisation of these faces.

2.3. *Microstructural and Elemental Characterisation (OM, SEM, and EDX)*

2.3.1. Optical Microscopy (OM)

Optical microscopy (OM) was utilised to examine melt pool morphology on the top, front, and side faces of the samples. Micrographs were captured at magnifications of 100×, 200×, and 400× to identify differences in melt pool geometry indicative of variations in thermal conditions and porosity distribution. Qualitative analysis concentrated on distinguishing melt pool shapes, grain structure features, and larger porosity defects. Due to OM resolution constraints, detailed grain structure analysis was subsequently conducted via SEM. Quantitative measurements of melt pool depth were conducted following ASTM E112-13 [28]. Melt pool depths and porosity analyses were carried out using ImageJ software (Version 1.54d) on micrographs captured at 100× magnification. A total of 18 pores were identified in 7 analysed micrographs. These pores were within the resolution limits of OM and clearly distinguishable from other microstructural features or potential contamination. Within ImageJ, the micrographs were set to 8-bit and manually thresholded using a threshold value that allowed the binary image to accurately represent the morphology of the pore. The "Analyse Particles" function provided the shape descriptors for each pore. Pores with an area below 15 μm^2 were omitted from the analysis as they were difficult to distinguish from other microstructural features.

Porosity was classified by measuring circularity, aspect ratio, and Feret diameter of individual pores. Each pore was categorised as either gas porosity or laser-induced porosity based on established shape descriptors from existing literature [29,30]. Gas pores were defined as having high circularity (>0.7), low aspect ratios (1–1.3), and smaller Feret diameters, indicating spherical shapes arising from gas entrapment. Laser-induced pores exhibited larger variations in shape descriptors, with low-to-moderate circularity (<0.7), moderate-to-high aspect ratios (1.5–12.2), and larger Feret diameters due to the formation mechanisms of the pores. The laser-induced porosity category includes both types of irregular pore—lack-of-fusion and keyhole porosity. They have been grouped together as they share morphological characteristics that may impact fatigue response—larger size and irregularly shaped with sharp corners and edges, facilitating a heightened stress concentration effect. Of the 18 pores analysed, 8 were identified to be gas porosities, with the remaining 10 falling into the laser-induced pore category. Average values of these descriptors were calculated for each porosity type.

2.3.2. SEM and EDX

High-resolution microstructural characterisation was performed using a Tescan VEGA3 Scanning Electron Microscope (SEM, Tescan, Brno, Czech Republic). SEM imaging was conducted at magnifications ranging from 10,000× to 14,000× to analyse microporosity across the different sample faces. Additionally, energy-dispersive X-ray spectroscopy (EDX) was conducted to confirm the elemental composition, emphasising the key alloying elements (Fe, Cr, Ni, Mn, C, and O) present in the printed samples. The standard error (SE) reported in the EDX elemental analysis was calculated based on the average of more than 10 individual measurements ($n > 10$) taken from each sample face.

2.4. *Crystallographic Structure and Phase Analysis (XRD and EBSD)*

X-ray diffraction (XRD) analysis was performed using a Rigaku SmartLab X-ray Diffractometer (Rigaku, Tokyo, Japan) with Cu Kα radiation (λ = 1.5406 Å) operated at 45 kV and 200 mA. XRD scans were conducted at a speed of 2°/min, covering a 2θ diffraction angle range from 20° to 80°, to verify the crystalline phases and confirm the face-centred cubic (FCC) structure characteristic of austenitic stainless steel. Electron backscatter diffraction (EBSD, Oxford Instruments, Abingdon, UK) analysis was conducted at an accelerating voltage of 20 kV with a specimen tilt angle of 70° to characterise crystallographic orientations and grain structures on the top, front, and side faces of the PBF-LB/M samples. Euler orientation maps, inverse pole figure (IPF) maps, and phase distributions were generated. Data collection was performed at approximately 40 Hz with a hit rate exceeding 98% for all faces. Crystallographic orientations were defined using the Bunge rotation convention.

2.5. *Four-Point Bending Tests and Fatigue Testing*

2.5.1. Sample Preparation and Geometry

Single-edge notched bend (SENB) specimens of PBF-LB/M SS316L were used for quasi-static bending and fatigue tests. A four-point bend test was selected to simulate a bone plate fracture fixation, as a tibia fracture fixation was shown to be subject to bending loads imparting simultaneous tensile and compressive stresses [31]. A V-notch was introduced via Electrical Discharge Machining (EDM) wire cutting following ISO 12108 [32] guidelines to localise crack initiation and facilitate crack monitoring.

Figure 3a,b illustrate the four-point bending test setup, loading and support node arrangement, and SENB specimen geometry. Figure 3c provides details of the V-notch geometry, which has a notch width and depth of 1 mm. Due to inherent surface roughness from the PBF-LB/M printing process, samples were initially printed oversized (3.5 mm thick) and machined down to a final thickness of 2 mm using EDM wire cutting, resulting in a final geometry of approximately 82 mm × 20 mm × 2 mm.

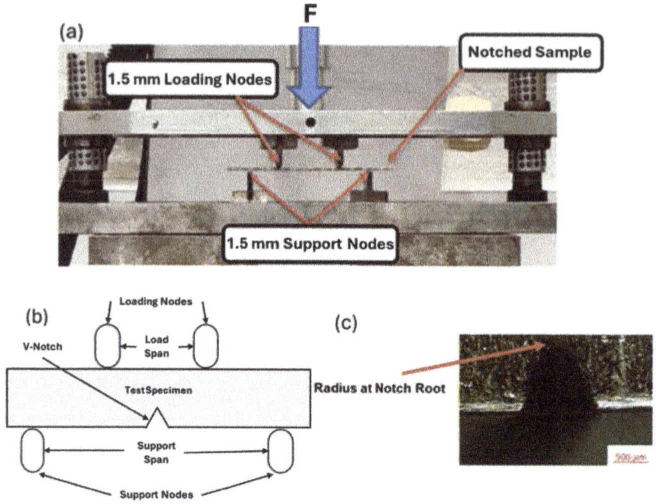

Figure 3. (**a**) Photograph of the four-point bending test setup showing loading and support nodes, (**b**) schematic illustration highlighting loading and support span positions and SENB sample geometry (Adapted from Ref. [28]), and (**c**) optical microscopy image showing the V-notch geometry and radius at the notch root produced by EDM wire cutting.

2.5.2. Quasi-Static Four-Point Bending Tests

Quasi-static four-point bending tests were performed to determine the elastic and early plastic deformation regimes used for fatigue testing. Tests were carried out using an Instron test frame at a loading rate of 45 mm/min. Load and displacement data were continuously recorded at intervals of 0.01 s. Elastic and early plastic deformation regimes were defined from the recorded load–displacement data, as illustrated in Figure 4. Note that the displacement offset observed at approximately 0.1 mm resulted from initial preloading to ensure consistent sample–fixture contact before testing.

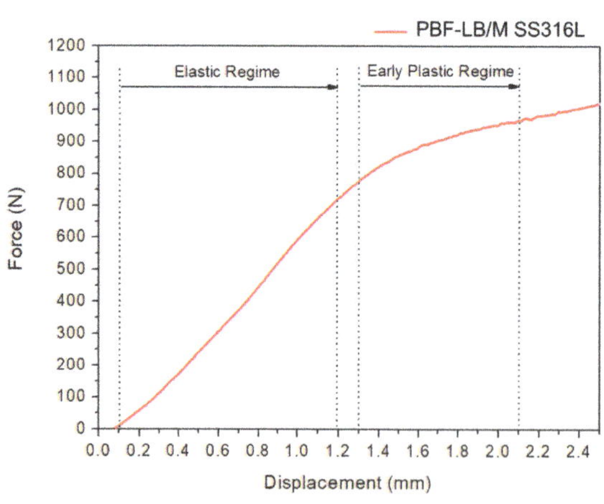

Figure 4. Load–displacement curve from quasi-static four-point bending tests of PBF-LB/M SS316L, indicating the elastic (0.1–1.2 mm) and early plastic (1.3–2.1 mm) deformation regimes used to determine displacement ranges for fatigue tests.

2.5.3. Fatigue Testing Parameters

Displacement-controlled fatigue tests were conducted using elastic and early plastic deformation regimes derived from quasi-static tests. Elastic-regime tests employed displacement cycles of 0.1–1.2 mm, cycling within the elastic region of the quasi-static test, while early-plastic-regime tests were cycled from 1.3 to 2.1 mm, cycling within the early plastic region of the quasi-static test. In displacement-controlled fatigue testing, the stress ratio (R) constantly changes as the sample deforms under successive cycles. In this case, the initial stress ratios are presented. For the elastic regime, initially R = 0.01 and for the early plastic regime, initially R = 0.8. Frequencies of 5 Hz, 10 Hz, 15 Hz, and 20 Hz were evaluated; a frequency of 10 Hz was identified as optimal for accurate displacement control and efficient test duration. Fatigue tests were performed on an Instron servo-hydraulic fatigue testing machine at room temperature and ambient humidity. Peak load and displacement values for each fatigue cycle were recorded. A minimum of three samples were tested for each condition (elastic and early plastic regimes) to ensure reliability.

2.5.4. Real Time Crack Monitoring

Real-time monitoring of crack initiation and propagation was performed using an AmScope 3MP High-Speed digital microscope (AmScope, Irvine, CA, USA). The experimental arrangement used for real-time crack monitoring is illustrated in Figure 5. Due to challenges in obtaining continuous high-quality images, tests were periodically paused to manually capture detailed images of crack progression. Tests automatically terminated either upon specimen fracture or after completing 1 million cycles. Post-testing fracture surface morphologies and crack paths were examined via SEM, as detailed in Section 2.5.5.

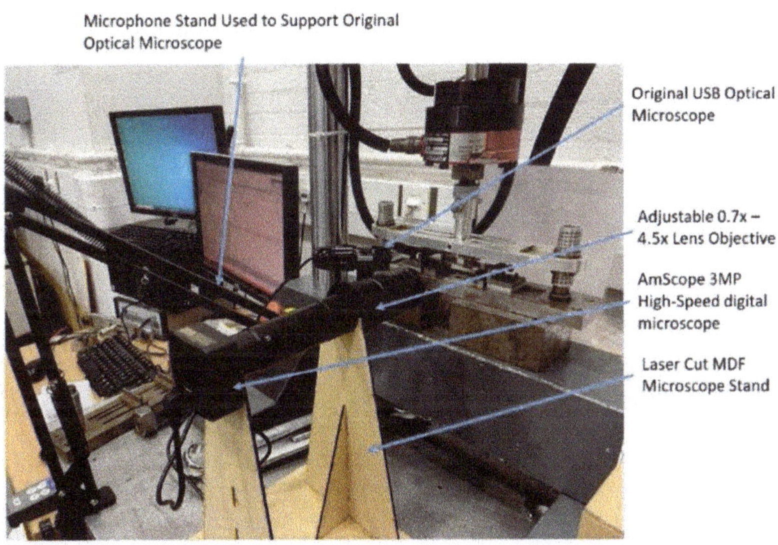

Figure 5. Experimental setup for real-time crack monitoring during fatigue testing, showing the positioning of the high-speed digital microscope and the USB optical microscope used for illumination and image capturing.

2.5.5. SEM Fractography Analysis

Fracture surfaces of samples tested under elastic and early plastic fatigue conditions were analysed using SEM. Low-magnification images (50×) were first taken to identify crack initiation points, propagation paths, and final fracture zones. High-magnification images (1000×) were then captured to closely examine specific fracture features, such as fatigue striations, ductile fracture characteristics, and defects influencing crack growth.

3. Results

3.1. OM Melt Pool Morphology and Measurements

Analysis of OM images for the PBF-LB/M samples showed distinct variations in melt pool characteristics on the top, front, and side faces. The top face exhibited a "chessboard" morphology with visible laser scan paths (Figure 6a), characterised by discontinuities along scan tracks, limited overlap, and clear scanning patterns. Gas porosity was also observed. In contrast, the side and front faces displayed a "fish-scale" morphology, presenting cross-sectional views of vertically aligned melt pools (Figure 6b,c). These faces showed continuous melt pool boundaries with minimal porosity. Columnar grains aligned parallel to the build direction were distinctly visible on the front face (Figure 6b). The fish-scale morphology appeared more defined and organised on the front face, while the side face displayed elongated melt pools interspersed with fish-scale features.

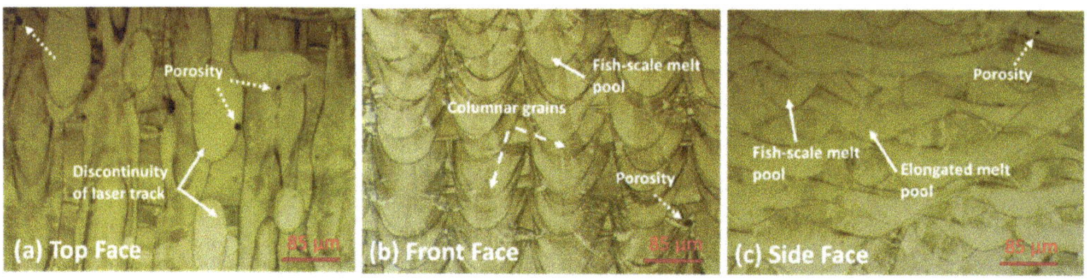

Figure 6. OM images at 200× magnification illustrating characteristic morphologies of each face: (**a**) top, (**b**) front, and (**c**) side. Unique features specific to each face are highlighted by white arrows.

Quantitative analysis results of melt pool depth are summarised in Figure 7. Depth measurements were assessed on the side and front faces due to clear cross-sectional visibility. Depth measurements on the front face included 121 intercepts, resulting in an average depth of 48 μm ± 5.6 μm (Figure 7a). On the side face, 234 intercepts were measured across the bottom (71 intercepts), middle (76 intercepts), and top (87 intercepts) regions, averaging 41 μm ± 6.4 μm (Figure 7b). Figure 7b shows a linear correlation between melt pool depth and distance from the build plate. Figure 7c provides a visual reference for the specific regions (bottom, middle, top) used in these measurements on the side face. The deepest melt pools were found at the bottom region near the support structure, which had lower variability. These findings align with observations by Leary et al. [33], suggesting enhanced heat dissipation facilitated using a support structure. The average melt pool depth on the front face was 14% larger than on the side face, indicating distinct melt pool formation dynamics between these faces. Additionally, the front face exhibited lower variability (±5.6 μm) compared to the side face (±6.4 μm), reflecting more uniform melt pool formation.

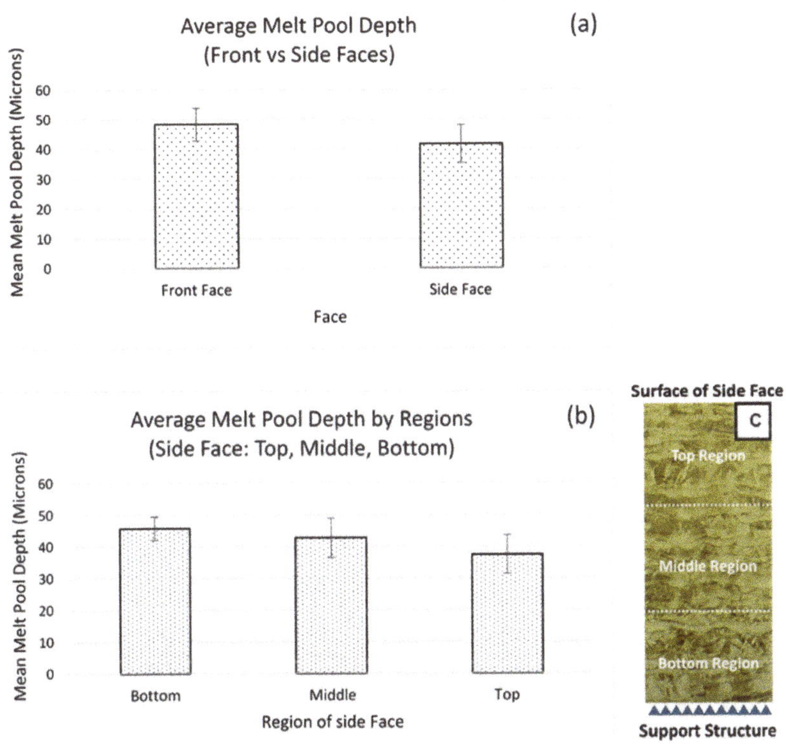

Figure 7. Quantitative assessment of melt pool dimensions: (**a**) average melt pool depth on the front and side faces, (**b**) average melt pool depth in different regions of the side face, and (**c**) OM image regions used for measurements.

3.2. OM Porosity Morphology and Quantification

3.2.1. Porosity Morphology and Distribution

Following the evaluation of microstructural attributes, the next step is to examine defects identified within the samples. Understanding defect types within the sample assists in assessing the fatigue behaviour of specimens. The analysis of porosity focuses on morphology, position, and distribution of pores. OM images indicate significant densification on all faces. However, defects persist on each face. Pores smaller than 15 μm^2 were excluded from the analysis, as they were difficult to categorise and distinguish from darker regions related to microstructure or melt pool boundaries. Figure 8 displays OM images of sample faces, each showing different degrees of porosity. On the top face (Figure 8a), numerous pores of varying sizes and shapes were identified. Energy-induced pores, indicated by dashed red arrows, are predominantly located at melt pool edges and exhibit irregular shapes. Gas pores, marked with solid blue arrows, were observed within melt pools and exhibit higher circularity.

The front face (Figure 8b, 400× magnification) showed near-total densification, with only one gas pore detected (solid blue arrow). The front face image required higher magnification (400×) due to minimal porosity, enabling clearer visual identification of defects. The side face (Figure 8c) exhibited a more complex porosity pattern. Gas porosities (solid blue arrows) within melt pools showed high circularity and smaller Feret diameters. Irregularly shaped energy-induced pores (dashed red arrows), located at melt pool edges, were elongated and related to fluctuations in energy input. The distribution and morphology of porosity indicate that despite high densification, process-related defects remain, potentially affecting mechanical properties and fatigue resistance.

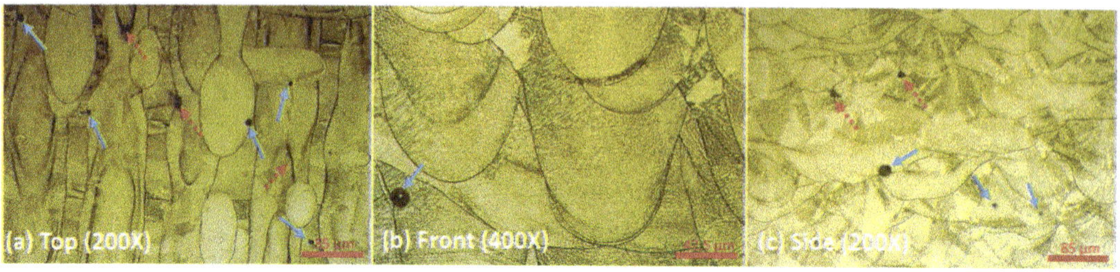

Figure 8. OM images of various faces: (**a**) top (200×), (**b**) front (400×), and (**c**) side (200×), exhibiting porosity. Gas pores are marked with solid blue arrows; energy-induced pores are indicated by dashed red arrows.

3.2.2. Quantitative Analysis of Pore Shape Descriptors

In a comprehensive analysis of pore characteristics, several key shape descriptors, including Feret diameter, circularity, aspect ratio, and $\sqrt{Area}$, were evaluated to identify correlations critical for understanding defect morphology. Each plot in Figure 9 includes a trend line and corresponding equation with an R^2 value indicating the strength of the correlation.

Certain parameter relationships, such as Feret diameter versus aspect ratio, $\sqrt{Area}$ versus aspect ratio, and circularity versus $\sqrt{Area}$, showed negligible statistical significance, with R^2 values under 0.35. In contrast, moderate correlations were observed for circularity versus Feret diameter (Figure 9a) and circularity versus aspect ratio (Figure 9b), exhibiting R^2 values of 0.63 and 0.52, respectively. The negative correlation in Figure 9a indicates larger pores typically have lower circularity, suggesting they are more irregularly shaped. Similarly, the negative correlation in Figure 9b implies that pores with higher circularity have lower aspect ratios, further confirming their rounded morphology.

Additionally, a strong positive correlation (R^2 = 0.82) was identified between $\sqrt{Area}$ and Feret diameter (Figure 9c), confirming the intuitive relationship between pore size and linear measurements. Although general trends were identified, five individual pores with similar sizes (Feret diameters ranging from 34 to 37 μm) exhibited substantial differences in other shape descriptors. Aspect ratios varied significantly from 1.6 to 12.3, circularity ranged widely from 0.15 to 0.52, and $\sqrt{Area}$ varied between 8.5 and 22.6 μm. This substantial variation highlights the necessity of utilising multiple shape descriptors rather than relying solely on size when characterising pores. The multi-parameter approach provides a more accurate assessment of the potential impact of porosity on material fatigue performance.

Figure 10a–c present average values of key shape descriptors for each porosity type, validating classification criteria used in this study and quantifying typical pore morphologies. Gas porosities exhibit high circularity (>0.7), low aspect ratios (1–1.3), and the smallest Feret diameters (9.2 μm), aligning closely with descriptions from Wang et al. [29], confirming their spherical shape and small size. Conversely, laser-induced porosities exhibit low circularity (<0.45), significantly higher aspect ratios (3.91), and larger Feret diameters (32.6 μm), confirming their irregular morphology. Additionally, an analysis of defect locations revealed distinct distribution patterns: gas porosities were predominantly found within the melt pool boundaries, whereas laser-induced defects were commonly located along melt pool edges. This pattern is consistent with observations reported by Sanaei and Fatemi [34], further supporting established knowledge regarding porosity formation mechanisms in PBF-LB/M processes.

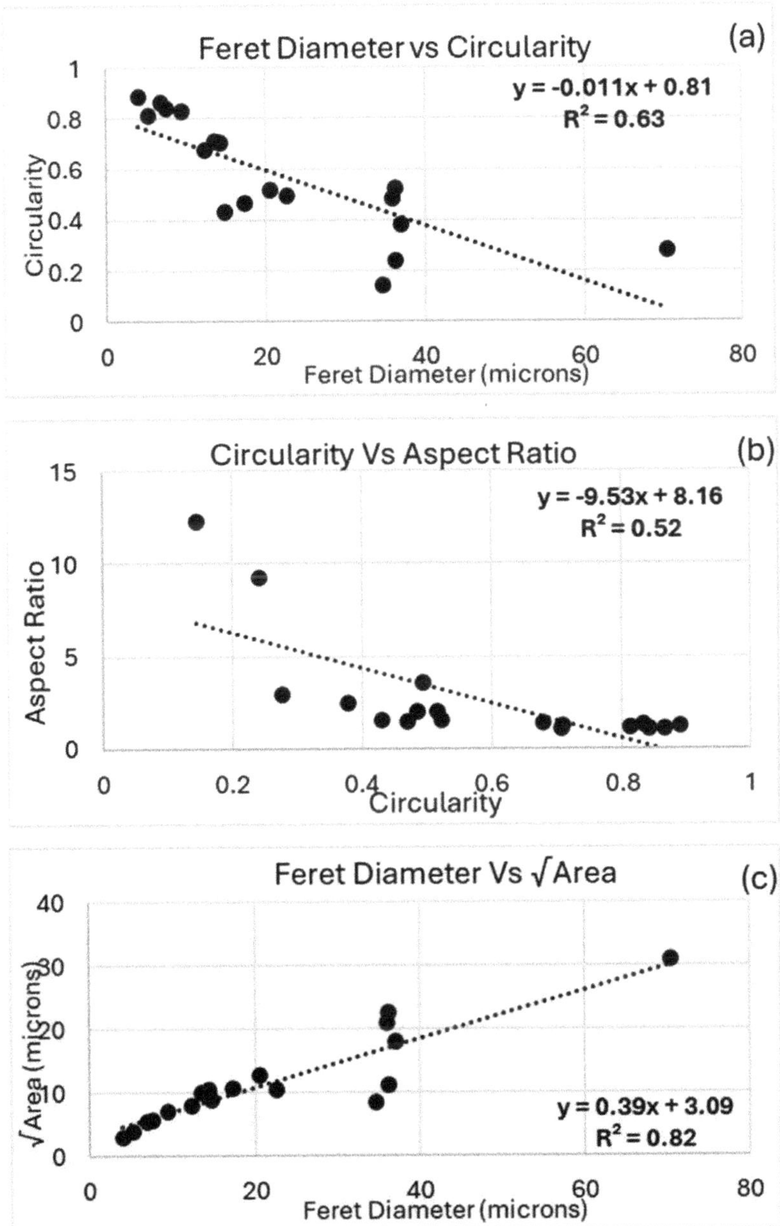

Figure 9. Comparisons between key shape descriptors to determine relationships: (**a**) circularity vs. Feret diameter, (**b**) circularity vs. aspect ratio, and (**c**) $\sqrt{\text{Area}}$ vs. Feret diameter.

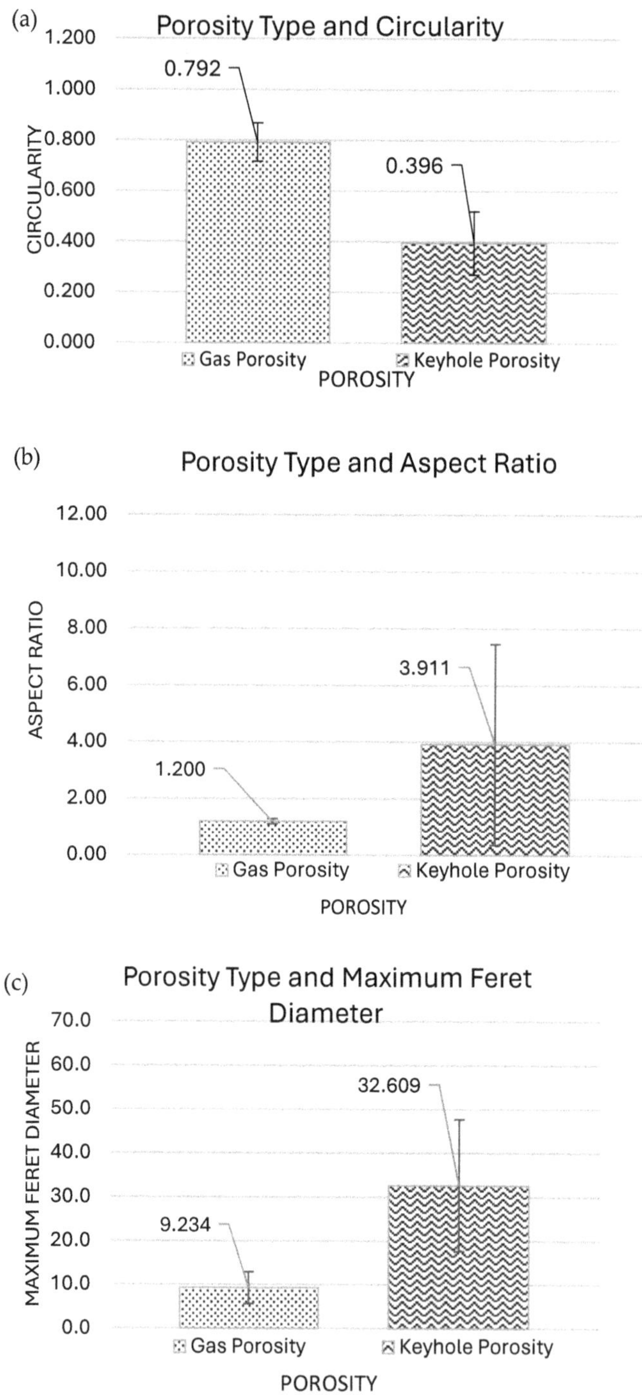

Figure 10. Comparison of (**a**) circularity, (**b**) aspect ratio, and (**c**) maximum Feret diameter among porosity types.

3.3. SEM Microstructure and Microporosity

SEM analysis provided detailed insights into the microstructural characteristics observed on different faces of the PBF-LB/M samples (Figure 11).

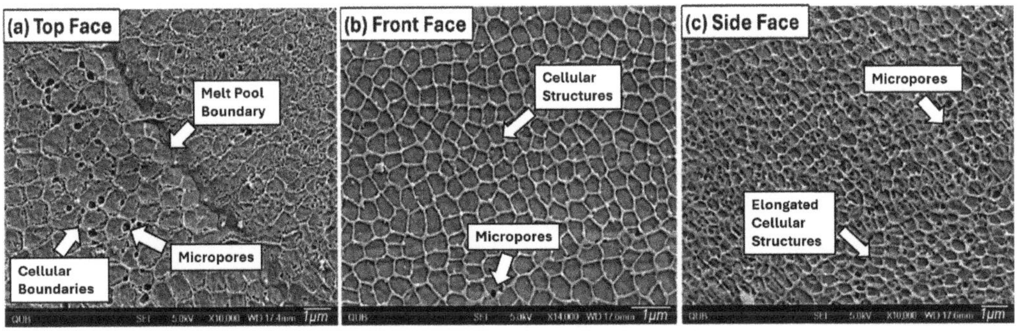

Figure 11. SEM images showing the microstructural features from different faces: (**a**) top face, (**b**) front face, and (**c**) side face.

The top face (Figure 11a) exhibited substantial microporosity, mostly smaller than 1 μm, clearly concentrated at cellular boundaries. These micropores appear as distinct, dark voids located at interfaces between cellular structures, which are formed by rapid solidification due to constitutional supercooling and solute partitioning [35]. Melt pool boundaries were also visible but relatively indistinct. In contrast, the front face (Figure 11b) showed minimal microporosity and highly uniform, clearly defined cellular structures, suggesting improved local solidification. The side face (Figure 11c) presented moderately elongated cellular structures aligned with the build direction. Microporosity was observed here but appeared smaller and less extensive compared to the top face, again primarily concentrated along cellular boundaries. The SEM images confirm that microporosity is closely associated with cellular boundaries and reveal clear differences in cellular morphologies across different build orientations.

3.4. EDX Elemental Composition Analysis

EDX was conducted to determine the elemental composition of the PBF-LB/M samples. Table 1 presents the average composition (wt%) of the front, side, and top faces of the samples.

Table 1. Average elemental composition (wt%) of different faces measured by EDX analysis (n > 10).

Face	Fe (wt%)	Cr (wt%)	Ni (wt%)	Mn (wt%)	C (wt%)	O (wt%)	Others: Mo, Si, S (wt%)
Top	66.0 ± 0.17	16.0 ± 0.04	10.1 ± 0.07	1.2 ± 0.02	1.2 ± 0.02	0.8 ± 0.11	4.7 ± 0.18
Front	65.0 ± 0.28	16.1 ± 0.04	10.0 ± 0.07	1.3 ± 0.07	5.1 ± 0.08	0.9 ± 0.15	1.6 ± 0.13
Side	66.2 ± 0.22	16.3 ± 0.06	10.0 ± 0.06	1.2 ± 0.05	3.8 ± 0.17	0.8 ± 0.08	1.7 ± 0.20

The primary constituents, such as Fe, Cr, and Ni, were consistent across all faces, indicating uniformity in the main alloying elements and conformance to typical compositional specifications for standard SS316L. However, EDX analysis has inherent limitations in accurately detecting and quantifying lighter elements, such as carbon and oxygen. Accurate determination of these elements would require specialised analytical methods (e.g., XPS), which are beyond the scope of this study. Therefore, the reported results primarily represent the heavier metallic alloying constituents.

3.5. XRD Phase Structure Analysis

XRD analysis was conducted to identify and compare crystalline phases in the PBF-LB/M and conventional SS316L samples (Figure 12).

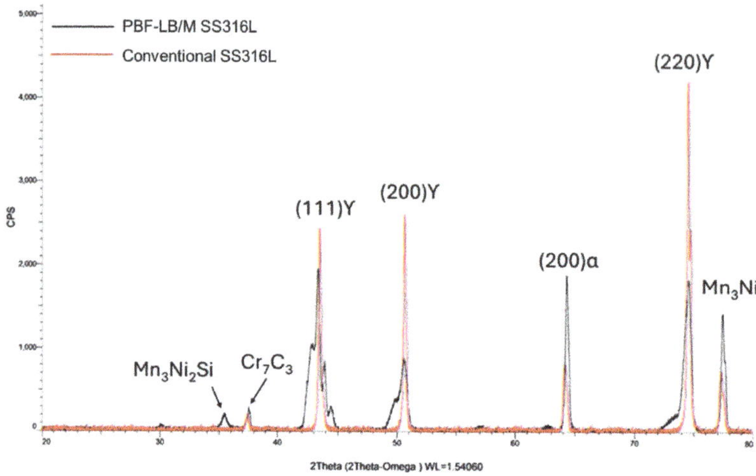

Figure 12. XRD patterns comparing phase structures of the PBF-LB/M and conventional SS316L samples. The black curve represents the top face of the PBF-LB/M sample.

Prominent peaks were identified at approximately 43.6°, 50.7°, and 74.7°, corresponding to the (111)γ, (200)γ, and (220)γ austenitic reflections, respectively, confirming a predominantly austenitic (FCC) structure in both samples. Peak broadening observed in the PBF-LB/M sample is attributed to residual stresses and grain refinement resulting from the rapid cooling rates inherent in the PBF-LB/M process. A ferritic (BCC) phase was also identified by the (200) α-ferrite diffraction peak around 65°, with higher intensity observed in the PBF-LB/M sample. Additionally, minor peaks indicated the presence of chromium carbide (Cr_7C_3) and the intermetallic compound Mn_3Ni_2Si. The carbide peak detected in the PBF-LB/M sample aligns with elevated carbon levels identified by EDX. Particularly, the Mn_3Ni_2Si intermetallic peak exhibited higher intensity in the PBF-LB/M sample, suggesting localised segregation conditions favourable for intermetallic formation, despite the rapid solidification conditions typically associated with PBF-LB/M fabrication.

3.6. EBSD Grain Orientation and Morphology Analysis

EBSD analyses were conducted to examine grain orientations and morphological characteristics of PBF-LB/M SS316L samples on the top, front, and side faces. Figure 13 presents EBSD Euler maps illustrating grain morphology and crystallographic orientations on the (a) top face, (b) front face, and (c) side face. Columnar grains aligned with the build direction were clearly observed on the front and side faces, marked by dashed lines, resulting from directional heat flow and thermal gradients during the printing process. Columnar grains were absent on the top face, consistent with typical PBF-LB/M processes in which grain elongation occurs primarily perpendicularly to the build plate.

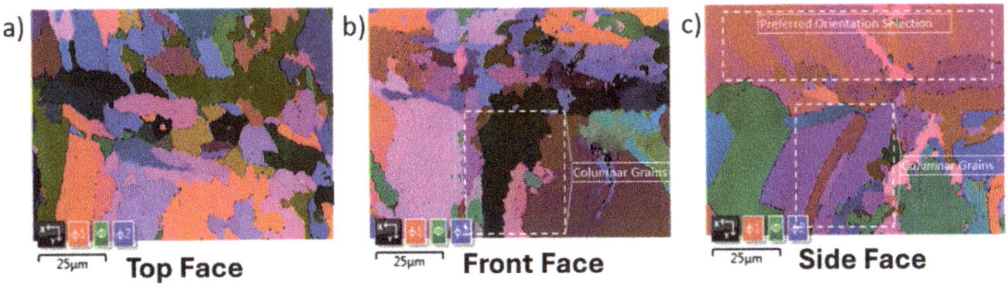

Figure 13. EBSD Euler maps illustrating grain morphology and crystallographic orientations on the (**a**) top face, (**b**) front face, and (**c**) side face of PBF-LB/M samples. White dashed lines mark columnar grain structures aligned with the build direction. The highlighted area on the side face indicates a region of preferred crystallographic orientation.

Additionally, the side face displayed a region with distinct preferred crystallographic orientation, highlighted by a red-dominant region. This observation aligns with previous research by Andreau et al. [36], indicating shifts toward preferred orientations with increased distance from the build plate. Grain size was noted to decrease with increasing distance from the build plate on the front face, a trend not observed on the side face.

Further EBSD analysis using IPF colour orientation maps (Figure 14) provided additional details on crystallographic textures. IPF-X, IPF-Y, and IPF-Z maps for the top face showed mixed grain orientations moderately favouring (001) and (101) planes, consistent with the absence of columnar grains. The front face exhibited moderate textures primarily oriented towards (001)-(111), confirming vertical grain elongation along the build direction as previously observed. The side face displayed the strongest crystallographic texture, predominantly aligned along the (001) direction vertically, with horizontal alignment toward the (101)-(111) planes, clearly reflecting pronounced directional grain growth.

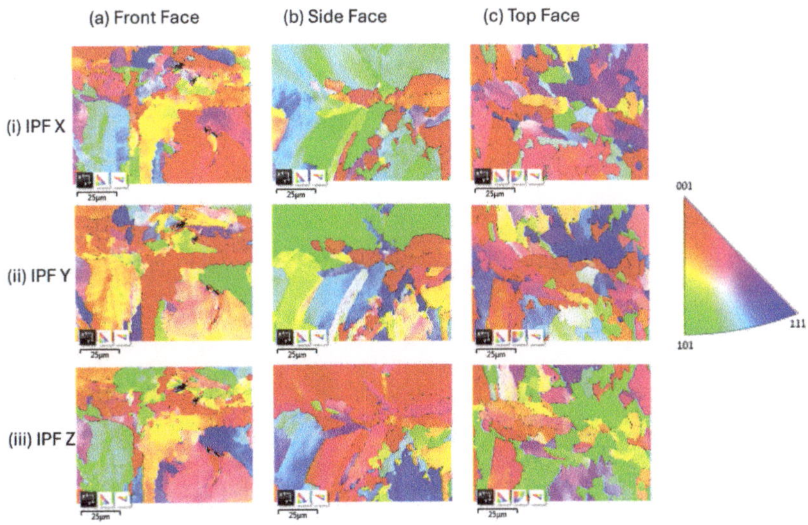

Figure 14. EBSD IPF orientation maps for the (**a**) front face (XZ-plane, parallel to the build direction), (**b**) side face (YZ-plane, parallel to the build direction, orthogonal to scanning direction), and (**c**) top face (XY-plane, perpendicular to the build direction) of PBF-LB/M SS316L. Rows (i–iii) represent crystallographic orientations relative to the X, Y, and Z directions, respectively. The inset FCC colour triangle indicates the corresponding crystallographic directions associated with each colour.

Phase fraction analysis, summarised in Table 2, indicates that the sample predominantly comprises Fe-FCC (>98%) with minor phases, including Fe-BCC, Cr-based carbides ($Cr_{23}C_6$ and Cr_7C_3), and indexing ambiguities classified as zero solutions, possibly representing grain boundaries, local segregations, or fine grains below EBSD resolution.

Table 2. Phase fraction percentages for the top, front, and side faces. The predominant phase is Fe-FCC, with minor presence of Fe-BCC, Cr-based carbides ($Cr_{23}C_6$ and Cr_7C_3), and indexing ambiguities represented by zero solutions.

Phase Name	Top	Front	Side
Fe-BCC	0.04	0.03	0.03
Fe-FCC	98.35	98.71	99.07
$Cr_{23}C_6$	0.07	0.07	0.05
Cr_7C_3	0.03	0.02	0.02
Zero Solutions	1.51	1.17	0.83

The dominance of the Fe-FCC phase confirms stable and consistent microstructure formation under the selected PBF-LB/M parameters.

The EBSD analysis confirms and complements the XRD findings by mapping the dominant FCC structure along with minor BCC and carbide phases on the top face of the PBF-LB/M sample. While EBSD directly identifies these phases and provides spatial distribution, certain microstructural details, such as grain refinement and residual stresses indicated by peak broadening, are better captured by XRD. It is noted that XRD detected a minor Mn_3Ni_2Si intermetallic phase that was not included in the EBSD phase configuration, as this phase is uncommon and not typically available in standard EBSD databases. Given its unexpected occurrence, it was not targeted during the initial EBSD setup. This highlights the importance of combining EBSD with XRD for comprehensive microstructural analysis under rapid solidification conditions.

When considered alongside the previously reported PBF-LB/M scanning strategy for this sample [27], the mixed orientation texture observed on the top face can be attributed to alternating horizontal and vertical scanning patterns, which encourage multiple grain nucleation sites and isotropic grain growth within individual layers. The moderate columnar growth on the front face and pronounced crystallographic texture on the side face result directly from the significant thermal gradients and epitaxial grain growth inherent to the PBF-LB/M process. Such microstructural anisotropy is likely to affect mechanical properties, particularly fatigue performance.

3.7. *Bending Fatigue Behaviour and Crack Propagation*

Bending fatigue tests were conducted to evaluate the fatigue behaviour of PBF-LB/M SS316L samples. Static bending tests defined baseline deformation characteristics, identifying elastic and early plastic deformation regimes. The bending fatigue curves (Figure 15) illustrate the differences between samples tested in the elastic regime (0.1–1.2 mm) and those tested in the early plastic regime (1.3–2.1 mm).

During static bending tests (Figure 4), the elastic limit occurred at approximately 700 N with a displacement of 1.2 mm, while the early plastic deformation regime extended up to about 900 N at 2.1 mm displacement. In cyclic fatigue tests conducted under displacement control, the measured peak cyclic forces were slightly higher, approximately 750 N in the elastic regime and 1000 N in the early plastic deformation regime, compared to static bending tests. This increase in cyclic peak force arises due to higher instantaneous strain rates associated with cyclic loading, demonstrating the strain-rate sensitivity of SS316L. Additionally, repeated cyclic loading induced localised strain-hardening, temporarily increasing the material stiffness.

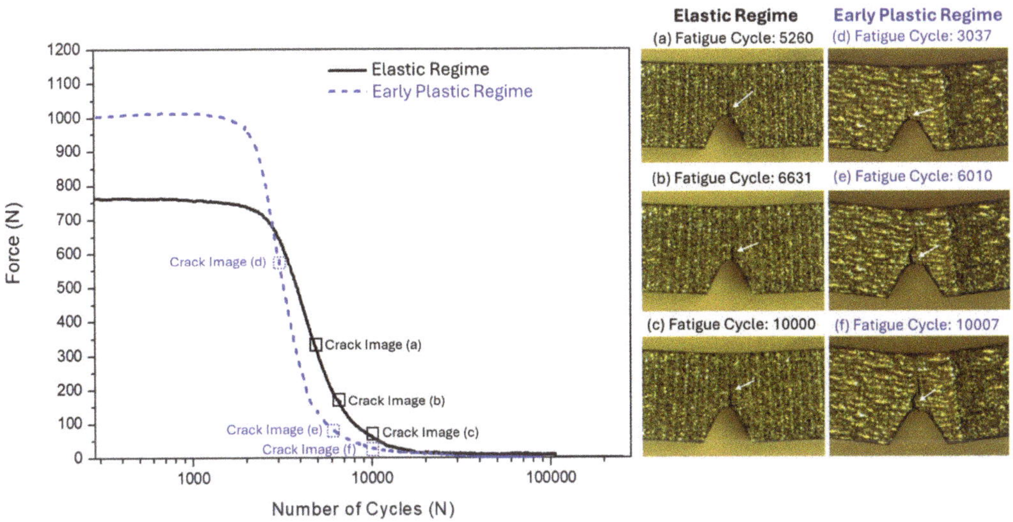

Figure 15. Bending fatigue curves of PBF-LB/M SS316L tested under displacement control in the elastic and early plastic deformation regimes, showing force reduction with increasing fatigue cycles. Insets (a–c) and (d–f) correspond to specific cycle counts, illustrating the progression of macroscopic fatigue cracks (white arrows) for samples tested in elastic and early plastic regions, respectively.

Within the elastic regime, stable cyclic forces were maintained up to approximately 1800 cycles, followed by a decline indicating damage accumulation. Macroscopic cracks were visually detected at about 5260 cycles, with subsequent progressive crack propagation evident at higher cycles. Fatigue cracks propagating in this regime exhibited straighter and more consistent paths due to smaller, more uniform plastic zones at the crack tip, limiting crack tip deformation and stress redistribution, resulting in stable and predictable crack paths. The accompanying images (Figure 15a–c) show the progression of external macroscopic cracks in the elastic regime corresponding to specific cycle counts. In contrast, tests within the early plastic deformation regime showed stable cyclic forces only up to approximately 1200 cycles, with a rapid decline thereafter, indicating more profound internal damage. Macroscopic cracks became externally visible around 3037 cycles and continued to propagate rapidly. Crack propagation in this regime followed irregular or deflected paths due to larger, more heterogeneous plastic zones around the crack tip, causing uneven stress redistribution and irregular propagation patterns. The progression of external macroscopic cracks in the early plastic regime is illustrated in Figure 15d–f.

The delayed external visibility of macroscopic cracks in both regimes suggests internal microstructural damage occurred earlier. Under cyclic loading, internal micro-cracks accumulated and grew progressively due to localised stress concentrations and material imperfections (e.g., porosity), becoming externally visible only after substantial internal damage. Consequently, the measured force began to decline before macroscopic cracks became visible, indicating that internal damage had progressed significantly ahead of external crack appearance.

Using a failure criterion defined as a drop to 10% of the initial measured cyclic peak force, samples in the elastic regime failed at approximately 2850 cycles, while those tested in the early plastic regime failed earlier, at around 2280 cycles. The shorter fatigue life in the early plastic deformation regime was mainly due to increased plastic deformation, causing higher local stress concentrations, increased strain amplitudes, and accelerated internal crack initiation. Fatigue performance was relatively limited in both regimes, mainly due to stress concentrations from the V-notched samples, designed to facilitate stress concentration effects and crack monitoring.

It is important to note that direct comparisons of fatigue strength with previously published studies remain challenging due to substantial differences in experimental conditions [37], including sample geometry, surface finishing, processing parameters (e.g., laser energy density and scanning strategies), and fatigue loading conditions (e.g., load levels, load ratios, frequency, and testing methods like axial, rotating–bending, or four-point bending fatigue). Nonetheless, the literature generally indicates lower fatigue strength for additively manufactured SS316L compared to conventionally manufactured counterparts [38]. This reduced fatigue performance primarily arises from inherent defects (as detailed in the following sections). Therefore, to achieve fatigue strengths comparable to conventionally manufactured components, optimised processing parameters combined with appropriate post-processing treatments, such as heat treatments, hot isostatic pressing (HIP), or surface finishing, are typically required [37].

3.8. SEM Fracture Surfaces After Bending Fatigue Tests

Figure 16 shows merged cross-sectional SEM images of fracture surfaces from samples tested under elastic and early plastic bending fatigue conditions. The images highlight four key regions examined, namely, surface, top, middle, and bottom, extending from the notched tensile side toward the compression side. During bending fatigue tests, tensile stresses at the notch initiate cracks, whereas compressive stresses on the opposite side inhibit crack growth.

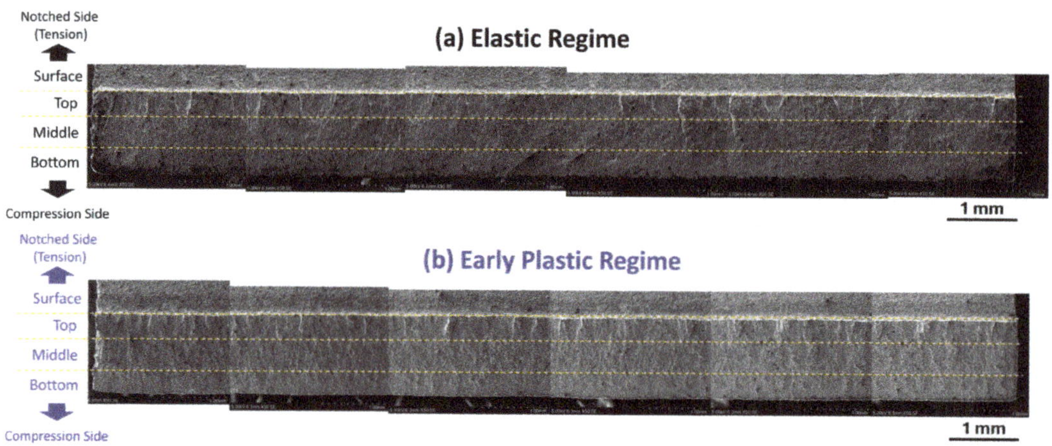

Figure 16. Merged cross-sectional SEM images (50×) of fracture surfaces showing clearly defined regions (surface, top, middle, and bottom) for samples tested under (**a**) elastic and (**b**) early plastic loading regimes.

Figure 17 presents lower-magnification SEM images (1000×) illustrating fracture morphologies at various depths within samples tested under elastic and early plastic bending fatigue regimes. These images clearly show variations in fracture morphology across the sample thickness, from the outer surface (initially under tension) toward the inner surface (initially under compression). Key fracture features, including surface irregularities, micropores, unmelted powder particles, cracks, and areas of flat, brittle fracture surfaces, are indicated by arrows.

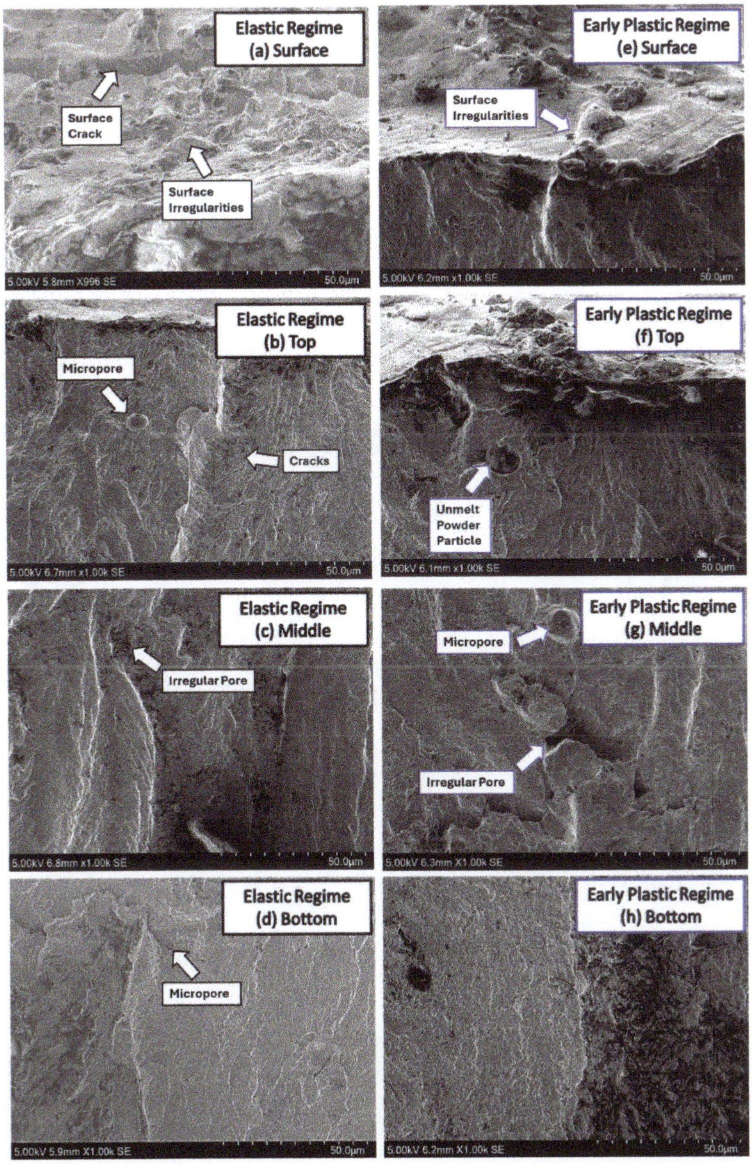

Figure 17. Lower-magnification SEM fractographs at the surface, top, middle, and bottom regions for samples tested under elastic (**a–d**) and early plastic (**e–h**) loading regimes.

For more detailed insights into specific fatigue-related features, Figure 18 provides higher-magnification SEM fractographs (2000× to 5000×). Under the elastic regime (below the yield point), fracture surfaces exhibited micropores (Figure 18a), unmelted powder particles (Figure 18b), and clear fatigue striations associated with micropores (Figure 18c). Samples tested under the early plastic regime (just beyond the yield point) similarly revealed irregular pores (Figure 18d), internal cracks within unmelted powder particles (Figure 18e), and distinct fatigue striations accompanied by micropores (Figure 18f). Both loading regimes exhibited comparable fracture morphologies dominated by fatigue processes. These observations indicate that fatigue crack initiation was primarily influenced by manufacturing-induced defects, particularly micropores and unmelted powder particles.

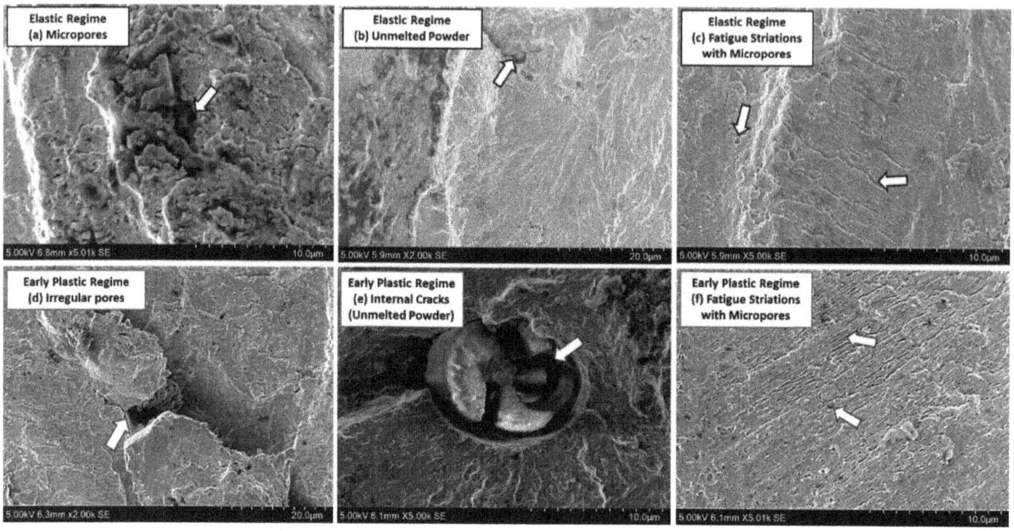

Figure 18. High-magnification SEM fractographs illustrating fatigue crack features under elastic (**a–c**) and early plastic (**d–f**) loading regimes: (**a**) crack initiation from micropores; (**b**) cracks propagating around unmelted powder particle; (**c**) clear fatigue striations accompanied by micropores; (**d**) crack propagation around irregular pores; (**e**) internal cracks within an unmelted powder particle; and (**f**) distinct fatigue striations with micropores. White arrows highlight the described features, indicating fatigue-dominated fracture morphology.

3.9. *Comprehensive Microstructure and Chemical Analyses*

Comprehensive microstructural analyses (Sections 3.1–3.6) were conducted to interpret the fatigue behaviour observed in bending fatigue tests (Sections 3.7 and 3.8).

3.9.1. Microstructural and Porosity Findings

PBF-LB/M as a manufacturing method produces a hierarchical structure of melt pools, grains, and the solidification structure. OM images indicated face-dependent melt pool morphologies: a chessboard pattern on the top face and fish-scale patterns on the front and side faces. Variations in melt pool morphology and cooling conditions influenced the resulting grain structure and the quality of consolidation, subsequently determining the locations of fatigue crack initiation and influencing crack propagation paths. The solidification structure is unique to PBF-LB/M SS316L when compared to conventional SS316L, and may improve fatigue resistance, offering resistance to slip and leading to crack branching. Porosity observed at melt pool boundaries, clearly identified in OM images, acted as preferential sites for fatigue crack initiation. SEM analysis further revealed micropores concentrated along grain boundaries, especially on the top surface, providing additional sites for potential fatigue crack initiation.

EDX analysis confirmed the primary alloying elements (Fe, Cr, Ni) and indicated elevated carbon content, particularly on the front face. However, due to limitations such as surface contamination, detection sensitivity, and spectral peak overlap, EDX alone cannot reliably quantify carbon or confirm carbide formation without additional supporting evidence. XRD analysis provided such additional evidence by identifying a predominantly FCC austenitic structure with minor ferritic phases and chromium carbides (Cr_7C_3). However, carbides were not directly visible in SEM images, likely because their size was below the resolution limits at the employed magnifications. As a result, despite XRD confirmation, the direct role of carbides in influencing fatigue fracture remains unclear and requires further investigation.

3.9.2. Influence of Microstructure on Bending Fatigue Behaviour

The fatigue behaviour observed in this study was closely related to microstructural defects such as unmelted powder particles and micropores formed during the PBF-LB/M process. Under elastic bending conditions (below yield point), fatigue cracks mainly initiated at micropores due to local stress concentrations around these defects. Previous studies [6,39] showed that the shape and size of pores significantly affect local stress levels. Larger, irregular pores, often formed by the laser melting process, typically cause higher stress concentrations due to their sharp edges compared to smaller, spherical gas pores. Because bending creates the highest tensile stress near surfaces farthest from the neutral axis, defects located in these areas became critical sites for fatigue crack initiation [6]. Additionally, grain structures formed by directional solidification during PBF-LB/M could also affect fatigue performance [40–42]. Columnar grains aligned with the build direction may influence how easily cracks initiate and propagate during cyclic loading. The fine cellular structures created by rapid cooling in PBF-LB/M can help slow crack growth by limiting the movement of defects (dislocations) at lower stresses.

3.9.3. Relative Influence of Microstructural Factors

A comparative ranking of microstructural factors influencing the bending fatigue behaviour is proposed. Micropores, particularly irregular pores, along with unmelted powder particles were identified as the most influential factors under both elastic and early plastic loading conditions, clearly supported by observed fatigue crack initiation at these defects due to local stress concentrations. Grain anisotropy, characterised by columnar grains aligned with the build direction (confirmed by EBSD), was also identified as a contributing factor. However, its direct impact on fatigue fracture was not conclusively observable in SEM analysis, and thus further experimental confirmation is necessary. Other microstructural factors, such as chromium carbides (detected by XRD), refined cellular solidification structures, and melt pool morphology, remain speculative. The exact roles of these features require further targeted investigation to confirm their influence on fatigue performance.

4. Conclusions

Microstructural and porosity analyses of PBF-LB/M SS316L were performed to evaluate bending fatigue performance under two distinct loading conditions: an elastic regime (below yield stress) and an early plastic regime (slightly above yield stress). Despite achieving high densification (>99%), residual defects, including micropores and unmelted powder particles, persisted due to inherent limitations of the PBF-LB/M process. SEM analyses demonstrated that these defects, particularly irregular pores, significantly influenced fatigue crack initiation. XRD confirmed a predominantly FCC-austenitic structure with minor ferritic phases and chromium carbides (Cr_7C_3), although carbides were not directly visible in SEM observations. Chemical composition of the primary alloying elements (Fe, Cr, Ni) remained consistent across all sample faces. Grain structures varied by sample orientation, exhibiting columnar grains parallel to the build direction on the front and side faces, and isotropic grains on the top face. However, explicit evidence of grain orientation influencing crack propagation paths was not conclusively observed, highlighting areas requiring further investigation. Fatigue life was found to be shorter in the early plastic regime, with irregular crack propagation observed on external surfaces. Although direct fatigue strength comparisons remain challenging due to varied experimental conditions reported in the literature, achieving fatigue performance comparable to conventional manufacturing generally requires optimised processing and appropriate post-processing treatments to mitigate inherent additive manufacturing defects.

Author Contributions: Conceptualisation, C.-W.C.; Methodology, C.K., W.X., B.F., C.W. and C.-W.C.; Validation, C.K., W.X., B.F., C.W. and C.-W.C.; Formal analysis, C.K., W.X., B.F. and S.D.; Investigation, C.K., W.X., B.F. and C.W.; Resources, H.-C.M. and C.-W.C.; Data curation, C.K., W.X., B.F. and C.-W.C.; Writing—original draft preparation, C.K., W.X. and S.D.; Writing—review and editing, H.-C.M. and C.-W.C.; Visualisation, C.K., W.X. and B.F.; Supervision, H.-C.M. and C.-W.C.; Project administration, C.-W.C.; Funding acquisition, H.-C.M. and C.-W.C. All authors have read and agreed to the published version of the manuscript.

Funding: This research was funded by the Research Grants Council of the Hong Kong Special Administrative Region (grant number 152131/18E) and Queen's University Belfast (grant number R8448MEE). Support from the infrastructure of Queen's University Belfast, The Hong Kong Polytechnic University, and the University Research Facility in 3D Printing (U3DP) is also acknowledged.

Data Availability Statement: The raw data supporting the conclusions of this article will be made available by the authors on request.

Conflicts of Interest: The authors declare no conflicts of interest.

References

1. Lee, J.-A.; Koh, Y.-G.; Kang, K.-T. Biomechanical and Clinical Effect of Patient-Specific or Customized Knee Implants: A Review. *J. Clin. Med.* **2020**, *9*, 1559. [CrossRef] [PubMed]
2. Moret, C.S.; Schelker, B.L.; Hirschmann, M.T. Clinical and Radiological Outcomes after Knee Arthroplasty with Patient-Specific versus Off-the-Shelf Knee Implants: A Systematic Review. *J. Pers. Med.* **2021**, *11*, 590. [CrossRef] [PubMed]
3. O'Connor, M.I.; Kransdorf, M.J. Customized Knee Arthroplasty and the Role of Preoperative Imaging. *Am. J. Roentgenol.* **2013**, *201*, W443–W450. [CrossRef] [PubMed]
4. Nagpure, D.; Asutkar, S. 3D Printing in Surgery: Transforming Patient-Specific Solutions. *Multidiscip. Rev.* **2024**, *8*, 2025154. [CrossRef]
5. Stendal, J.; Fergani, O.; Yamaguchi, H.; Espallargas, N. A Comparative Tribocorrosion Study of Additive Manufactured and Wrought 316L Stainless Steel in Simulated Body Fluids. *J. Bio Tribo Corros.* **2018**, *4*, s40735-s017. [CrossRef]
6. Liang, X.; Hor, A.; Robert, C.; Salem, M.; Lin, F.; Morel, F. High. Cycle Fatigue Behavior of 316L Steel Fabricated by Laser Powder Bed Fusion: Effects of Surface Defect and Loading Mode. *Int. J. Fatigue* **2022**, *160*, 106843. [CrossRef]
7. Kotrych, D.; Angelini, A.; Bohatyrewicz, A.; Ruggieri, P. 3D Printing for Patient-Specific Implants in Musculoskeletal Oncology. *EFORT Open Rev.* **2023**, *8*, 331–339. [CrossRef]
8. Pilliar, R.M.; Cameron, H.U.; Binnington, A.G.; Szivek, J.; Macnab, I. Bone Ingrowth and Stress Shielding with a Porous Surface Coated Fracture Fixation Plate. *J. Biomed. Mater. Res.* **1979**, *13*, 799–810. [CrossRef]
9. Chua, M.; Chui, C.-K. Optimization of Patient-Specific Design of Medical Implants for Manufacturing. *Procedia CIRP* **2016**, *40*, 402–406. [CrossRef]
10. Yadav, D.; Garg, R.; Ahlawat, A.; Chhabra, D. 3D Printable Biomaterials for Orthopedic Implants: Solution for Sustainable and Circular Economy. *Resour. Policy* **2020**, *68*, 101767. [CrossRef]
11. Fantin, L. de B.; Moreira, M.F.; de Souza, J.M.P. Failure Analysis of a Fractured Stainless Steel Femoral Bone Plate. *Eng. Fail. Anal.* **2025**, *173*, 109466. [CrossRef]
12. Guerra-Fuentes, L.; Garcia-Sanchez, E.; Juarez-Hernandez, A.; Hernandez-Rodriguez, M.A.L. Failure Analysis in 316L Stainless Steel Supracondylar Blade Plate. *Eng. Fail. Anal.* **2015**, *57*, 243–247. [CrossRef]
13. Sudhakar, K.V. Metallurgical Investigation of a Failure in 316L Stainless Steel Orthopaedic Implant. *Eng. Fail. Anal.* **2005**, *12*, 249–256. [CrossRef]
14. Polak, S.; Beever, L.; Wade, A.; Fukuoka, M.; Worth, A.J. Biomechanical Comparison of Titanium Alloy Additively Manufactured and Conventionally Manufactured Plate-Screw Constructs. *N. Zealand Vet. J.* **2024**, *72*, 17–27. [CrossRef]
15. Wu, K.; Li, B.; Guo, J. Fatigue Crack Growth and Fracture of Internal Fixation Materials in In Vivo Environments—A Review. *Materials* **2021**, *14*, 176. [CrossRef]

16. Zhang, H.; Chiang, R.; Qin, H.; Ren, Z.; Hou, X.; Doll, G.; Vasudevan, V.; Dong, Y.; Ye, C. The Effects of Ultrasonic Nanocrystal Surface Modification on the Fatigue Performance of 3D-Printed Ti64. *Int. J. Fatigue* **2017**, *103*. [CrossRef]
17. Rivolta, B.; Gerosa, R.; Panzeri, D. Selective Laser Melted 316L Stainless Steel: Influence of Surface and Inner Defects on Fatigue Behavior. *Int. J. Fatigue* **2023**, *172*, 107664. [CrossRef]
18. Concli, F.; Gerosa, R.; Panzeri, D.; Fraccaroli, L. High and Low Cycle Fatigue Properties of Selective Laser Melted AISI 316L and AlSi10Mg. *Int. J. Fatigue* **2023**, *177*, 107931. [CrossRef]
19. Blinn, B.; Krebs, F.; Ley, M.; Gläßner, C.; Smaga, M.; Aurich, J.C.; Teutsch, R.; Beck, T. Influence of the Chemical Composition of the Used Powder on the Fatigue Behavior of Additively Manufactured Materials. *Metals* **2019**, *9*, 1285. [CrossRef]
20. Kong, D.; Ni, X.; Dong, C.; Zhang, L.; Man, C.; Cheng, X.; Li, X. Anisotropy in the Microstructure and Mechanical Property for the Bulk and Porous 316L Stainless Steel Fabricated via Selective Laser Melting. *Mater. Lett.* **2019**, *235*, 1–5. [CrossRef]
21. Marattukalam, J.J.; Karlsson, D.; Pacheco, V.; Beran, P.; Wiklund, U.; Jansson, U.; Hjörvarsson, B.; Sahlberg, M. The Effect of Laser Scanning Strategies on Texture, Mechanical Properties, and Site-Specific Grain Orientation in Selective Laser Melted 316L SS. *Mater. Des.* **2020**, *193*, 108852. [CrossRef]
22. Dixit, S.; Liu, S.; Murdoch, H.A.; Smith, P.M. Investigating Build Orientation-Induced Mechanical Anisotropy in Additive Manufacturing 316L Stainless Steel. *Mater. Sci. Eng. A* **2023**, *880*, 145308. [CrossRef]
23. Ganesh, P.; Kaul, R.; Sasikala, G.; Kumar, H.; Venugopal, S.; Tiwari, P.; Rai, S.; Prasad, R.C.; Kukreja, L.M. Fatigue Crack Propagation and Fracture Toughness of Laser Rapid Manufactured Structures of AISI 316L Stainless Steel. *Metallogr. Microstruct. Anal.* **2014**, *3*, 36–45. [CrossRef]
24. Jaskari, M.; Mäkikangas, J.; Järvenpää, A.; Mäntyjärvi, K.; Karjalainen, P. Effect of High Porosity on Bending Fatigue Properties of 3D Printed AISI 316L Steel. *Procedia Manuf.* **2019**, *36*, 33–41. [CrossRef]
25. Chen, Y.; Wang, X.; Shen, J.; Peng, Y.; Jiang, Y.; Yang, X.; Leen, S.B.; Gong, J. Deformation Mechanisms of Selective Laser Melted 316L Austenitic Stainless Steel in High Temperature Low Cycle Fatigue. *Mater. Sci. Eng. A* **2022**, *843*, 143123. [CrossRef]
26. Winwood, K.; Zioupos, P.; Currey, J.D.; Cotton, J.R.; Taylor, M. The Importance of the Elastic and Plastic Components of Strain in Tensile and Compressive Fatigue of Human Cortical Bone in Relation to Orthopaedic Biomechanics. *J. Musculoskelet. Neuronal Interact.* **2006**, *6*, 134–141.
27. Xie, W.; Hussain, I.; Man, H.-C.; Chan, C.-W. Adoption of Appropriate Scanning Strategy on Selective Laser Melting of Stainless Steel 316L to Enable Healthy Mesenchymal Stem Cells Response. *J. Laser Appl.* **2023**, *35*, 022019. [CrossRef]
28. *ASTM E112-13(2021)*; Standard Test Methods for Determining Average Grain Size. ASTM International: West Conshohocken, PA, USA, 2021. Available online: https://compass.astm.org/document/?contentCode=ASTM%7CE0112-13R21%7Cen-US&proxycl=https://secure.astm.org&fromLogin=true (accessed on 18 October 2023).
29. Wang, S.; Ning, J.; Zhu, L.; Yang, Z.; Yan, W.; Dun, Y.; Xue, P.; Xu, P.; Bose, S.; Bandyopadhyay, A. Role of Porosity Defects in Metal 3D Printing: Formation Mechanisms, Impacts on Properties and Mitigation Strategies. *Mater. Today* **2022**, *59*, 133–160. [CrossRef]
30. Liu, J.; Song, Y.; Chen, C.; Wang, X.; Li, H.; Zhou, C.; Wang, J.; Guo, K.; Sun, J. Effect of Scanning Speed on the Microstructure and Mechanical Behavior of 316L Stainless Steel Fabricated by Selective Laser Melting. *Mater. Des.* **2020**, *186*, 108355. [CrossRef]
31. Minns, R.J.; Bremble, G.R.; Campbell, J. A Biomechanical Study of Internal Fixation of the Tibial Shaft. *J. Biomech.* **1977**, *10*, 569–579. [CrossRef]
32. *ISO 12108:2018*; Metallic Materials—Fatigue Testing—Fatigue Crack Growth Method. International Organization for Standardization: Geneva, Switzerland, 2018. Available online: https://www.iso.org/standard/73809.html (accessed on 24 March 2025).
33. Leary, M.; Maconachie, T.; Sarker, A.; Faruque, O.; Brandt, M. Mechanical and Thermal Characterisation of AlSi10Mg SLM Block Support Structures. *Mater. Des.* **2019**, *183*, 108138. [CrossRef]
34. Sanaei, N.; Fatemi, A. Defects in Additive Manufactured Metals and Their Effect on Fatigue Performance: A State-of-the-Art Review. *Prog. Mater. Sci.* **2021**, *117*, 100724. [CrossRef]
35. Qu, H.; Li, J.; Zhang, F.; Bai, J. Anisotropic Cellular Structure and Texture Microstructure of 316L Stainless Steel Fabricated by Selective Laser Melting via Rotation Scanning Strategy. *Mater. Des.* **2022**, *215*, 110454. [CrossRef]
36. Andreau, O.; Koutiri, I.; Peyre, P.; Penot, J.-D.; Saintier, N.; Pessard, E.; De Terris, T.; Dupuy, C.; Baudin, T. Texture Control of 316L Parts by Modulation of the Melt Pool Morphology in Selective Laser Melting. *J. Mater. Process. Technol.* **2019**, *264*, 21–31. [CrossRef]
37. Avanzini, A. Fatigue Behavior of Additively Manufactured Stainless Steel 316L. *Materials* **2022**, *16*, 65. [CrossRef]
38. Chepkoech, M.; Omoniyi, P.; Owolabi, G. Fatigue Response of Additive-Manufactured 316L Stainless Steel. *Metals* **2024**, *14*, 988. [CrossRef]
39. Zhang, W.; Pommier, S.; Curtit, F.; Léopold, G.; Courtin, S. Mode I Crack Propagation under High Cyclic Loading in 316L Stainless Steel. *Procedia Mater. Sci.* **2014**, *3*, 1197–1203. [CrossRef]
40. Deev, A.A.; Kuznetcov, P.A.; Petrov, S.N. Anisotropy of Mechanical Properties and Its Correlation with the Structure of the Stainless Steel 316L Produced by the SLM Method. *Phys. Procedia* **2016**, *83*, 789–796. [CrossRef]

41. Carassus, H.; Guérin, J.D.; Morvan, H.; Haugou, G.; Sadat, T.; Guérard, S.; Markiewicz, E. An Experimental Investigation into Influences of Build Orientation and Specimen Thickness on Quasi-Static and Dynamic Mechanical Responses of Selective Laser Melting 316L Stainless Steel. *Mater. Sci. Eng. A* **2022**, *835*, 142683. [CrossRef]
42. Aiza, I.; Baldi, C.; de la Vega, F.M.; Sebastiani, S.; Veronese, N.E.; Yousefi, M.; Mosallanejad, M.H.; Maleki, E.; Guagliano, M.; Iuliano, L.; et al. Effects of Build Orientation and Inclined Features on Physical, Microstructural and Mechanical Properties of Powder Bed Fusion Additively Manufactured Metallic Parts. *Prog. Mater. Sci.* **2025**, *147*, 101357. [CrossRef]

metals

MDPI

Article

The Effect of Drag Finishing on Additively Manufactured Customized Dental Crowns

Cosmin Cosma [1,*], Martin Melichar [2], Stelian Libu [3], Alexandru Popan [1], Glad Contiu [1], Cristina Teusan [1], Petru Berce [1] and Nicolae Balc [1,*]

1 Department of Manufacturing Engineering, Technical University of Cluj-Napoca, 400641 Cluj-Napoca, Romania; ioan.popan@tcm.utcluj.ro (A.P.); glad.contiu@tcm.utcluj.ro (G.C.); christina-valentina.teusan@man.eu (C.T.); petru.berce@tcm.utcluj.ro (P.B.)
2 Regional Technological Institute, University of West Bohemia, 30614 Pilsen, Czech Republic; mech@fst.zcu.cz
3 Department of Digital Technology, Dental Design Lab, 18121 Athens, Greece; dentaldesignlab.gr@gmail.com
* Correspondence: cosmin.cosma@tcm.utcluj.ro (C.C.); nicolae.balc@tcm.utcluj.ro (N.B.)

Abstract: Cobalt–chromium (CoCr) alloys are frequently used to produce customized dental applications such as crowns, bridges, or prostheses. These medical products have anatomical forms, and can be effectively manufactured using the laser-based powder bed fusion (PBF-LB/M) technique. A major disadvantage of this approach is the extended time required to refine the resultant surface. The purpose of this research is to reduce the surface roughness of PBF-LB/M/CoCr dental crowns by adopting drag finishing (DF) technology. To evaluate the impact of this automatic post-processing, surface roughness measurements and geometrical investigations were undertaken. The microstructure was characterized using scanning electron microscopy (SEM), and the chemical composition was verified by energy-dispersive X-ray spectroscopy (EDAX). On outside surfaces, the DF post-processing decreased the initial surface roughness by 70–90%. The dental crown's surface roughness value after DF post-processing was comparable to that of the basic form (cylinder). The lowest roughness was obtained with DF3 post-processing (Ra~0.60 μm). The inner surfaces were limitedly finished. The 3D surface texture showed that the DF method reduced the height of peaks, uniformizing the surfaces. CMM work compared the deviations between the virtual model and the printed samples before and after DF post-processing. This analysis revealed that dimensional deviations were reduced on the outside crown walls, ranging from +0.01 to +0.05 mm. The laser parameters and the heat treatment applied increased the hardness of CoCr crowns to 520 HV, but the proper DF conditions identified reduced the surface roughness and improved the accuracy.

Keywords: complex surfaces; CoCr; drag finishing; surface roughness; accuracy evaluation

1. Introduction

Nowadays, laser-based powder bed fusion (PBF-LB/M) is one of the most frequently used additive manufacturing (AM) methods, able to produce directly metallic prototypes from virtual models. The PBF-LB/M process rapidly penetrated the dental field, where for each patient, there is a need to manufacture a custom and anatomic prosthetic restoration [1]. CoCr alloys are frequently applied in dental applications to develop porcelain-fused-to-metal prostheses [2]. The main customized products made of CoCr powder via the PBF-LB/M process are dental crowns, bridges, prostheses, implant bars for over-dentures, and removable frames [3–7]. The advantages of applying CoCr alloys in prosthetic restorations as a substructure are the mechanical properties (e.g., high elastic modulus) that make it possible to reduce the dimensions of the framework, and the adequate bond strength between the porcelain and metal, providing corrosion resistance [8–10].

From a manufacturing point of view, a major disadvantage of PBF-LB/M parts is the extended time required to refine the resultant surface roughness of complex dental applications [11,12]. It is well known that PBF-LB/M components have a rough and uneven surface, influenced by the stair-step effect, semi-melted grains attached to surfaces, micropores, and adverse residual stresses [13–15]. In general, the surface roughness ranges from 5 to 15 μm in the case of the Ra parameter [16–18]. Often, the post-processing procedure of CoCr crowns or prostheses is performed manually by dental technicians using pneumatic straight grinders. A concern regarding this procedure relates to the fact that the dental technicians are susceptible to inhaling grinding dust during adjustments and polishing of the CoCr parts [19,20]. Moreover, the final surface roughness and accuracy of CoCr frameworks depend on dental technicians' skills and abilities. These issues motivated the authors to explore the possibility of using an automatic system for finishing PBF-LB/M-processed dental applications.

Various researchers describe how the surfaces of PBF-LB/M-processed components can be finished by automatic techniques that are commonly used in industrial applications. The main methods that can be applied to metal AM parts are vibratory bowl abrasion, hot cutter machining, chemical and electromechanical polishing, and laser micromachining [13,21–24]. For example, Lober et al. present four post-processing technologies and combinations of them applied to stainless PBF-LB/M-fabricated steel parts (grinding, sand blasting, electrolytic polishing, and plasma polishing) [17]. To reduce the roughness of titanium (Ti) specimens processed by AM, Ma et al. and Mohammad et al. tested a laser polishing technology [25,26]. However, most AM studies are based on basic shapes. Significant challenges remain in this field, because dental applications or customized implants frequently have high complexity and anatomical shapes, making it difficult to improve the surface quality.

Drag finishing (DF) is a mass finishing method, and a review of the literature reveals that only limited scientific research has been published on PBF-LB/M-processed components post-processed using this technology. Previous studies have focused mainly on Ti alloys [27] and stainless steel [28,29]. This emphasizes the need to evaluate the use of DF technology on hard metals like CoCr. Compared with other biocompatible dental alloys, such as Ti6Al4V [30], Ni-Cr [31], or stainless steel [32], CoCr is the hardest and most laborious for post-processing. To obtain a lower surface roughness on complex CoCr specimens, adapted DF conditions are still required. Moreover, it is also necessary to determine the influence of DF on dimensional deviations.

The main purpose of this research was to reduce the surface roughness of PBF-LB/M dental crowns by adapting the DF technology for use on CoCr alloys. Surface roughness measurements and geometrical investigations were undertaken to evaluate the dimensional deviations that occurred before and after DF post-processing. The microstructure was characterized by scanning electron microscopy (SEM), and the chemical composition was verified by energy-dispersive X-ray spectroscopy (EDAX). Additionally, 3D surface topography analyses and hardness measurements were considered. To evaluate the impact of this automatic post-processing, the PBF-LB/M specimen had a complex shape (dental crown) and a basic form (cylinder).

2. Experimental Section

2.1. Material

The present investigation was carried out using CoCr alloy powder (commercially named Starbond CoS 30) produced by S & S Scheftner (Mainz, Germany). In general, the diameter of the grains was between 10 and 30 μm. SEM investigations revealed that the grains of this powder had a spherical shape, with a very limited amount of satellites (Figure 1a). Some deformed particles were also found in the powder feedstock. Because this powder was obtained through a gas atomization process, the topology of CoCr grain was rougher, with a moon-like surface (Figure 1b). The powder flowability corresponded to laser bed fusion requirements. The solidus–liquidus interval of this alloy was between 1305 °C and 1400 °C, and the printed samples were highly corrosion resistant. According to EDAX spectrometry, this powder had the chemical composition detailed in Table 1, which respects ISO 22674 for dental materials [33]. Beryllium, cadmium, and nickel were absent from the feedstock.

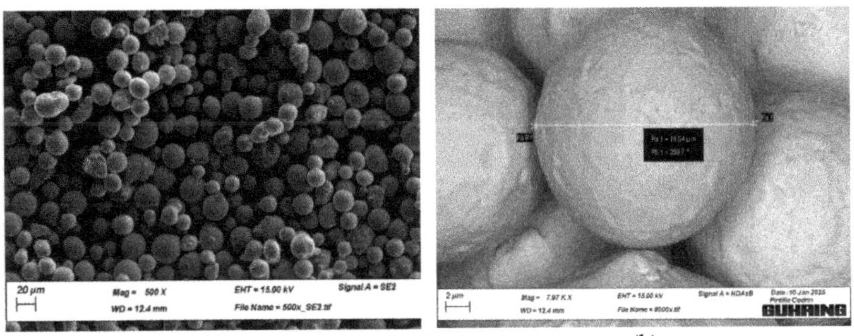

(a) (b)

Figure 1. SEM investigation of CoCr powder: (**a**) feedstock; (**b**) grain morphology.

Table 1. Chemical composition of CoCr powder.

Chemical Element	Co	Cr	W	Mo	Si	Mn	O	Other Elements (C, Fe, N)
Maximum weight percentage [%]	59.3	25.1	9.4	3.4	1.2	0.9	0.5	Max. 1

2.2. Specimen Design

The complex model considered was a customized dental crown containing multiple anatomical shapes, sharp internal corners, and uneven thicknesses (Figure 2a). The main steps undertaken to design this part are specific to the restorative dentistry field. Initially, intraoral digital scanning of a prepared maxillary first molar (known as "16") was conducted using the Dental Wings IntraOral (DWIO) system (Dental Wings GmbH, Chemnitz, Germany). The resulting scan was modeled using DWOS Designing Dental Restorations software (version 8.1, Dental Wings GmbH, Chemnitz, Germany) to obtain the virtual metallic substructure of the dental crown. The virtual model was designed with thin walls, with a thickness between 0.50 and 0.70 mm (Figure 2). These surfaces were modeled according to patient tooth anatomy prepared by a dentist, tilted at angles normally varying between 55° and 88°. An example is illustrated in Figure 2b, where the wall angles are measured as 65° on the left side and 55° on the right side. Because crown margins are a critical factor in restoration fit, a chamfer finishing line was applied (see Figure 2b). This type of final margin is recommended for metal–ceramic restorations, and it has been demonstrated to exert the least stress on tooth dentine. Moreover, the dental crown was fixed on a cylindrical holder with a 12.00 mm diameter (see Figure 2a). The cylindrical holder was added to analyze the influence of the drag finishing process on complex shapes vs. basic shapes.

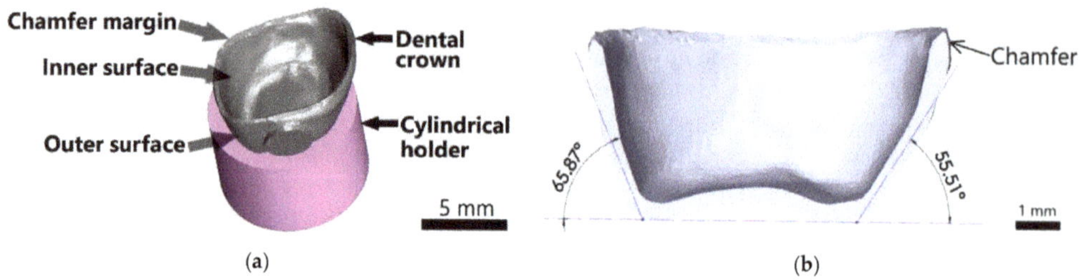

Figure 2. (**a**) Virtual design of dental crown fixed on cylindrical holder and its main surfaces (inner, outer, and chamfer margin); (**b**) typical angled walls of dental crown designed with chamfer margin.

2.3. PBF-LB/M Manufacturing

Sample fabrication was carried out using a MySint 100 system (Sisma S.p.A., Vicenza, Italy), a PBF-LB/M machine equipped with a 150 W continuous-wave Nd:YAG fiber laser. Under the slice-by-slice principle, this laser metal fusion technology is capable of manufacturing complex and anatomical parts. The working parameters used, as shown in Table 2, were selected based on our PBF-LB/M manufacturing expertise [2,34,35] and relevant guidance from the literature [14,36–38]. These laser parameters were programmed to scan the outer and inner boundaries, respectively, of the hatch area. The scanning strategy chosen was a chessboard pattern (rotated islands), and the specimens were built up vertically (build orientation is shown in Figure 2b, cavities facing up). To sustain the specimens during the AM process, they were anchored with conic supports, modeled in AutoFab software according to best practices from the literature [39]. Under a high-purity nitrogen atmosphere and a limited oxygen level (under 1%), twenty dental crowns were manufactured at the same time. Hence, homogeneity of the test was ensured. To avoid any unfavorable effects of residual stress, the specimens were exposed to a stress relief treatment. The heat treatment was conducted in an electric oven (type N 31/H, Nabertherm GmbH, Lilienthal, Germany) in an air atmosphere. Firstly, the specimens were heated up to 860 °C at a rate of 6 °C/min. They were held at this temperature for one hour, followed by a cooling rate of 10 °C/min. After the temperature decreased to 300 °C, the oven door was opened for natural cooling. Finally, the specimens were cleaned off supports and randomly divided into 4 groups, one for control (as-built surfaces) and three for the drag finishing process (5 samples per group).

Table 2. Technological parameters of PBF-LB/M/CoCr manufacturing.

Laser Power (W)	Laser Scanning Speed (mm/s)	Hatch Spacing (μm)	Layer Thickness (μm)	Laser Spot Diameter (μm)	Scanning Strategy
85	960	60	25	30	Chessboard

2.4. Drag Finishing Process

The equipment used for DF post-processing was the OTEC DF Series 3 (OTEC Präzisionsfinish GmbH, Straubenhardt, Germany), and it is illustrated in Figure 3a. To adapt the DF technology for hard metals such as CoCr alloys, a proper abrasive medium was identified and tested. The actual DF post-processing was carried out using the HSC 1/300 product impregnated with a type OTEC P28 polishing paste. HSC 1/300 is a mixture of 70% walnut shell granulate, 0.80–1.30 mm in size, and 30% silicon carbide (SiC) grains, with an approx. 200 μm diameter. This abrasive medium is illustrated in Figure 3b. The DF procedure was carried out in dry conditions. The workpieces were clamped in a rotary head sustained by a rod, and they were gradually immersed into the abrasive medium, which exerted pressure on them. To release the dental cavity from the grains during the DF process, they were quickly ejected (7 times). Varying the machining time and keeping the other parameters constant, the samples were post-processed. Table 3 details the DF process conditions considered for each group (DF1, DF2, and DF3). The actual DF parameters and abrasive medium were employed based on initial research carried out by the Regional Technological Institute (University of West Bohemia) in the Czech Republic [40,41].

(**a**) (**b**)

Figure 3. (**a**) Drag finishing equipment (OTEC DF Series 3); (**b**) HSC 1/300 abrasive medium containing walnut shells and SiC grains.

Table 3. Process conditions of DF method.

Code of DF Condition	Main Heat Speed [rpm]	Spindle Speed [rpm]	Immersion Depth [mm]	Clockwise Rotation [min]	Counter-Clockwise Rotation [min]	Total Finishing Time [min]
DF1				22.5	22.5	45
DF2	35	60	420	30	30	60
DF3				37.5	37.5	75

2.5. Examination Methods

The roughness evaluation was conducted using an Alicona G4 Infinite Focus (Alicona Imaging, Graz, Austria) microscope. This optical profilometry system uses focus-variation technology to extract 3D morphology and depth information from curved surfaces [42,43]. In addition to the profile roughness parameters calculated according to ISO 4287 [44], this instrument offered topographical information about surface texture. The objective magnification was 20×, the cut-off wavelength was 250 μm (λc), the vertical resolution was 50 μm, and lateral resolution setup was 3.5 μm. Optical micrographs were recorded, and twelve roughness parameters were calculated. More than 10 measurements were carried out on each surface, with a profile length of approximately 3.5 mm, at 1 mm distance. The surface roughness profiles were measured perpendicular to the build layers, on outer surfaces.

The geometrical inspection was carried out with a coordinate measuring machine (CMM) type Carl Zeiss Navigator 7 (Carl Zeiss IQS Deutschland GmbH, Oberkochen, Germany) and Calypso 5.8 software (Zeiss, Maple Grove, MN, USA). This CMM made accurate determinations, comparing the deviations between the virtual model and the PBF-LB/M samples. The dimensions were evaluated using point-to-3D-model distance, according to ISO 10360 [45]. On the virtual model of the sample, 180 points were defined as follows: 80 points on the inner surfaces and 100 points on the outer surfaces (Figure 4a). The CNC travel paths were automatically generated. To position all the samples, a small cylinder was considered as a reference (horizontal cylinder, Figure 4a, left image). All these points were measured on PBF-LB/M samples using a star stylus with a Ruby ball tip (Ø 0.5 mm, Figure 4b). This scanning head measures with approx. 1 μm (0.9 + L/350 μm) precision, applying a light force of 0.1 N

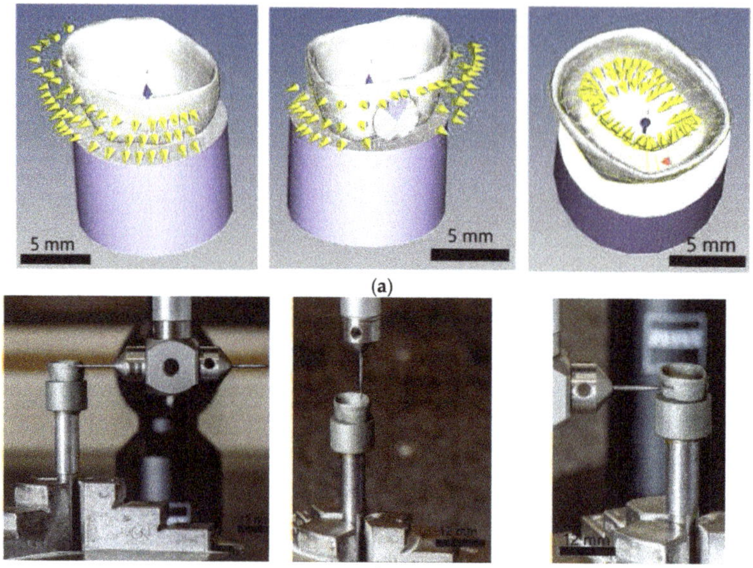

(a)

(b)

Figure 4. (**a**) Measured points on dental crowns; (**b**) CMM investigation on initial surface of CoCr dental crown.

Vickers hardness tests were completed with a Wilson Tukon 1102 (Buehler, Lake Bluff, IL, USA) instrument, applying a force of 1 kg across the CoCr surfaces, and the dwell time was 15 s. Several trials were conducted on each sample, in accordance with ISO 6507 (n = 15) [46], and the mean microhardness was calculated. All these examinations were conducted on the initial surface and after the DF post-processing.

Microstructural aspects, as well as the chemical composition, were investigated by scanning electron microscopy (SEM) and energy dispersive X-ray analysis (EDAX). The analyses were performed using a Sigma Gemini Zeiss microscope (Carl Zeiss Microscopy Deutschland GmbH, Oberkochen, Germany). Before the SEM investigations, the specimens are cleaned by ultrasound, while sunken into isopropyl alcohol at 35 °C for 30 min at a frequency of 17 kHz. Finally, the samples were washed with distilled water and dried with hot air. During the SEM investigations, the morphological characteristics of the surfaces were recorded and analyzed. EDAX spectrometry was used for elemental mapping of the CoCr surfaces.

2.6. Statistical Analysis

To test the significance of the data obtained, statistical analysis was performed in TIBCO Statistica software solution (version 14.1.0, StatSoft GmbH, Hamburg, Germany). Consistency of the samples was achieved by using uniform production technology under constant conditions. Analysis of variance (ANOVA) was used to verify data homogeneity. ANOVA is a mathematical–statistical method that allows for verification of whether the value of a random variable for a particular individual is statistically significantly affected by the value of any feature that can be observed individually. This feature must have only a finite number of possible values (at least two), and is used to divide individuals into groups of comparison. However, the quantitative value of a feature does not have the character of a measurement. If a specific quantitative value must also be considered as a measurement of a particular feature, a linear model must be used instead of variance analysis. The ANOVA was based on a regression function. A mathematical model was created to evaluate the significance level of the parameter ($p < \alpha$, where $\alpha = 0.05$ or 5%).

3. Results

3.1. Surface Roughness

Figure 5 shows the sample surfaces before and after they were post-processed with DF for 45–75 min. The initial CoCr samples were covered by a thin oxide layer, as can be seen in Figure 5. The oxidation layer appeared because the stress relief treatment was performed without a protective gas environment. The inner surface roughness was reduced by less than 25%, indicating a reduced impact of DF post-processing on the inside walls (see Table 5). Since DF post-processing had a limited effect on the inner surfaces, the outer surfaces were analyzed in greater detail. The DF post-processing was able to deburr, smooth, and polish the outer surfaces in one operation. On the external surfaces, the DF post-processing firstly deburred the CoCr surfaces, secondly cleaned the surfaces of attached partially melted grains, and thirdly polished the surfaces. All the DF conditions completely removed the oxide layer and the undesirable grains attached to the outer surfaces (Figure 5). Differences in polishing were seen according to the DF time (45, 60, and 75 min).

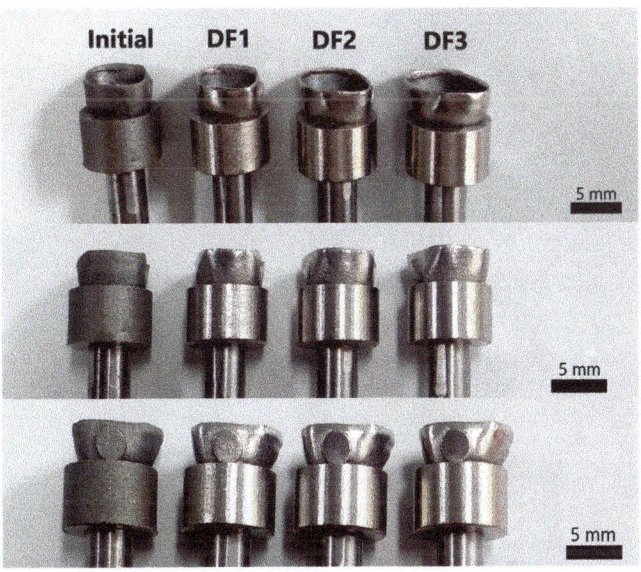

Figure 5. Dental crowns fabricated by PBF-LB/M/CoCr, before and after drag finishing of different durations.

To quantify the effect of DF parameters on the CoCr outer surfaces, Tables 4 and 5 summarize the Ra and Rz values measured on dental crowns with a basic cylindrical form and with anatomical shapes (inner, outer surfaces, and chamfer margin). The as-built Ra roughness varies between 5.20 and 6.95 μm, and in the case of Rz, between 23.80 and 53.32 μm (Tables 4 and 5). Analyzing the profiles recorded, they are non-periodic, with specific peaks and valleys. Significant differences were recorded in the surface roughness before and after the DF process. Overall, DF post-processing reduced both Ra and Rz roughness values by 73–91% for basic shapes (cylinders), 70–90% for outer surfaces, and 76–93% for the chamfer margin. The lowest roughness was obtained on samples post-processed with the DF3 condition (see Tables 4 and 5). These results indicate that the DF surface is more suitable for future applications, because the peaks were removed from the surface. These peaks can concentrate stress, so they could cause the dental crown to crack.

Table 4. Surface roughness of cylindrical surfaces before and after different DF post-processing (mean values).

Code of DF Condition	Ra Surface Roughness [μm]			Rz Surface Roughness [μm]		
	As-Built	After DF	Difference [%]	As-Built	After DF	Difference [%]
DF1	5.20	1.52	73.19	23.80	7.65	67.86
DF2	5.54	1.05	82.84	34.09	5.38	84.22
DF3	6.05	0.51	91.57	29.81	4.05	86.41

Table 5. Roughness of outer and inner surfaces and chamfer margins of dental crowns before and after DF post-processing (mean values).

	Code of DF Condition	Ra Surface Roughness [μm]			Rz Surface Roughness [μm]		
		As-Built	After DF	Difference [%]	As-Built	After DF	Difference [%]
Outer surfaces	DF1	5.95	1.60	70.45	27.83	8.20	70.54
	DF2	6.25	1.10	84.38	53.32	6.59	87.64
	DF3	6.14	0.61	90.07	32.49	4.53	86.06
Chamfer	DF1	6.29	1.48	76.47	36.48	7.40	79.72
	DF2	6.01	0.98	83.69	35.46	5.39	84.80
	DF3	6.67	0.49	92.65	40.02	2.79	93.02
Inner surfaces	DF1	6.95	5.24	24.60	40.31	31.65	21.48
	DF2	5.85	4.97	15.04	29.87	23.07	22.77
	DF3	6.77	5.48	19.05	35.38	31.11	12.07

To establish the impact of DF post-processing on outer surfaces made of CoCr powder, the present study details various roughness parameters, as follows: Ra, Rq, Rku, Rsk, Rz, Rp, Rv, Rc, Rt, Rsm, Rdq, and Rmax. To gain a complete perspective of the DF surfaces, Table 6 presents the mean values of these roughness parameters. Non-contact 3D surface characterization allowed us to calculate various roughness profile parameters, grouped into average amplitude parameters, amplitude parameters (peak and valley), and others. The average amplitude parameters included the arithmetical mean height (Ra), the root mean square deviation (Rq), the kurtosis (Rku), and the skewness roughness profiles (Rsk). The amplitude parameters included the maximum height of the peak and valley (Rz), the maximum profile peak height (Rp), the maximum profile valley depth (Rv), the mean height (Rc), and the total height (Rt). Other parameters considered were the mean spacing of profile irregularities of the roughness profile (Rsm), the root mean square slope (Rdq), the maximum peak-to-valley height of the roughness profile within a sampling length (Rmax), and the ratio of Rt/Rz extreme scratch/peak values. All the roughness parameters were calculated using the specific profile recorded and according to ISO 4287 equations and filters [44]. Their significance has also been well explained previously [47]. Table 6 details all the roughness parameters measured on each drag-finished surface. Because the Rsk values are less than 0, the height distribution is above the mean line. The Rku relates to the tip geometry of peaks and valleys, and the DF1 surfaces have an even height distribution (Rku < 3). DF2 and DF3 surfaces have a sharp height distribution (Rku > 3). Regarding the amplitude parameters, Table 6 provides evidence of the fact that the DF3 conditions significantly limited the peak's height (Rp) and the valley's depth (Rv), making a uniform surface. The hybrid Rdq parameter represents the numerical steepness, which was reduced to 0.08 µm for the DF3 surface. It is confirmed that after DF3, the lowest surface roughness value among all conditions was achieved.

Table 6. Mean values of profile roughness measurements on CoCr outer surfaces after different drag finishing conditions.

Code of DF Condition	Amplitude Average Parameters				Amplitude Parameters (Peak and Valley)				Others				
	Ra [µm]	Rq [µm]	Rsk [µm]	Rku [µm]	Rz [µm]	Rp [µm]	Rv [µm]	Rc [µm]	Rt [µm]	Rsm [µm]	Rmax [µm]	Rdq [µm]	Rt/Rz Ratio
DF1	1.60	1.99	−0.19	2.87	8.20	5.31	6.67	6.07	11.98	178.14	11.23	0.12	1.46
DF2	1.10	1.52	−0.68	6.57	6.59	4.71	7.76	5.75	12.47	280.08	12.47	0.11	1.89
DF3	0.61	1.04	−1.29	7.97	4.53	2.41	5.79	3.96	8.20	292.47	8.20	0.08	1.80

In addition to roughness measurements, 3D images of the surface topography were collected using height-based coloring, both on the cylinder shape and on the dental crown outer surfaces. The depth information from curved surfaces was recorded, and height map visualizations are shown in Figures 6–9. The initial surface is presented in Figure 6, where the peaks and valleys are between −30 µm and +30 µm. In some areas, they were beyond this scale. The typical topography for each DF surface is illustrated in Figures 7–9 for basic and complex shapes. The map scale was reduced to ±14 µm for outer surfaces. As can be seen from the 3D surface texture, the DF method reduced the height of peaks, uniformizing the surfaces. Compared with as-built surfaces, the parameters of DF3 finishing could significantly improve the quality of basic and anatomical shapes (i.e., Figure 6b vs. Figure 8c). The chamfer area of the dental crown was smoother, and the surface texture has value of ±8 µm on the map scale (Figure 9). The results confirm the feasibility of employing the DF3 conditions for finishing dental applications made from CoCr.

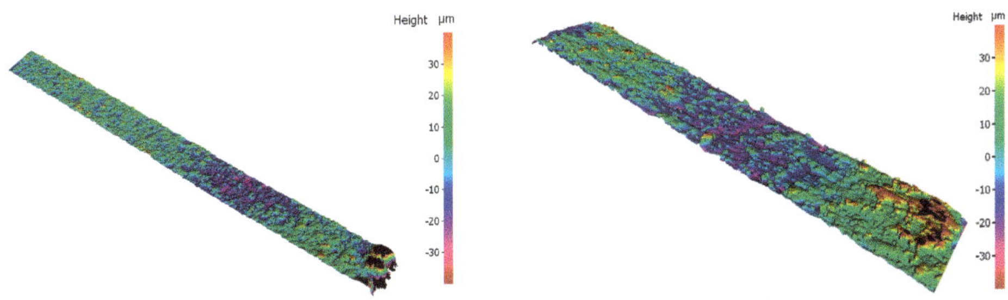

(a) (b)

Figure 6. Height map visualization of initial surface topography: (a) cylinder; (b) crown outer surfaces.

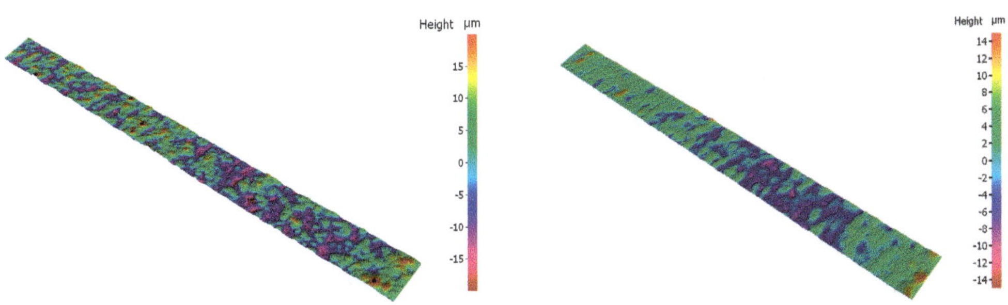

(a) (b)

Figure 7. *Cont.*

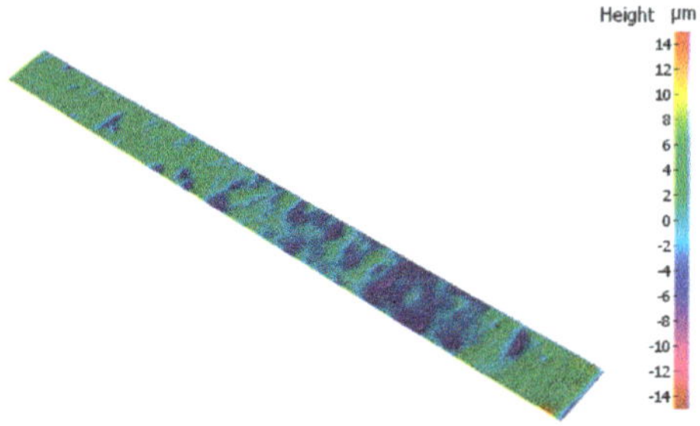

(**c**)

Figure 7. Height map visualization of cylinder surface topography after DF post-processing: (**a**) DF1, (**b**) DF2, (**c**) DF3.

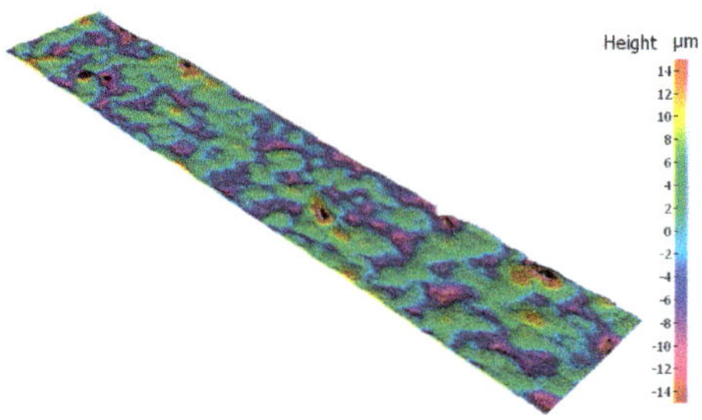

(**a**)

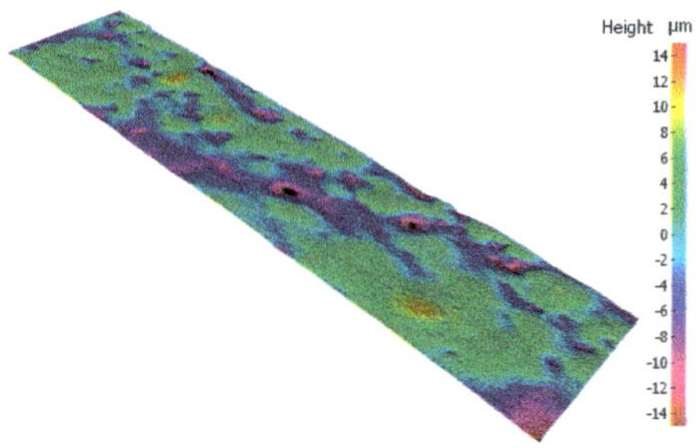

(**b**)

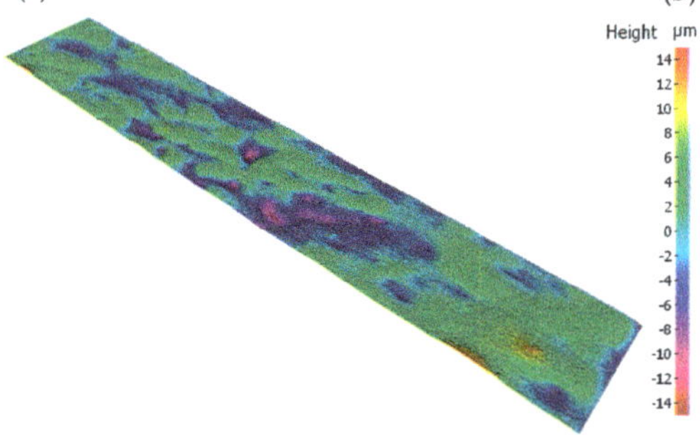

(**c**)

Figure 8. Height map visualization of crown outer surface topography after DF post-processing: (**a**) DF1, (**b**) DF2, (**c**) DF3.

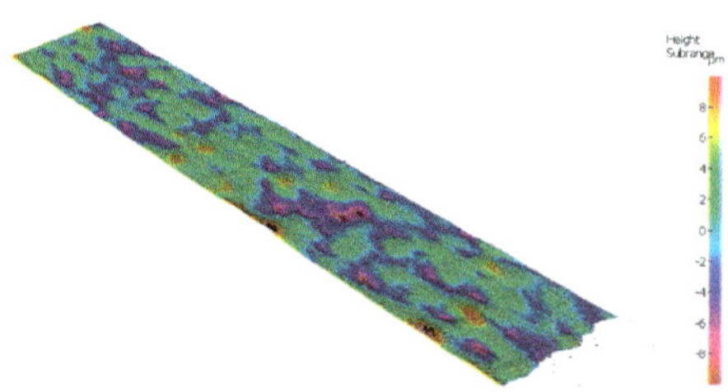

(**a**)

Figure 9. *Cont.*

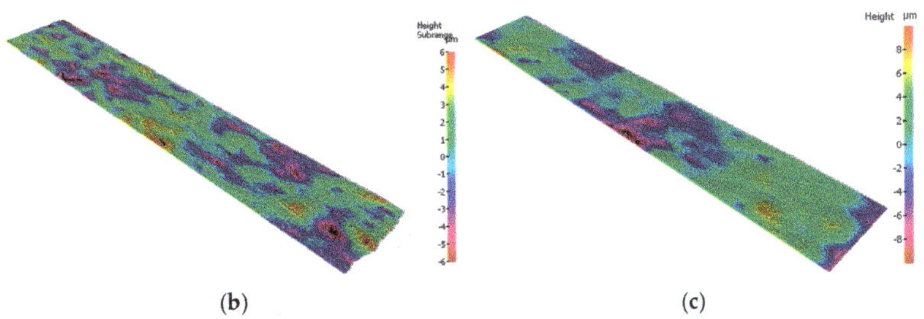

Figure 9. Height map visualization of chamfer surface topography after DF post-processing: (**a**) DF1, (**b**) DF2, (**c**) DF3.

3.2. Dimensional Deviations—Accuracy

The graphs shown in Figure 10 summarize the CMM work, where the difference values between the virtual model points and the actual ones located on DF surfaces are depicted (point-to-3D-model distance measurements). In Figure 10, the measured points considered are from the outside of the dental crown. To gain a deeper comprehension, values around the 0.00 difference level mean that the DF parameters improve the accuracy of the PBF-LB/M crown, and dimensional deviations are very limited. For example, DF3 finishing reduced the dimensional deviations on the outside crown walls, and the difference values between the virtual model and the surfaces range from +0.010 to +0.050 mm. Because in all the cases, the difference in distance has predominantly positive values, this means that the DF process did not polish the surfaces too much and affect the thin walls of the crown. After DF post-processing, each sample had reduced geometrical deviations.

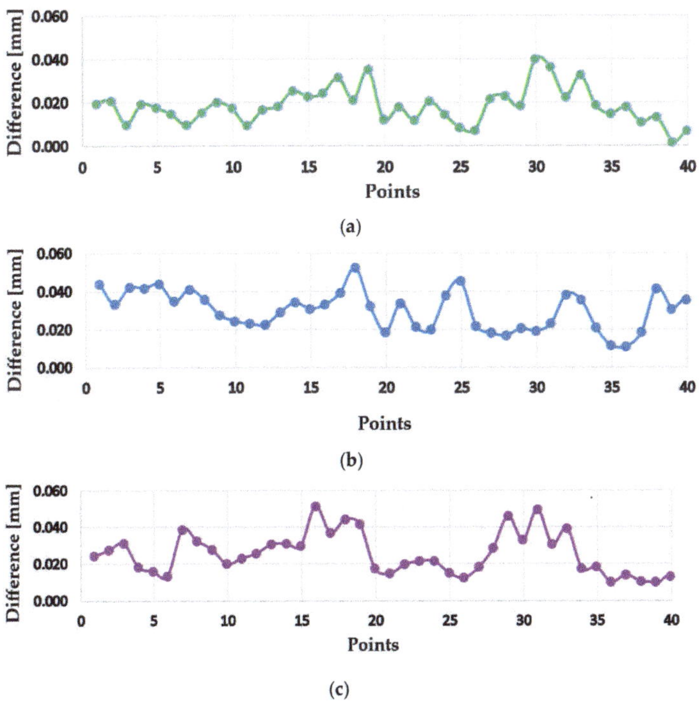

Figure 10. Typical graphs regarding the difference between the virtual model and the PBF-LB/M/CoCr crown after DF post-processing. The point-to-3D-model distance on outer surface after (**a**) DF1, (**b**) DF2, and (**c**) DF3.

Figure 11 shows a representative CMM graph obtained from the measurements recorded inside of the crown cavity (intaglio surface). Forty points are considered, and the difference between them, before and after DF, is marked with a green line (Figure 11). In general, the green line of difference is ±0.005 mm from the 0.000 level, which means that the post-processing did not remove material from the inner surfaces of the crown. To release abrasive grains from the cavity during the DF process, the crowns were ejected seven times. When the crowns were immersed in abrasive media, their cavity was very rapidly filled with abrasive grains, which limited the removal mechanism. Moreover, the reduced size of the crown cavity restricted the movement of the abrasive (11.40 × 8.60 mm).

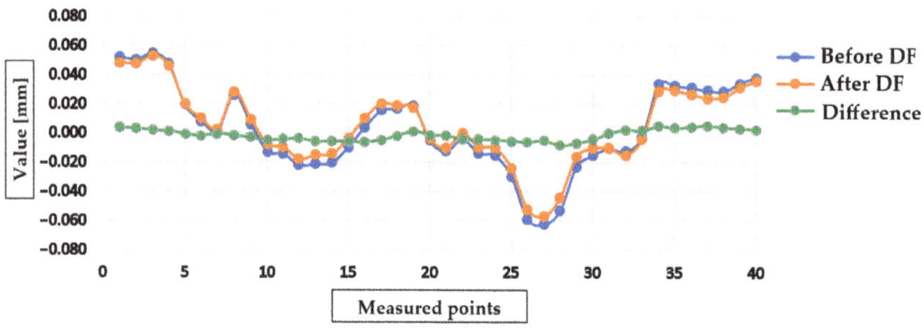

Figure 11. Typical graphs regarding the difference before and after DF post-processing.

Point-to-3D-model distance measurements were performed on the inner surfaces of crown (as-built vs. DF surface).

3.3. Surface Hardness

Figure 12 displays representative micrographs of Vickers indentations. The indentation impression shape was a little distorted because of elastic recovery, which is common for anisotropic materials [2]. This effect is more pronounced on as-built surfaces (Figure 12a). The Vickers hardness scores are depicted in Table 7. The CoCr specimens post-processed with DF had a hardness between 524 HV1 and 530 HV1. The DF3 parameters reduced the standard deviation of the mean value to ±27 HV1.

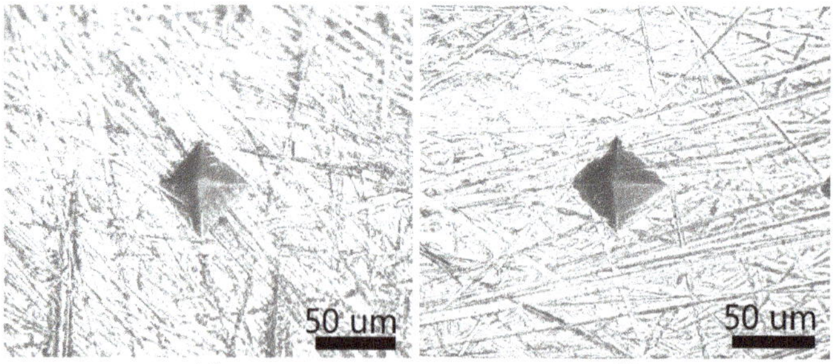

Figure 12. Micrographs of Vickers hardness indentation: testing performed on (**a**) initial surface and (**b**) after DF3 post-processing.

Table 7. Vickers hardness measurements of CoCr surfaces after applications of different DF conditions (mean value ± standard deviation).

Surface Condition	Initial	DF1	DF2	DF3
Vickers hardness [HV1]	520 ± 56	524 ± 42	527 ± 38	530 ± 27

3.4. SEM Analysis of Microstructure

Representative SEM images for the analysis of surface morphology prior to and following DF post-processing are presented in Figure 13. On as-built surfaces, a large amount of unmelted or partially melted grains were adhered to the surfaces, and tightly spaced straight contours were found (Figure 13a). Due to layering, the microtopography of surfaces contains specific waviness (Figure 13a). The initial surface was rough. However, after DF post-processing, a significant improvement on the outer surfaces was achieved, as can be seen in Figure 13b. Large droplets and sintered particles were removed. A smooth, uniform, and dense geometry can be observed across the entire outer surfaces. Grinding tracks are visible, caused by the solid abrasive particles (HSC 1/300). It can also be observed that the underlying surface is free of cracks. The chamfer margins were polished on both the intaglio (inner) and external (outer) surfaces (Figure 13b). An accurate and polished chamfer is beneficial for the long-term stability of a dental crown. The inner surfaces were limitedly polished, as can be seen in Figure 13b (top-right image). To reduce the surface roughness of the inner surfaces, a finer abrasive powder should be considered. On the other hand, micro-roughness of the inner surfaces can increase the adhesion of luting cement. Luting cement is used to bond a CoCr dental crown onto a prepared tooth (dentine tissue), and its thickness is between 25 µm and 40 µm. For clinical application, future in vitro studies will evaluate the shear bond strength of different cements applied to CoCr parts fabricated by the PBF-LB/M method and post-processed with DF.

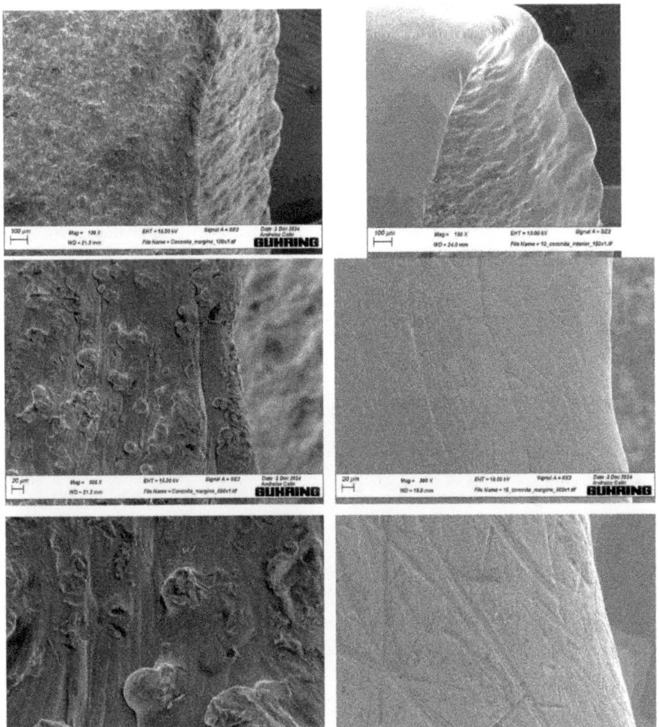

(a) (b)

Figure 13. Representative surface morphology of CoCr crn outer surfaces and chamfer margin: (**a**) as-built surfaces with unmelted or partially melted grains attached; (**b**) after DF3 post-processing with grinding tracks. Magnification ×150, ×500, and ×2500.

In the as-built state, the PBF-LB/M microstructures exhibited specific fine cellular or columnar dendritic structures [48]. Compared with conventional casting, finer grains were obtained through the PBF-LB/M process, which contributed to excellent mechanical properties. Because the samples were exposed to heat treatment at 860 °C for 1 h to release residual stress, the specific fine dendritic structures were found to disappear, and grain boundaries became decorated with precipitates. A similar microstructure was reported for CoCr samples heat-treated at 750–900 °C for stress relief [48,49]. Moreover, the highest mechanical resistance was obtained when the heat treatment was performed at 850 °C for 1 h [49]. Previous research has shown that after heat treatment, the composition segregations caused by non-equilibrium solidification are gradually eliminated, so both the molten pool and cellular boundaries vanish [49]. After heat treatment, intermetallic compounds gradually precipitated. As the SEM images in Figure 14 show, many fine intermetallic compounds were densely precipitated in the CoCr specimens.

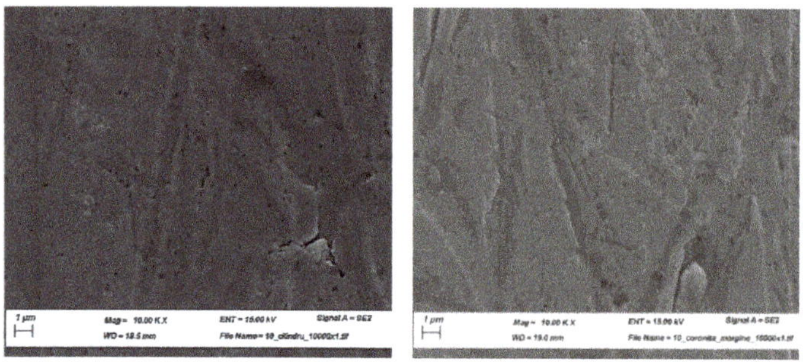

Figure 14. SEM micrographs of CoCr crown walls post-processed with DF; magnification ×10,000.

To test whether the DF method contaminated or changed the chemical composition of surfaces, EDAX spectrometry was used to compare the original powder and post-processed specimens. Within the limitations of the EDAX analyses, no variation in elemental composition was identified (Figure 15).

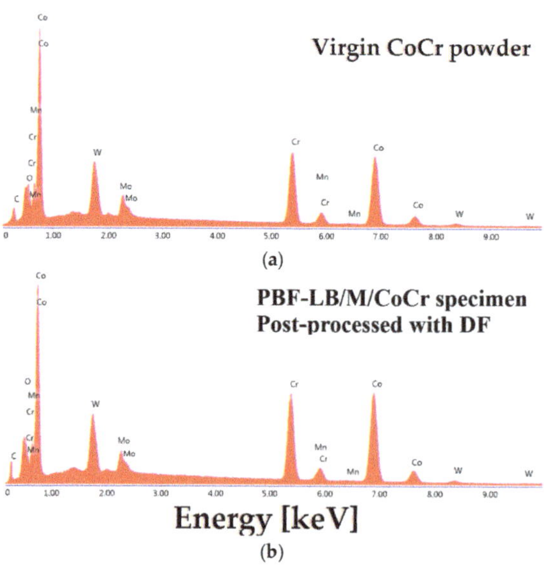

Figure 15. EDAX spectrum of original CoCr powder and PBF-LB/M/CoCr specimens.

3.5. Statistical Analysis

A model was created to evaluate the significance level ($p < \alpha$, $\alpha = 0.05$). The model was created according to the methodology described. Using this model, the effects of initial surface roughness and DF duration were tested. The *p*-value was less than 0.05 (or 5%). This means that no other influences or selected parameters had any influence on the tests performed. By confirming the homogeneity of the tested samples, it was possible to construct a linear model. The model was created to estimate the Ra roughness of the outer surface after the drag finishing procedure. In this linear model, the following parameters were substituted into the Regression Equation (1):

$$\mathrm{ER} = 2.695 + 0.0512 \times \mathrm{RAP} - 0.03214 \times \mathrm{DFT} \quad (1)$$

where ER is the estimated Ra roughness of outer surfaces after drag finishing (μm), RAP is the initial roughness after printing (μm), and DFT is the drag finishing time (min).

This equation suggests that for every 1 μm increase in initial roughness, the final Ra increases by 0.0512 μm, assuming the same drag finishing time. On the other hand, if the DF time increases by 1 min, the Ra decreases by 0.03214 μm. Another way to predict the surface roughness is by using the graph created from this regression function. To construct a graphical model, it is necessary to know the input values—in our case, the initial surface roughness after the printing process (RAP) and the drag finishing time (DFT). The intersection of these values is the resulting surface roughness. The graphical model is shown in Figure 16.

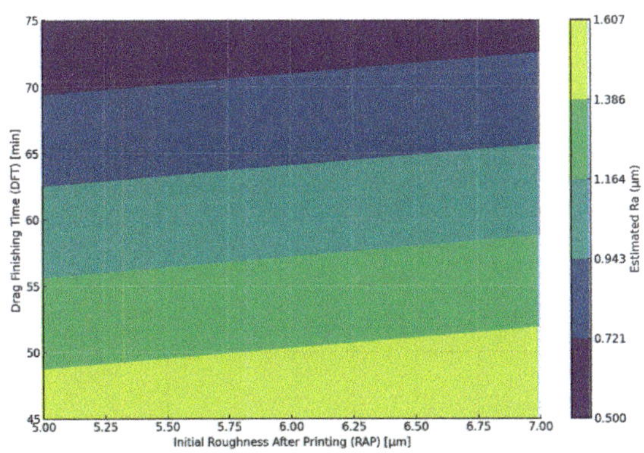

Figure 16. A graphical model derived from the regression function illustrating the effect of initial roughness (post-printing) and drag finishing duration on the outer surface roughness.

In the case of statistical processing of data from the measurement of the point-to-model distance, the same procedure was used. The linear model was calculated, and the following parameters were substituted into Equation (2):

$$\mathrm{ED} = 0.0151 + 0.1669 \times \mathrm{MD} - 0.0000861 \times \mathrm{DFT} \quad (2)$$

where ED is the estimated distance between the outer CMM point and the 3D model after the DF procedure (μm); MD is the point-to-model distance, or the measured distance between the determinate CMM point and the 3D model, measured before the DF procedure (μm); and DFT is the drag finishing time (min).

MD is the dominant factor influencing the ED value, and the DF time can help to reduce the distance between the CMM points and the 3D model, but its effect is much smaller. In other words, ED values could predict material loss on outer surfaces after the

DF procedure. Next, a graphical model was created, where based on the point-to-model value of the printed crown and the drag finishing time, the expected material removed is determined by the color spectrum in the graph (Figure 17). For example, if the CoCr crowns are post-processed for 75 min, their accuracy could be improved, reducing the dimensional deviations from 0.07 mm to 0.02 mm.

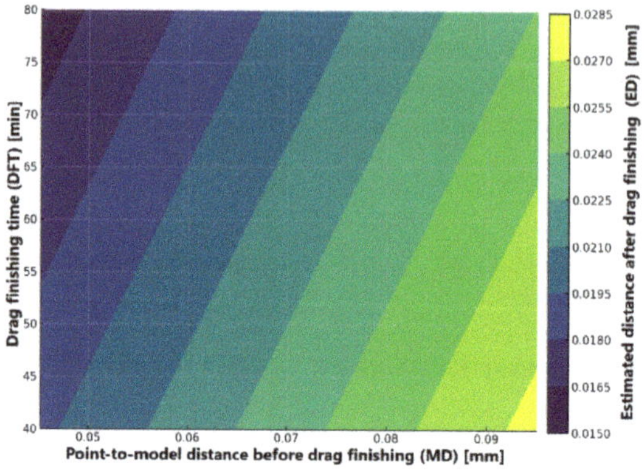

Figure 17. Effect of DF time on estimated distance after DF procedure (thickness of material removed). Initial point-to-model CMM measurements were considered.

It should be noted that these two equations are valid only when respecting the DF conditions detailed in Table 2 and for CoCr dental applications.

To test and validate Equations (1) and (2), five other samples were PBF-LB/M-fabricated and post-processed using drag finishing, as described in Section 2. Table 8 below summarizes the results, with analysis focused on the outer surfaces. The first model can accurately predict the Ra surface roughness after the DF procedure, considering the initial roughness and the drag finishing time. The difference between the estimated and the measured Ra is lower than 0.05 µm. The results confirm the close match between the estimated and measured Ra values, supporting the validation of Equation (1). The coefficient of determination (R^2) is 0.991, indicating a strong fit of the predicted outcome. For the second equation, the model's average prediction error is around 2–5 microns, which may be acceptable, depending on required manufacturing tolerance. The model has a good fit for practical use, with a high R^2 (0.947). It can predict material loss following the DF procedure and how the dimensional deviations of external surfaces may be minimized.

Table 8. Testing trails to validate Equations (1) and (2) for CoCr dental crowns after DF post-processing.

Sample No.	Ra After Printing (µm)	Drag Finishing Time (min)	Ra After DF, Estimated [a] (µm)	Ra After DF, Measured (µm)	Difference [b] (µm)	Point-to-Model Distance Before DF (mm)	ED After DF [c] (mm)	Measured Distance After DF (mm)	Difference (mm) [d]
1	5.92	45	1.55	1.59	0.04	0.070	0.023	0.025	0.002
2	5.75	45	1.54	1.58	0.04	0.081	0.025	0.028	0.003
3	6.34	60	1.09	1.12	0.03	0.052	0.018	0.022	0.004
4	6.18	75	0.60	0.64	0.04	0.062	0.019	0.024	0.005
5	5.97	75	0.59	0.64	0.05	0.091	0.024	0.027	0.003

[a] The Ra surface roughness after DF post-processing was predicted with Equation (1), considering the initial Ra and DF timing (outer surfaces were considered). [b] The difference between the measured Ra and the estimated value predicted by Equation (1). [c] The ED after DF post-processing was predicted by Equation (2), considering the initial point-to-model distance and DF timing. [d] The difference between the distance from the measured CMM point to the 3D model and the estimated one (ED), predicted by Equation (2).

4. Discussion

Every dental application produced by the PBF-LB/M process requires two essential steps—finishing and polishing. The quality of polishing largely depends on the dental technician's experience and manual skill, as these tasks are still performed freehand. While numerous studies have investigated the efficiency of finishing and polishing CoCr surfaces, research on fully automated post-processing methods remains limited. To reduce the workload on dental technicians and enhance surface quality, DF post-processing was considered in this study.

Unfinished surfaces of components processed by PBF-LB/M not only present appearance issues, but can also lead to problems such as reduced mechanical properties. Rough surfaces can be the source of crack initiation, and can decrease fatigue strength. The DF post-processing method was adapted to reduce surface roughness and to create uniform height distribution. In this work, the DF3 conditions significantly reduced the roughness of outer surfaces (see Tables 3–6). The DF method improved the quality of CoCr dental applications, providing reliable, repeatable, and controlled results. On the other hand, the DF1 conditions provided a micro-rough surface that may be suitable for medical implants (Ra 1.52–1.60 μm). If the surface of a PBF-LB/M-processed implant is cleaned and has micro-roughness, this can positively influence its biocompatibility. Herrero-Climen et al. demonstrate that micro-rough surfaces can promote the biological response and improve the osseointegration process [50].

The inner surfaces were limitedly polished, but this micro-rough surface of the dental crown can increase the adhesion of luting cement. Future in vitro studies should evaluate the shear bond strength of different cements applied to CoCr dental crowns fabricated by the PBF-LB/M method and DF3 post-processed.

Table 9 summarizes some reported works focused on how post-processing methods can reduce the roughness of CoCr, Ti, and PBF-LB/M-fabricated stainless steel parts. In general, the as-built surfaces have a roughness ranging from 6 to 15 μm in the case of the Ra parameter. The main post-processing methods used to reduce the PBF-LB/M surface roughness are drag finishing, vibratory finishing, sandblasting, electro-polishing, and laser polishing. Because medical applications fabricated by PBF-LB/M are complex, studies also focused on complex surfaces, not just on basic shapes, were identified (see Table 9). Jamal et al. demonstrate that if the Ti6Al4V parts are DF post-processed, the Ra surface roughness can be reduced from 7.5 μm to 0.6 μm (see Table 9). This value was obtained on angled surfaces, and the parts were maintained for 120 min in a wet environment (SC15), immersed in a ceramic medium (universal compound SC15). Due to the abrasive medium (mixture of walnut shells and SiC) and polishing paste used, the current study shows a similar value for surface roughness, but a reduced time for the DF process (75 min). Other PBF-LB/M studies demonstrate that the surface roughness of 316L samples can be reduced by 60–70% if DF post-treatment is applied for 120–240 min (see results of Kaynak et al. [29] and Behjat et al. [51], depicted in Table 9). The findings of the present study demonstrate that it is feasible to reduce the surface roughness by 87% in 75 min.

The effect of the operating parameters on the quality of the laser polishing process has been studied by several researchers [22,25,53]. A recent review focused on this topic highlights a significant reduction in Ra, down to 0.10 μm, obtained for printed Inconel 718 parts with an initial surface roughness of 7.5 μm [54]. After laser polishing, the hardness of the polished surface could increase by at least 27.5%. Ma et al. studied the execution of a polishing process with a pulsed fiber laser on a Ti alloy, and found that the surface roughness was reduced from 7.2 to 0.7 μm (Table 9), but the study was limited to the basic shape [25]. Gora et al. [22] performed laser polishing on CoCr parts manufactured using PBF-LB/M, and reported a reduction in surface roughness from 12.8 μm to 0.6 μm. This study was also limited to the basic shape. Furthermore, the Fraunhofer Institute for Laser Technology conducted extensive research on PBF-LB/M processing of medical implants made from CoCr [53]. The fabricated knee implants were subsequently post-processed using laser polishing, resulting in an average surface roughness of Ra 0.3 μm (see Table 9). However, laser polishing of custom complex parts is expensive because the equipment needs to be programmed, an action that will generate supplementary costs for each part. Laser polishing can change the microstructure phase on polished surfaces [25], which may contribute to different biocompatibility behavior, and it can also affect corrosion resistance. The actual DF process provides a fine surface finish with a high degree of uniformity across the CoCr samples, but the final result may not always match the ultra-smooth finish achievable with laser polishing (Ra 0.10 μm, [54]) or plasma electrolytic polishing (Ra 0.10–0.20 nm [52]).

Table 9. Comparison regarding the surface roughness obtained on PBF-LB/M-fabricated parts after different post-processing techniques.

Material	Specimen Shape [Basic/Complex]	Post-Processing Method	Ra of as-Built Specimen [μm]	Ra After Post-Processing [μm]	Vickers Hardness [HV]	Source
CoCr	Complex	DF, 75 min.	6.1	0.6	530	This study
Ti6Al4V	Complex	DF, 90–120 min.	7.5	0.6–0.8	~400	Jamal et al. [27]
316L	Basic	DF, 120–240 min.	6	2–3	280	Kaynak et al. [29]
316L	Basic	DF, 180 min.	12	5	~220	Behjat et al. [51]
Inconel 718	Basic	DF, 240 min.	4.3	2.2	-	Lee et al. [40]
Inconel 718	Basic	Grinding + DF, 240 min.	4.3	0.5	-	Lee et al. [40]
316L	Basic	Vibratory surface finishing	6	4	280	Kaynak et al. [29]
CoCr	Complex	Electro-polishing	9.1	1.4	-	Demir et al. [16]
CoCr	Basic	Electro-polishing	20	0.02	-	Valentinčič et al. [52]
316L *	Basic	Electro-polishing	15.1	9.2	~250	Lober et al. [17]
316L *	Basic	Plasma-polishing	15.1	8.5	~250	Lober et al. [17]
316L *	Basic	Blasted with sand (90–250 μm)	15.1	3.8	~250	Lober et al. [17]
CoCr	Complex	Laser polishing	7	0.3	-	Fraunhofer Inst. Laser Tech. [53]
CoCr	Basic	Laser polishing	12.8	0.6	-	Gora et al. [22]
Ti6Al4V	Basic	Laser polishing	5.2	0.4	345	Ma et al. [25]
Inconel 718	Basic	Laser polishing	7.5	0.1	440	Gisario et al. [54]
316L *	Basic	Laser polishing	4.6	0.3	~302	Gisario et al. [54]
15-5PH *	Complex	Abrasive flow machining (150 cycles)	13.9	2.5	650	Duval-Chaneac et al. [23]

* Stainless steel alloy; DF—drag finishing post-processing method.

Depending on the material type, post-processing method, and shape complexity, the Ra surface roughness of PBF-LB/M parts can be reduced to less than 1.60 μm. Some examples are detailed in Table 9. This Ra value is typical for CNC-machined parts fabricated with finishing cutting parameters. Moreover, the combination of two automated post-processing techniques has emerged as a promising approach. In a recent study, Mizokoshi et al. employed a combination of barrel finishing (25 min) followed by magnetic polishing using stainless steel needles (2 h), achieving an Ra surface roughness of approximately 1.76 μm. The lowest Ra value of 0.05 μm was attained by combining barrel finishing (25 min) with dry electrolyte polishing using plastic resin media (15 min) [55]. This level of surface smoothness was comparable to results obtained through freehand finishing (10 min) and polishing (10 min) by a highly experienced dental technician with over 30 years of laboratory practice [55]. However, a significant drawback of combining two automated post-processing machines is the increased labor and equipment cost, which may limit their widespread adoption in routine dental manufacturing.

The present study indicates that the finishing approach contributed to surface quality, and the DF method is feasible for reducing the roughness of CoCr dental applications, being similar to mechanical polishing (i.e., grinding with SiC papers P300, Ra 0.52 μm) [17]. The DF parameters and abrasive materials tested improved the surface quality of PBF-LB/M-fabricated parts, meeting the surface requirements (Ra < 0.8 μm).

On the other hand, the DF process has some disadvantages that must be considered. These include limitations in processing large parts and holes, a slow speed compared with laser polishing or electro-polishing, a relatively low material removal rate, and medium-induced wear. For example, Valentinčič et al. showed that additively manufactured CoCr parts polished with plasma electrolytic polishing using a 0.3 M $(NH_4)_2SO_4$ aqueous solution could achieve a final Ra roughness of 20 nm [52]. This post-processing result was obtained in just 8 min. Moreover, DF post-processing is not feasible for lattice scaffolds or cardiovascular stents, because it is impossible to polish the thin struts, and electro-polishing is a suitable solution. Understanding these limitations is crucial when determining whether the DF process is the most suitable option for a given application, or whether alternative finishing processes might provide better results.

Compared with other biocompatible dental alloys such as Ti6Al4V, Ni-Cr, and stainless steel, CoCr is the hardest and most laborious for post-processing. For example, in Rockwell tests, CoCr had a hardness between 43 and 52 HRC [56,57], the Ti6Al4V alloy had 34 HRC [30], the Ni-Cr alloy had around 41 HRC [31], and 316L stainless steel had approx. 30 HRC [17,32]. In the literature, various PBF-LB/M studies have been described, providing information about the Vickers hardness of CoCr. Table 10 summarizes the Vickers hardness results of CoCr, measured on both PBF-LB/M-fabricated and cast specimens. Depending on the laser parameter setup, the scanning strategy adopted, and the heat treatment conditions, the CoCr specimens could possess a Vickers hardness between 325 HV and 550 HV (see Table 10). Only a limited number of PBF-LB/M studies have reported Vickers hardness values above 550 HV for CoCr alloys [2,9], indicating exceptionally high hardness. From a manufacturing perspective, the hardness of cast CoCr ranges from 280 HV to 385 HV (Table 10). Compared with the hardness obtained with the casting method, the hardness of the present PBF-LB/M surfaces was superior by 36% (Table 10). According to ASTM F75 [58] and ASTM F90 requirements [59], the hardness of the present specimens increased by 33% (Table 10).

In this work, the laser parameters and the heat treatment applied increased the hardness to 530 HV, higher than that achieved in other PBF-LB/M studies [57,60,61]. The PBF-LB/M process was able to provide this improved hardness because the melt pool cooled down rapidly after the laser beam had passed, and this procedure was repeated in cycles, affecting the previous layer. As has been reported, the higher microhardness encountered can be attributed to a finer microstructure (small cell size) [2,64–66]. Moreover, Ayyıldız et al. demonstrated that the atmospheric conditions of heat treatment could increase the hardness of PBF-LB/M surfaces [57]. Because the heat treatment in the present study was performed in a normal atmosphere with oxygen, it could contribute to this increased hardness. Improvement in the hardness of PBF-LB/M surfaces could improve the service life of metal–ceramic restorations, but future studies are needed to evaluate this achievement.

Table 10. Comparison of CoCr Vickers hardness obtained on PBF-LB/M and cast specimens.

Manufacturing Method	Vickers Hardness [HV]	Source
PBF-LB/M	524–530	Present study after DF post-processing
	400–460	[60]
	325–482	[57]
	440	[61]
	564–570	[9]
	498–618 *	[2]
Casting	280–385	[5]
	323	[62]
	287–295 **	[63]
	266–345	Standard of CoCr alloy for surgical implants [58]
	300–400	Standard of CoCr alloy for surgical implants [59]

* Laser remelting of boundary surfaces; ** different Si concentrations (0.1–1%).

Porcelain-fused-to-metal crowns are frequently used in dental prosthetics to protect the remaining tooth structure (dentine tissue). Due to their good clinical performance, with low failure rates, these metal–ceramic restorations are developed using CoCr as a metallic substructure (see ISO 22764, [33]). Many countries, such as the United States of America, have decided to replace NiCr alloys with CoCr to prepare porcelain-fused-to-metal dentures [67]. This alloy is considered a highly corrosion-resistant and biocompatible material in dentistry [1–5,67]. To prepare metal–ceramic restorations, the outer surfaces of CoCr are covered by veneering ceramic materials. The inside of the CoCr-ceramic crown is luted with cement (thin layer between 50 μm and 120 μm) onto host dentine. Based on the EDAX results, the post-processed CoCr crowns may be suitable for use in fixed metal–ceramic prostheses. Future studies should analyze corrosion potential, pitting potential, and electrochemical impedance to confirm the biocompatibility of CoCr crowns post-processed with DF.

From a clinical implication perspective, the reduced roughness achieved when crowns are post-processed using DF may allow this to be considered a promising alternative to deburring, smoothing, and polishing the outer side of PBF-LB/M-fabricated surfaces. It is a feasible solution to automate the post-processing of PBF-LB/M/CoCr components. Currently, CoCr crowns manufactured using this technology are manually post-processed by dental technicians using pneumatic straight grinders.

Moreover, the current work demonstrates that even if the surface hardness has high values up to 530 HV, it is possible to identify the proper DF condition to reduce the Ra roughness of CoCr dental crowns from approx. 6 μm to 0.6 μm. The results regarding the post-processing of CoCr can contribute to expanding automatic post-processing of complex PBF-LB/M-fabricated parts. Future research is required to evaluate the fatigue resistance and metal–ceramic bonding of CoCr dental crowns that have been PBF-LB/M-manufactured and post-processed via DF3. According to the ISO 9693 standard, the minimum adhesion strength should be 25 MPa [68].

5. Conclusions

The findings of this study support the following conclusions:

- DF is an effective and feasible method for improving both the surface quality and dimensional accuracy of CoCr dental crowns manufactured via PBF-LB/M, offering a viable pathway for automating post-processing procedures.
- The outer surface roughness of CoCr crowns could be reduced by up to 90% when the DF processing time was increased from 45 to 75 min, achieving values of Ra 0.6 μm, Rz 4.5 μm, and Rq 1.0 μm. However, the effectiveness of the post-processing was limited on the inner surfaces.
- Three-dimensional surface topography analyses after DF revealed a notable reduction in peak heights, resulting in smoother, more uniform surfaces and effective removal of residual particles.
- The DF method enhanced dimensional accuracy, with minimal deviations (+0.010 to +0.050 mm) observed on the outer surfaces, indicating that delicate features, such as thin crown walls, were preserved and not over-polished during the process.
- The empirical equations developed were validated through experimental results, demonstrating their ability to accurately predict both surface roughness and material loss following the DF process.
- Despite the high hardness of CoCr crowns (530 HV), the DF3 method produced high-quality surface finishes within a shorter processing time, offering a more time-efficient alternative compared to other DF works that recommend longer times.

- To fully understand the clinical significance of these results, further studies are necessary to evaluate the bonding strength between ceramic veneering materials applied to CoCr crowns that have undergone DF3 post-processing.

Author Contributions: Conceptualization, C.C. and N.B.; methodology, M.M. and A.P.; validation, M.M., S.L., G.C. and C.T.; formal analysis, C.T., M.M. and A.P.; investigation, C.C., M.M., S.L., G.C. and C.T.; resources, S.L., A.P. and P.B.; data curation, A.P. and P.B.; writing—original draft preparation, C.C., M.M. and A.P.; writing—review and editing, C.C. and N.B.; visualization, C.T., G.C. and S.L.; supervision, N.B. and P.B.; project administration, N.B. and G.C.; funding acquisition, C.C. and P.B. All authors have read and agreed to the published version of the manuscript.

Funding: This work was supported by the Central European Exchange Program for University Studies (CEEPUS) and by the OpTi-DeP Project (no. BG101/2016), financed by the Executive Unit for Financing Higher Education, Research, Development, and Innovation (UEFISCDI) of the Romanian Government.

Data Availability Statement: The original contributions presented in this study are included in the article. Further inquiries can be directed to the corresponding author.

Acknowledgments: The authors thank Guhring Company for helping with the scanning electron microscopy.

Conflicts of Interest: The authors declare no conflicts of interest.

Abbreviations

The following abbreviations are used in this manuscript:

15-5PH	Precipitation-hardening stainless steel alloy
316L	Stainless steel alloy
AM	Additive manufacturing
ANOVA	Analysis of variance, statistical method
ASTM	American Society for Testing and Material standard
CMM	Coordinate measuring machine
CNC	Computer numerical control
CoCr	Cobalt–chromium alloy
DF	Drag finishing process
DF1, DF2, DF3	Drag finishing process with different finishing times
DFT	Drag finishing total time
DWIO	Dental Wings IntraOral system
ED	Estimated distance between CMM point and 3D model after DF procedure
EDAX	Energy-dispersive X-ray spectroscopy (EDAX)
HRC	Rockwell hardness scale C
HSC	Mixture of 70% walnut shell granulates and 30% silicon carbide
HV	Vickers surface hardness
Inconel 718	Nickel-based superalloy
ISO	International organization for standardization
MD	Point-to-model distance or measured distance between determinate CMM point and 3D model, measured before DF procedure
NiCr	Nickel–chromium alloy
Nd:YAG	Neodymium-doped Yttrium Aluminum Garnet, solid-state laser
R^2	R squared or coefficient of determination
Ra	Surface roughness parameter
Rp, Rv, Rc, Rt	Amplitude parameters, peak and valley
Rq, Rku, Rsk	Average amplitude parameters
Rsm, Rdq, Rmax	Other surface roughness parameters
Rz	Surface roughness parameter
PBF-LB/M	Scanning electron microscopy

SiC	Silicon carbide
Ti	Titanium
Ti6Al4V	Titanium alloy
W	Watts

References

1. Dawood, A.; Marti, B.; Sauret-Jackson, V.; Darwood, A. 3D printing in dentistry. *Nat. Br. Dent. J.* **2015**, *219*, 521–529. [CrossRef] [PubMed]
2. Cosma, C.; Moldovan, M.; Simion, M.; Balc, N. Impact of laser parameters on additively manufactured cobalt-chromium restorations. *J Prosthet. Dent.* **2022**, *128*, 421–429. [CrossRef] [PubMed]
3. Yager, S.; Ma, J.; Ozcan, H.; Kilinc, H.I.; Elwany, A.; Karaman, I. Mechanical Properties and Microstructure of Removable Partial Denture Clasps Manufactured Using Selective Laser Melting. *Addit. Manuf.* **2015**, *8*, 117–123. [CrossRef]
4. Hedberg, Y.; Hedberg, Y.; Qian, B.; Shen, Z.; Virtanen, S.; Odnevall Wallinder, I. In Vitro Biocompatibility of CoCrMo Dental Alloys Fabricated by Selective Laser Melting. *Dent. Mater.* **2014**, *30*, 525–534. [CrossRef]
5. Al Jabbari, Y.S. Physico-Mechanical Properties and Prosthodontic Applications of Co-Cr Dental Alloys: A Review of the Literature. *J. Adv. Prosthodont.* **2014**, *6*, 138–145. [CrossRef]
6. Oltean-Dan, D.; Dogaru, G.-B.; Tomoaia-Cotisel, M.; Apostu, D.; Mester, A.; Benea, H.-R.-C.; Paiusan, M.-G.; Jianu, E.-M.; Mocanu, A.; Balint, R.; et al. Enhancement of Bone Consolidation Using High-Frequency Pulsed Electromagnetic Short-Waves and Titanium Implants Coated with Biomimetic Composite Embedded into PLA Matrix: In Vivo Evaluation. *Int. J. Nanomed.* **2019**, *14*, 5799–5816. [CrossRef]
7. Munteanu, L.; Iancu, I.; Breazu, C.; Cioltean, C.; Brânzilă, S.; Odainii, A.; Furda, P.; Bocşe, H.; Herdean, A.; Bartoş, D.; et al. Rare Causes of Gastrointestinal Bleeding: Focus on Pancreatic Pathology and Visceral Artery Aneurysms. *Chirurgia* **2021**, *116*, S5–S15.
8. Kassapidou, M.; Franke Stenport, V.; Hjalmarsson, L.; Johansson, C.B. Cobalt-Chromium Alloys in Fixed Prosthodontics in Sweden. *Acta Biomater Odontol. Scand.* **2017**, *3*, 53–62. [CrossRef]
9. Lu, Y.; Wu, S.; Gan, Y.; Li, J.; Zhao, C.; Zhuo, D.; Lin, J. Investigation on the microstructure, mechanical property and corrosion behavior of the selective laser melted CoCrW alloy for dental application. *Mater. Sci. Eng. C Mater. Biol. Appl.* **2015**, *49*, 517–525. [CrossRef]
10. Soporan, V.S.; Lehene, T.R.; Pădureţu, S.; Gabor, T. Engineering of circular economy and good management of industrial material resources. *IOP Conf. Ser. Mater. Sci. Eng.* **2019**, *572*, 012086. [CrossRef]
11. Mitić, J.; Vitković, N.; Trajanović, M.; Górski, F.; Păcurar, A.; Borzan, C.; Sabău, E.; Păcurar, R. Utilizing Artificial Neural Networks for Geometric Bone Model Reconstruction in Mandibular Prognathism Patients. *Mathematics* **2024**, *12*, 1577. [CrossRef]
12. Charles, A.; Elkaseer, A.; Thijs, L.; Hagenmeyer, V.; Scholz, S. Effect of Process Parameters on the Generated Surface Roughness of Down-Facing Surfaces in Selective Laser Melting. *Appl. Sci.* **2019**, *9*, 1256. [CrossRef]
13. Băilă, D.I.; Doicin, C.V.; Cotruț, C.M.; Ulmeanu, M.E.; Ghionea, I.G.; Tarbă, C.I. Sintering the beaks of the elevator manufactured by direct metal laser sintering (DMLS) process from Co-Cr alloy. *Metalurgija* **2016**, *55*, 663–666.
14. Hong, M.-H.; Min, B.; Kwon, T.-Y. The Influence of Process Parameters on the Surface Roughness of a 3D-Printed Co–Cr Dental Alloy Produced via Selective Laser Melting. *Appl. Sci.* **2016**, *6*, 401. [CrossRef]
15. Martin, A.A.; Calta, N.P.; Khairallah, S.A.; Wang, J.; Depond, P.J.; Fong, A.Y.; Thampy, V.; Guss, G.M.; Kiss, A.M.; Stone, K.H.; et al. Dynamics of pore formation during laser powder bed fusion additive manufacturing. *Nat. Commun.* **2019**, *10*, 1987. [CrossRef]
16. Demir, A.G.; Previtali, B. Additive manufacturing of cardiovascular CoCr stents by selective laser melting. *Mater. Des.* **2017**, *119*, 338–350. [CrossRef]
17. Löber, L.; Flache, C.; Petters, R.; Kühn, U.; Eckert, J. Comparison of different post processing technologies for SLM generated 316l steel parts. *Rapid Prototyp. J.* **2013**, *19*, 173–179. [CrossRef]
18. Buican, G.R.; Oancea, G.; Martins, R.F. Study on SLM manufacturing of teeth used for dental tools testing. *MATEC Web Conf.* **2017**, *94*, 03002. [CrossRef]
19. Luca, A.; Balc, N.; Popan, A.; Ceclan, V.; Panc, N. Improving the quality of the parts made by rapid metal casting process. *Acad. J. Manuf. Eng.* **2014**, *12*, 82–86.
20. Selden, A.I.; Persson, B.; Bornberger-Dankvardt, S.I.; Winstrom, L.E.; Bodin, L.S. Exposure to cobalt chromium dust and lung disorders in dental technicians. *Thorax* **1995**, *50*, 769–772. [CrossRef]
21. Kumbhar, N.N.; Mulay, A.V. Post Processing Methods used to Improve Surface Finish of Products which are Manufactured by Additive Manufacturing Technologies: A Review. *J. Inst. Eng. (India) Ser. C* **2016**, *99*, 481–487. [CrossRef]
22. Gora, W.S.; Tian, Y.; Cabo, A.P.; Ardron, M.; Maier, R.R.J.; Prangnell, P.; Weston, N.J.; Hand, D.P. Enhancing Surface Finish of Additively Manufactured Titanium and Cobalt Chrome Elements Using Laser Based Finishing. *Phys. Procedia* **2016**, *83*, 258–263. [CrossRef]

23. Duval-Chaneac; Han, S.; Claudin, C.; Salvatore, F.; Bajolet, J.; Rech, J. Experimental study on finishing of internal laser melting (SLM) surface with abrasive flow machining (AFM). *Precis. Eng.* **2018**, *54*, 1–6. [CrossRef]
24. Conțiu, G.; Popa, M.S.; Socaciu, L.; Pop, G. Fuzzy analytical hierarchy process applied to determine the material machinability in EDM process. *Acta Tech. Napoc. Ser. Appl. Math. Mech. Eng.* **2015**, *58*, 385–394.
25. Ma, C.P.; Guan, Y.C.; Zhou, W. Laser polishing of additive manufactured Ti alloys. *Opt. Lasers Eng.* **2017**, *93*, 171–177. [CrossRef]
26. Mohammad, A.; Mohammed, M.K.; Alahmari, A.M. Effect of laser ablation parameters on surface improvement of electron beam melted parts. *Int. J. Adv. Manuf. Technol.* **2016**, *87*, 1033–1044. [CrossRef]
27. Jamal, M.; Morgan, M. Design Process Control for Improved Surface Finish of Metal Additive Manufactured Parts of Complex Build Geometry. *Inventions* **2017**, *2*, 36. [CrossRef]
28. Zanger, F.; Kacaras, A.; Neuenfeldt, P.; Schulze, V. Optimization of the stream finishing process for mechanical surface treatment by numerical and experimental process analysis. *CIRP Ann.* **2019**, *68*, 373–376. [CrossRef]
29. Kaynak, Y.; Kitay, O. The effect of post-processing operations on surface characteristics of 316L stainless steel produced by selective laser melting. *Addit. Manuf.* **2018**, *26*, 84–93. [CrossRef]
30. Koike, M.; Greer, P.; Owen, K.; Lilly, G.; Murr, L.E.; Gaytan, S.M.; Martinez, E.; Okabe, T. Evaluation of Titanium Alloys Fabricated Using Rapid Prototyping Technologies—Electron Beam Melting and Laser Beam Melting. *Materials* **2011**, *4*, 1776–1792. [CrossRef]
31. Wang, X.; Gong, X.; Chou, K. Review on powder-bed laser additive manufacturing of Inconel 718 parts. *Proc. Inst. Mech. Eng. Part B J. Eng. Manuf.* **2016**, *231*, 1890–1903. [CrossRef]
32. Tan, C.; Zhou, K.; Kuang, M.; Ma, W.; Kuang, T. Microstructural characterization and properties of selective laser melted maraging steel with different build directions. *Sci. Technol. Adv. Mater.* **2018**, *19*, 746–758. [CrossRef]
33. *ISO 22674:2022*; Dentistry—Metallic Materials for Fixed and Removable Restorations and Appliances. ISO: Geneva, Switzerland, 2022.
34. Cuc, S.; Burde, A.; Cosma, C.; Leordean, D.; Rusu, M.; Balc, N.; Prodan, D.; Moldovan, M.; Ene, R. Adhesion between Biocomposites and Different Metallic Structures Additive Manufactured. *Coatings* **2021**, *11*, 483. [CrossRef]
35. Leordean, D.; Dudescu, C.; Marcu, T.; Berce, P.; Balc, N. Customized implants with specific properties, made by selective laser melting. *Rapid Prototyp. J.* **2015**, *21*, 98–104. [CrossRef]
36. Ucar, Y.; Ekren, O. Effect of layered manufacturing techniques, alloy powders, and layer thickness on mechanical properties of Co-Cr dental alloys. *J. Prosthet. Dent.* **2018**, *120*, 762–770. [CrossRef]
37. Cho, H.H.W.; Takaichi, A.; Kajima, Y.; Htat, H.L.; Kittikundecha, N.; Hanawa, T.; Wakabayashi, N. Effect of Post-Heat Treatment Cooling Conditions on Microstructures and Fatigue Properties of Cobalt Chromium Molybdenum Alloy Fabricated through Selective Laser Melting. *Metals* **2021**, *11*, 1005. [CrossRef]
38. Roudnicka, M.; Bigas, J.; Molnarova, O.; Palousek, D.; Vojtech, D. Different Response of Cast and 3D-Printed Co-Cr-Mo Alloy to Heat Treatment: A Thorough Microstructure Characterization. *Metals* **2021**, *11*, 687. [CrossRef]
39. Zhang, K.; Fu, G.; Zhang, P.; Ma, Z.; Mao, Z.; Zhang, D.Z. Study on the Geometric Design of Supports for Overhanging Structures Fabricated by Selective Laser Melting. *Materials* **2018**, *12*, 27. [CrossRef]
40. Lee, S.; Shao, S.; Wells, D.N.; Zetek, M.; Kepka, M.; Shamsaei, N. Fatigue behavior and modeling of additively manufactured IN718: The effect of surface treatments and surface measurement techniques. *J. Mater. Process. Technol.* **2022**, *302*, 117475. [CrossRef]
41. Pätoprstý, B.; Vozár, M.; Pokorný, P.; Vopát, T.; Buranský, I.; Zetek, M.; Cajthamlová, Š.; Laudát, V. Development of cutting edge radius size of solid carbide mills when drag finishing. In *Lecture Notes in Mechanical Engineering*; Springer: Singapore, 2020; pp. 95–100. [CrossRef]
42. Cook, R.B.; Shearwood-Porter, N.R.; Latham, J.M.; Wood, R.J.K. Volumetric assessment of material loss from retrieved cemented metal hip replacement stems. *Tribol. Int.* **2015**, *89*, 105–108. [CrossRef]
43. Panc, N.; Contiu, G.; Bocanet, V. Comparative Analysis of Surface Finishing for Different Cutting Strategies of Parts Made from POM C. In *Lecture Notes in Mechanical Engineering*; Springer: Singapore, 2018; pp. 324–332. [CrossRef]
44. *ISO 4278:1997*; Geometrical Product Specifications (GPS)—Surface Texture: Profile Method—Terms, Definitions and Surface Texture Parameters. ISO: Geneva, Switzerland, 1997.
45. *ISO 10360:2000*; Acceptance and Re-Verification Tests for Coordinate Measuring Machines. ISO: Geneva, Switzerland, 2000.
46. *ISO 6507-1:2018*; Metallic Materials—Vickers Hardness Test—Part 1: Test Method. ISO: Geneva, Switzerland, 2018.
47. Surface Roughness Measurement-Parameters, OLYMPUS, Tokyo, Japan. Available online: www.olympus-ims.com/en/metrology/surface-roughness-measurement-portal/parameters/ (accessed on 23 March 2024).
48. Takaichi, A.; Kajima, Y.; Jialin, D.; Wakabayashi, N. Metal Additive Manufacturing Technology for Digital Dentistry. *Health Care Curr Rev.* **2024**, *12*, 407. Available online: https://www.walshmedicalmedia.com/open-access/metal-additive-manufacturing-technology-for-digital-dentistry.pdf (accessed on 4 February 2025).

49. Wei, W.; Zhou, Y.; Sun, Q.; Li, N.; Yan, J.; Li, H.; Liu, W.; Huang, C. Microstructures and mechanical properties of dental Co-CR-MO-W alloys fabricated by selective laser melting at different subsequent heat treatment temperatures. *Metall. Mater. Trans. A* **2020**, *51*, 3205–3214. [CrossRef]
50. Herrero-Climent, M.; Lázaro, P.; Rios, J.V.; Lluch, S.; Marqués, M.; Guillem-Martí, J.; Gil, F.J. Influence of acid-etching after grit-blasted on osseointegration of titanium dental implants: In vitro and in vivo studies. *J. Mater. Sci. Mater. Med.* **2013**, *24*, 2047–2055. [CrossRef] [PubMed]
51. Behjat, A.; Lannunziata, E.; Gadalińska, E.; Iuliano, L.; Saboori, A. Improving the surface quality and mechanical properties of additively manufactured AISI 316L stainless steel by different surface post-treatment. *Procedia CIRP* **2023**, *118*, 771–776. [CrossRef]
52. Valentinčič, J.; Koroth, J.E.; Zeidler, H. Advancements in surface finish for additive manufacturing of metal parts: A comprehensive review of plasma electrolytic polishing (PEP). *Virtual Phys. Prototyp.* **2024**, *19*, e2364222. [CrossRef]
53. Shen, L.; Ross, I. Implant Manufacture out of CoCr by Means of SLM and Laser Polishing. Annual Report Fraunhofer Institute for Laser Technology ILT 2014. Available online: www.ilt.fraunhofer.de (accessed on 19 May 2020).
54. Gisario, A.; Barletta, M.; Veniali, F. Laser polishing: A review of a constantly growing technology in the surface finishing of components made by additive manufacturing. *Int. J. Adv. Manuf. Technol.* **2022**, *120*, 1433–1472. [CrossRef]
55. Mizokoshi, N.; Sakurai, T.; Shimpo, H.; Kawamura, N.; Ohkubo, C. Finishing efficiencies of additive-manufactured Co-Cr alloy and Ti alloy clasps. *Dent. Mater. J.* **2025**, 2024-356. [CrossRef]
56. Monroy, K.P.; Delgado, J.; Sereno, L.; Ciurana, J.; Hendrichs, N.J. Effects of the Selective Laser Melting manufacturing process on the properties of CoCrMo single tracks. *Met. Mater. Int.* **2014**, *20*, 873–884. [CrossRef]
57. Ayyıldız, S.; Soylu, E.H.; İde, S.; Kılıç, S.; Sipahi, C.; Pişkin, B.; Gökçe, H.S. Annealing of Co-Cr dental alloy: Effects on nanostructure and Rockwell hardness. *J. Adv. Prosthodont.* **2013**, *5*, 471. [CrossRef]
58. *ASTM F75-18*; Standard Specification for Cobalt-28 Chromium-6 Molybdenum Alloy Castings and Casting Alloy for Surgical Implants (UNS R30075). ASTM International: West Conshohocken, PA, USA, 2018.
59. *ASTM F90-14*; Standard Specification for Wrought Cobalt-20Chromium-15Tungsten-10Nickel Alloy for Surgical Implant Applications (UNS R30605). ASTM International: West Conshohocken, PA, USA, 2014.
60. Kotila, J.; Syvänen, T.; Hänninen, J.; Latikka, M.; Nyrhilä, O. Direct Metal Laser Sintering—New Possibilities in Biomedical Part Manufacturing. *Mater. Sci. Forum* **2007**, *534–536*, 461–464. [CrossRef]
61. Lapcevic, A.R.; Jevremovic, D.P.; Puskar, T.M.; Williams, R.J.; Eggbeer, D. Comparative analysis of structure and hardness of cast and direct metal laser sintering produced Co-Cr alloys used for dental devices. *Rapid Prototyp. J.* **2016**, *22*, 144–151. [CrossRef]
62. Zhou, Y.; Li, N.; Yan, J.; Zeng, Q. Comparative analysis of the microstructures and mechanical properties of Co-Cr dental alloys fabricated by different methods. *J. Prosthet. Dent.* **2018**, *120*, 617–623. [CrossRef] [PubMed]
63. Tunthawiroon, P.; Chiba, A. Effects of Si concentrations on microstructure and mechanical properties of as-cast CoCrMo alloys. *IOP Conf. Ser. Mater. Sci. Eng.* **2019**, *635*, 012006. [CrossRef]
64. Monkova, K.; Monka, P.P.; Pantazopoulos, G.A.; Toulfatzis, A.I.; Šmeringaiová, A.; Török, J.; Papadopoulou, S. Effect of Crosshead Speed and Volume Ratio on Compressive Mechanical Properties of Mono- and Double-Gyroid Structures Made of Inconel 718. *Materials* **2023**, *16*, 4973. [CrossRef]
65. Ibrahim, M.; Hulme, C.; Grasmo, G.; Aune, R.E. Preliminary Evaluation of Nickel Silicide (NiSi12-wt%) Laser Cladding for Enhancing Microhardness and Corrosion Resistance of S355 Steel. *Metals* **2024**, *14*, 1389. [CrossRef]
66. Grobelny, P.; Legutko, S.; Habrat, W.; Furmanski, L. Investigations of surface topography of titanium alloy manufactured with the use of 3D print. *IOP Conf. Ser. Mater. Sci. Eng.* **2018**, *393*, 012108. [CrossRef]
67. Fratila, A.; Jimenez-Marcos, C.; Mirza-Rosca, J.C.; Saceleanu, A. Mechanical properties and biocompatibility of various cobalt chromium dental alloys. *Mater. Chem. Phys.* **2023**, *304*, 127867. [CrossRef]
68. *ISO 9693:2019*; Dentistry—Compatibility Testing for Metal-Ceramic and Ceramic-Ceramic Systems. ISO: Geneva, Switzerland, 2019.

metals MDPI

Article

Optimized and Additively Manufactured Face Mills for Enhanced Cutting Performance

Ozgur Poyraz, Daniel Tomlinson, Anthony Molyneux, Marie E. Baxter, Evren Yasa * and James Hughes

The University of Sheffield Advanced Manufacturing Research Centre North-West, Blackburn BB2 7HP, UK; o.poyraz@amrc.co.uk (O.P.); d.tomlinson@amrc.co.uk (D.T.); a.molyneux@amrc.co.uk (A.M.); m.baxter@amrc.co.uk (M.E.B.); j.hughes@amrc.co.uk (J.H.)

* Correspondence: e.yasa@amrc.co.uk or e.yasa@sheffield.ac.uk

Abstract: With the growing acceptance of additive manufacturing (AM) across various sectors, laser-based powder bed fusion (PBF-LB) has widely been utilized to create intricately shaped parts from a range of metals alloys. The factors contributing to this reception are limited not only to the achievable geometrical complexity but also enhanced product functionality. Prominent sectors such as aerospace, defense, and biomedical have begun to leverage PBF-LB for a wide range of applications; its use in other industries, such as cutting tool manufacturing, remains more limited. Therefore, to address the potential of LPBF for higher performance in cutting tool applications, this study examines topology optimization and the laser powder bed fusion of face mills with experimental modal characteristics and cutting performance. The objectives in topology optimizations were to decrease the vibration magnitudes of face mills while keeping them stiff against deformation due to cutting forces. Three distinct designs were benchmarked, each with different weight reduction targets, and were fabricated using PBF-LB with M300 maraging steel. The optimized face mills demonstrated better performance with diminished vibration magnitudes and less tool wear patterns.

Keywords: laser-based powder bed fusion; maraging steel; topology optimization; face milling; tap testing; tool wear

1. Introduction

Additive manufacturing (AM) has revolutionized product development, design, and manufacturing by enabling unprecedented geometrical freedom. Unlike traditional manufacturing methods, which often rely on fixtures, molds, or cores, AM builds objects layer by layer, offering the flexibility to design without the constraints imposed by conventional production techniques.

Along with its design and manufacturing advantages, AM offers a diverse range of material options, each suited to specific process categories. Today, a wide range of materials, including metals, polymers, ceramics, and composites are used in various AM technologies, to manufacture products enabling customized solutions across different industries. As per international standards organizations, AM technologies are categorized into seven distinct groups which include binder jetting (BJ), directed energy deposition (DED), material extrusion (ME), material jetting (MJ), powder bed fusion (PBF), sheet lamination (SL), and vat photopolymerization (VP) [1].

Among these techniques, laser-based powder bed fusion (PBF-LB), a subset of the PBF group, stands out as the most technologically mature method for processing metals and their alloys [2]. Currently, PBF-LB can process a diverse selection of metal alloys, including aluminum, cobalt, copper, iron, nickel, and titanium alloys. Additionally, ongoing research is expanding to include materials previously unconsidered for PBF-LB, further enhancing the capabilities of the technique [3]. The process itself involves spreading a thin layer of fine metal powder over the build platform. This stage, influenced by different powder characteristics such as size, can impact process efficiency like spread ability and affect the part outcomes, including surface quality [4]. A laser, with a narrow focus and beam diameter of less than 100 μm, rapidly melts and solidifies the powder in the precise shape defined by the cross-section of the part. This layer-by-layer approach allows for the creation of complex geometries that would be challenging or even impossible to achieve using traditional manufacturing methods. With the combined application of a fine beam diameter and thin layer thicknesses, PBF-LB can create narrow melt pools to enable manufacturing of slender geometries [5], allowing the direct fabrication of intricate shapes such as turbine blades with internal cooling channels, impellers, or customized medical implants [6]. PBF-LB offers design flexibility through methods such as mass customization, topology optimization, lattice infilling for large volumes [7,8]. Industrial examples, such as the General Electric LEAP engine fuel nozzle, highlight the application of assembly consolidation, while topology optimization has been used to enhance the mechanical and dynamic properties of lightweight satellite structures [9,10].

Although the progress of topology optimization (TO) has been ongoing since 18th century with some considerable developments in theory and commercialization [11–14], it is the pairing of AM and TO that has shown to be an effective means for structural innovation, paving the path for broader applications [15]. As of today, density-based, hard-kill, boundary variation, and non-gradient topology optimization options are available under various commercial software packages, including Abaqus, Autodesk Fusion 360, Altair OptiStruct, Ansys Mechanical, Inspire, nTopology, and CAESS ProTOp [16]. Besides these, certain optimization software provides advanced features such as employing multiple objectives, integrating symmetrical solutions for effective computation, or executing calculations while taking various manufacturing constraints into account. Leveraging advancements in AM and TO, various industries such as aerospace, medicine, automotive industry, and machinery have launched numerous products, including gearboxes, engine brackets, landing gears, space frames, antenna enclosures, brake caliper casings, flywheels, and cam supports since the 2010s [16]. In contrast, the cutting tool sector has only begun its research and implementation efforts regarding the AM and TO of cutting tools and holders in recent years, mainly in the commercial domain.

Although metal tooling represents just 4% of total AM applications, a comprehensive literature review by Kelliger et al. (2024) revealed several possible advantages for cutting tools, categorized into cooling, light-weighting, damping, and functional integration through consolidation and topology optimization [17]. Figure 1 illustrates the categories of cutting tool materials and the potential enhancements achievable through the use of AM and TO [17].

Cutting tool cooling efficiency, which influences heat in the cutting zone, workpiece and tool temperatures, chip removal, and surface finish, can be significantly enhanced through AM. Lakner et al. (2019) demonstrated that an L-shaped nozzle design reduced flank wear when machining an AISI 4140 alloy [18]. Kugaevskii et al. (2019) further highlighted the effectiveness of AM in creating intricate shapes and efficient cooling holes [19]. Zäh et al. (2018) showcased the incorporation of spiral channels to improve cooling efficiency, made possible by the design flexibility of AM [20].

While vibration damping through lattice structures is a prevalent area of study, integrating these structures into cutting tools and minimizing the vibration amplitudes of the tools was only realized in 2019 by Vogel et al. [21]. They reengineered and produced a tool holder featuring an elliptical cavity filled with truncated octahedron lattices, and tap tests indicated a reduction in amplitudes from 2.75 to 0.75 μm/N [21]. In a second investigation, they created a milling holder designed to include truncated octahedron elements filled alongside denser $WCZrO_2$ particles and demonstrated the findings through tap tests and acoustic emissions throughout milling [22]. Hanzl et al. focused on face mills where a cutter was restructured into a shell filled with body-centered cubic (BCC) lattice structures and subsequently tap tested to demonstrate its damping enhancement [23,24]. In research conducted by Tomasoni et al. (2021), a face mill underwent topological optimization, and the enhanced conditions related to displacement and stresses were reported [25].

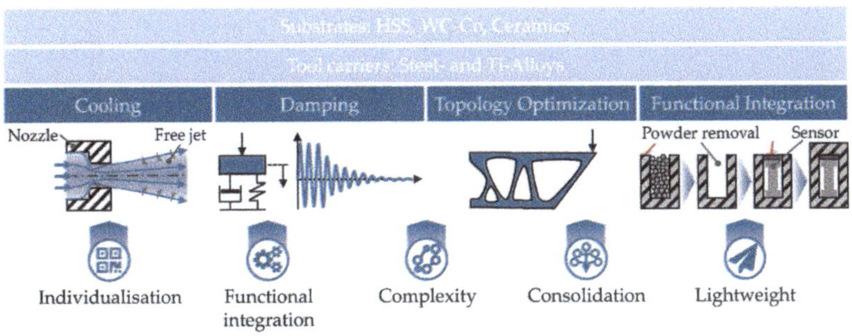

Figure 1. Potential cutting tool improvements via use of AM and TO by Kelliger et al. (licensed under CC BY 4.0) [17].

In addition to limited studies in academia, industrial cases have begun to emerge that leverage AM and integrate it with TO. These examples are provided by reputable tooling firms such as Kennametal, Sandvik Cormorant, and Seco. Kennametal's KENionic™ Technology provides lightweight poly-crystalline diamond (PCD) fine boring tools (Figure 2a), while Sandvik Coromant's Coromill lightweight Ti6Al4V face mill featuring silent mass-damped adaptors is sold commercially as an off-the-shelf product (Figure 2b). Furthermore, Seco Tools engages in hybrid AM tool production by building directly onto a HSK100 spindle cone (Figure 2c). Although industrial initiatives and advancements enhance the cutting-edge nature of AM tool design and production, access to detailed information from these case studies is limited due to IP considerations.

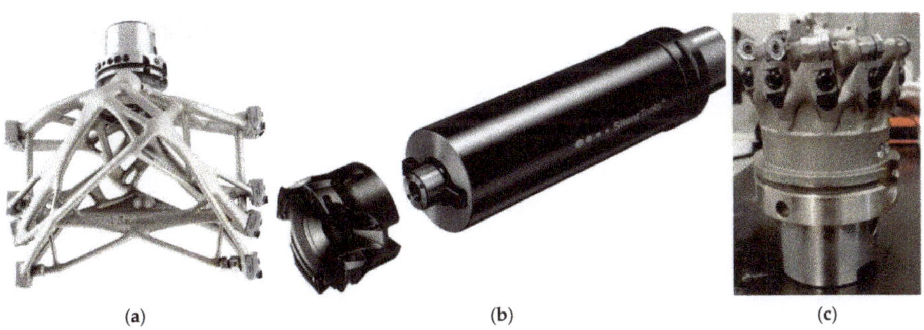

Figure 2. (**a**) Kennametal Lightweight PCD fine boring tool with steel body made with new KENionic ™ Technology (reprinted with permission from reference [26]); (**b**) Sandvik Coromill Light AM Tool with silent mass-damped adaptor (reprinted with permission from reference [27]); (**c**) Seco Hybrid AM tool for HSK interface (reprinted with permission from reference [28].

This article investigates the topology optimization of indexable face mills, considering modal characteristics into account as well as manufacturing constraints such as overhangs. Face mills, additively manufactured from M300 maraging steel via PBF-LB, were machined to final dimensions and underwent tap tests and cutting trials on Ti6Al4V wrought plates. Outcomes were benchmarked based on insert wear patterns following one hour of machining.

2. Materials and Methods

This study follows the research outline in Figure 3, starting with the selection of a commercially available indexable face mill, followed by preliminary finite element analysis (FEA) to assess the tool's initial state. After topology optimizations based on initial FEA, distinct tool designs were produced from M300 maraging steel powder by PBF-LB AM and heat treated in accordance with the material's specification. Wire electric discharge machining (WEDM) was then used to separate the face mills from the PBF-LB base plates, and their assembly surfaces, such as the screw holes or insert seatings, were finish machined. The face mills were then mounted to spindle extension arbors and adaptors for tap tests. Lastly, cutting trials on wrought Ti6Al4V were carried out using a fixed set of cutting parameters.

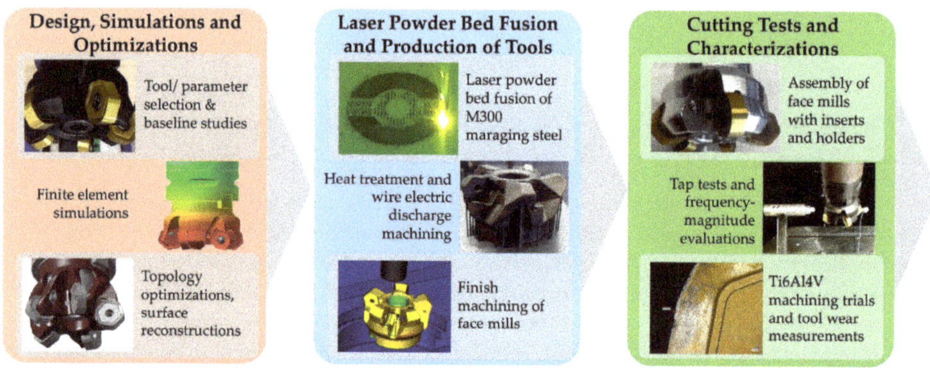

Figure 3. The research workflow for this study.

2.1. Design, Simulations and Optimizations

As a demonstrator, a 50 mm diameter indexable face mill with 5 cutting flutes suitable for cutting medium-to-high-strength materials was selected. The Kennametal Dodeka^{TM} 45° (Spain) indexable face mill was used with a compatible KCSM40 carbide-grade insert with the serial number HNGJ 0905-ANSNGD (United States of America). The Sandvik HSK-to-Coromant Capto® adaptor (C5-390.410-63 090C) (Germany) and the Coromant Capto®-to-arbor adaptor (C5-391.05C-22 025M) were included in the tool holding assembly, which was maintained at 115 mm length as an intermediate value (Figure 4a). Antivibration boring bars were intentionally omitted from the study.

FEA and TO were conducted with Altair Inspire© 2023.1 which invokes OptiStruct© 2023.1 for solving. Solid 3-dimensional (Tetra10/CTETRA) mesh elements were used for both the preliminary FEA and the latter TO. The reason for this preference is the ability of these elements to adapt to intricate geometric forms. Variable mesh sizes were used, and the ratio of meshes with a skew (distortion) of 0.999 was targeted to be below 0.1% for the whole assembly to maintain high quality.

Spindle extension, arbor, and inserts were chosen as non-design areas for optimizations, indicating that they cannot undergo any material removal or weight reduction (Figure 4a). Even though the face mill body was viewed as a design area, the mating features were divided and excluded from optimizations to maintain the assembly integrity (Figure 4d).

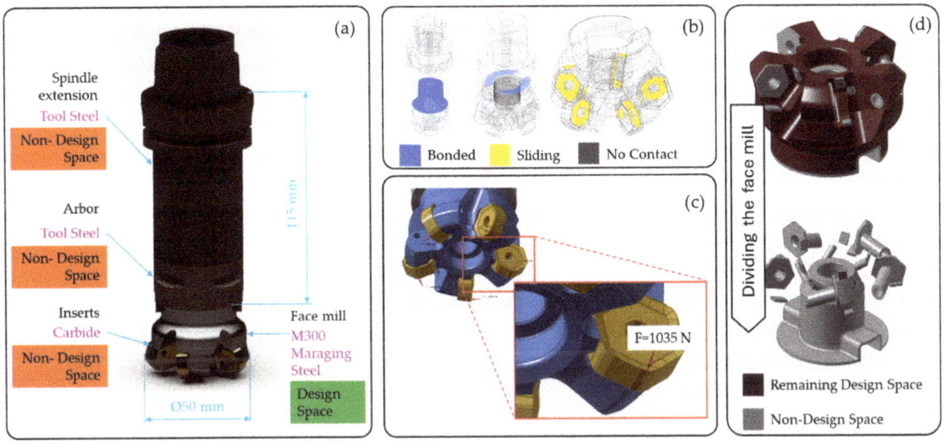

Figure 4. (**a**) Face mill assembly with component names, dimensions and materials, (**b**) contacts, (**c**) force boundary conditions, (**d**) design and non-design spaces of face mill body.

It was assumed that the contact between the inserts and the face mill was frictionless linear sliding. Bonded contacts were assigned between the remaining bodies, excluding the cylindrical face of the arbor that extends into the face mill in a loose state (Figure 4b).

To supply cutting forces as boundary conditions for FEA and TO, computer-aided manufacturing (CAM) and cutting force simulations were conducted. Toolpaths were created in Siemens NX 2412, utilizing a linear zig strategy to ensure climb cutting was maintained. For the FEA and TO, higher cutting parameters were considered to ensure safety and prevent the damage or deformation of the optimized face mills during testing. In this context, 50 m/min cutting speed (V_c), 0.31 mm feed per tooth (f_z), 2.25 mm axial depth of cut (a_p), and 30 mm radial depth of cut (a_e) were chosen. Cutting force simulations were performed with MACHPRO™ software, utilizing cutting force coefficients previously obtained by AMRC NW researchers, revealing a maximum force value of 1035 N in the tangential direction, which was used as a boundary condition for FEA and TO (Figure 4c).

A fixed constraint was applied to the spindle and machine interface, and 80% pretension loads were considered for the fasteners, including the inserts and the arbor.

The main goal for TO was to achieve weight reductions of at least 15%, 25%, and 35%, aiming to enhance the modal condition of the face mills by increasing frequencies or decreasing vibration magnitudes. Since reducing the weight and volume/thickness of the components could negatively affect the stiffness, the increases in deformation relative to the original face mill were also restricted to 20%, 25%, and 30%. Finally, rotational symmetry was incorporated into TO, and a manufacturing constraint of an overhang angle of 45° was implemented to facilitate support-free PBF-LB. The material properties used for FEA and TO are presented in Table 1 along with their corresponding references.

Table 1. Material properties used for FEA and TO.

Material	Density (kg/m^3)	Elastic Modulus (GPa)	Yield Strength (MPa)	Reference No.
M300	8100	199	1873	[29]
Tool Steel	7800	210	1650	[30]
Carbide	15,700	669	300	[31]

As the TO process reduces weight by discarding or voiding the mesh elements that are not necessary according to the analysis, faceted geometries and volumes were created. Due to the faceted shapes, these rough volumes need to be reconstructed to be utilized for subsequent operations. The majority of the surface reconstruction steps were performed using the Siemens NX 2412 Polygon Modeling module, incorporating the other components of assembly like the arbor and inserts. Additionally, the PolyNURBS Modeling module of Altair Inspire© 2023.1 was used for certain freeform surfaces that necessitated more complex forms and transitions. The final modifications on face mill designs were carried out by following Design for Additive Manufacturing (DfAM) principles. In this regard, the rotational axes of the face mills were adjusted to be parallel to the PBF-LB build direction, and their arbor mating faces were kept at the bottom, as these assembly faces would eventually require finish machining. Face mill geometries were then subjected to draft angle analysis based on a 45° overhanging limit, leading to a lower volume of supports. This achievement is also related to the overhang control constraint that was introduced for TO. As the last step, 0.5 mm stock envelopes were added to the up-facing mating surfaces like screw holes, insert seats and arbor interfaces.

2.2. Laser Powder Bed Fusion and Production of Face Mills

Carpenter Additive® gas atomized M300 maraging steel (United Kingdom) powders were employed for PBF-LB. M300 powder having iron as the balancing element contains the primary alloying elements of nickel at 18.2% weight, cobalt at 9.0% weight, molybdenum at 4.85% weight, and titanium at 0.87% weight. The M300 maraging steel used for this project exhibited an average powder size distribution of 35 μm with Dv (10) 19 μm and Dv (90) 52 μm levels.

PBF-LB production was conducted using a Renishaw® AM250 (United Kingdom) system located at the Royce Translational Centre (RTC) at the University of Sheffield. To ensure the reliability and repeatability of the additive manufacturing process, the machine underwent a comprehensive optical calibration and verification performed by the OEM. Routine servicing is conducted, including a thorough inspection to confirm optimal performance. This system features a 250 × 250 mm build area, a pulsed ytterbium fiber laser and an argon as an inert gas atmosphere. Volumes of both unoptimized and optimized face mill geometries were exported to a standard tessellation language (stl) file format, sliced with Materialise® Magics version 28, and prepared using Renishaw QuantAM software version 5.3. Pyramid type support structures were used where necessary, and density cubes were also included to validate the density of parts (Figure 5a).

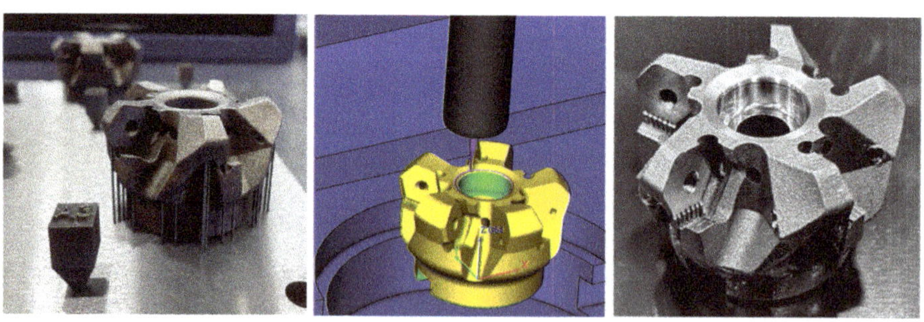

(a) (b) (c)

Figure 5. (**a**) Face mill and density cube and support structures following LPBF, (**b**) machining simulation using Vericut, (**c**) finished face mill.

The bulk material, or hatch region, was processed using a laser power of 200 W, a point distance of 65 μm, and an exposure time of 80 μs, with a hatch distance of 80 μm. The border was processed with a reduced laser power of 100 W, a finer point distance of 20 μm, and an exposure time of 50 μs to refine edge quality and minimize surface roughness. Both scanning strategies resulted in an equivalent scanning speed of 528 mm/s. The layer thickness was maintained at 40 μm to ensure a balance between build rate and part density.

The parts were heat treated in a TAV vacuum furnace (Model TH 30/30/30). The heat treatment recipe that was carried out on the samples is the standard stress-relief (SR) cycle recommended by the machine manufacturer [32]. The ramp up rate from room temperature was 10 °C per minute up to 850 °C, at which temperature the dwell time was 30 min, followed by an argon air cool. Following the SR, the parts were separated from the base plate using a GF AgieCharmilles CUT 550 P (Switzerland) wire electric discharge machine (WEDM).

Density measurements following Archimedes' principle were performed in accordance with ASTM 962-23 on a precision balance (Precisa Series 360 EP 125 SM) using distilled water. The results confirmed densities exceeding 99.9% [33].

The computer-aided manufacturing (CAM) for the face mills was conducted with Siemens NX 2412, and once the toolpaths were post-processed and transformed into G-codes, CGTech VERICUT 9.5.1 was employed for simulations to prevent any collisions on the machine. Finally, stock envelopes of 0.5 mm were finish machined on a Hermle C400 multi-axis milling machine located at AMRC NW (Figure 5b,c).

2.3. Tests and Investigations

Quality control of the produced tools was performed with respect to dimensional and surface quality factors. They underwent 3-dimensional (3D) scanning with the GOM ATOS-Q scanner, and stl models were generated using GOM Inspection software version 2021. Subsequently, these models were brought into Siemens NX 2412 along with nominal CAD files, and the software's automatic fitting option was utilized to best fit these two models. A 3D deviation gauge was utilized for a checking distance of 0.250 mm and an inner and outer tolerance of 0.02 mm. Surface quality inspections were conducted on specific surfaces of each produced face mill. These critical surfaces include the flat plane between the face mill and arbor, the insert seatings, the tips, and the radial outer diameter. These were measured using a Keyence VHX E-100 microscope (Japan) at 500× magnification. After taking measurements, VHX software was employed in 3D depth mode to obtain surface quality values. The arithmetic mean of the surface areas (Sa) was chosen to compare various face mills by setting the S-Filter to 2.5 μm and the L-Filter to 4 μm. After dimensional and surface quality assessments, each face mill was weighed using a Kern KB10000-1N balance, which has a readability increment of 0.1 mm.

As an assembly inspection, all face mills were assembled with spindle extensions, arbors, and inserts. Certain mating conditions, such as insert to face mill, were evaluated with a height gage and shims to identify any problems concerning their alignment or gaps between them. Additionally, the radial and axial runouts of insert tips were measured using a Nikken Elbo Controlli E46LA Tool Presetter and compared to those fitted on the original equipment manufacturer (OEM) face mill.

Final tap tests were carried out on the Mazak Variaxis I500 (Japan) utilizing the Productive Machines TapStarter kit (United Kingdom). This was utilized with the ICP 50,862 hammer, PCB Model 352C23 accelerometer, and the ICP Model 485B39 sensor signal conditioner, all of which were properly calibrated. Tap testing began solely with spindle extension, followed by the sequential addition of arbor, OEM face mill, and inserts (Figure 6a). This procedure was also carried out for additively manufactured face mills considering two axial directions as X and Y. The outputs were received as csv files in a columnar format, including frequency (1–10,000 Hz) and magnitude (m/N). In creating the frequency and magnitude graphs for this article, the resulting real and imaginary values were considered, and the averages of the X and Y directions are presented to provide a summary of numerous measurements. The frequency range of 300–1300 Hz was plotted to prevent any negligible values, emphasize the key window, and clearly present the benchmark within the focused range.

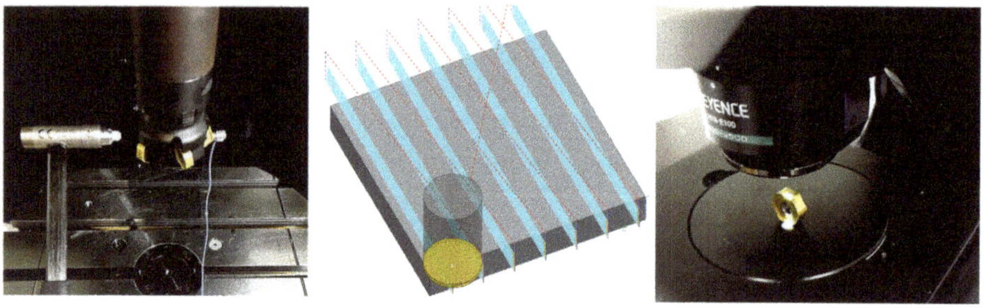

(**a**) (**b**) (**c**)

Figure 6. (**a**) Tap testing of face mills, (**b**) zig cutting strategy for machining tests, and (**c**) flank wear measurements under microscope.

Machining tests were performed on a Mazak Variaxis I500, located at AMRC NW, using four plates of Ti6Al4V (ASTM Grade 5 titanium) alloy with dimensions of $200 \times 200 \times 50$ mm^3 [34]. A zig cutting strategy was employed to keep each parameter and climb cutting the same throughout all machining (Figure 6b); 60 m/min cutting speed (Vc), 0.113 mm feed per tooth (fz), 2 mm axial depth of cut (ap), and 30 mm radial depth of cut (ae) were considered, and toolpath generation was accomplished using Siemens NX CAM module. Tests were deliberately paused at the 30 min mark during the cutting process to assess the progress of the tool wear and finished after 1 h of cutting.

International standards define tool deterioration as the changes observed on the cutting edge of tools resulting from the process of cutting [35]. Although this is a broad term, the standards encompass tool wear and brittle fractures or chipping as subcategories of deterioration. Among them, tool wear or flank wear refers to the loss of tool material from the flanks of the tool during the cutting process and it can be uniform (VB1), non-uniform (VB2), or localized (VB3) at certain locations. The standard suggests limits of 0.3 mm for uniform (VB1) wear and 0.5 mm for localized (VB3) wear [35]. Alongside the standards, earlier research on tool wear during Ti6Al4V machining has performed wear tests over specific time periods. In this context, prior researchers performed wear tests based on a 0.3 mm uniform tool wear limit, which lasted from thirty minutes to one hour [36,37]. Considering these factors, flank wear studies were conducted, and the flanks of 5 inserts from each face mill were examined using a Keyence VHX E-100 microscope at 100× magnification (Figure 6c). Inserts were positioned under the microscope at a right angle to the lens to observe the flank faces, and measurements were obtained using the length measurement feature of VHX Software.

Finally, the surface quality of the machined plates was measured along feed and cross-feed directions. The portable Mitutoyo Surftest SJ-210-Series 178 (Japan) was employed for roughness assessments utilizing constant sampling lengths and cut-off lengths ($\lambda s = 2.5$ µm, $\lambda c = 0.8$ µm) for each machined plate. The arithmetic mean values (Ra) were reported as the outcomes.

3. Results and Discussions

The preliminary FEA outcomes offered an in-depth understanding of the OEM tool's modal characteristics and deformation behavior under the stresses induced by cutting forces. The initial modal analysis of the tool assembly, consisting of the OEM face mill and inserts, revealed bending in various X and Y directions in the first modes, followed by torsional motion around the tool's rotational center, and culminating in axial displacement in the sixth mode. The results of the FEA simulations confirmed that the OEM face mill is already a stiffness-driven component, and it did not appear to be susceptible to high deformations under the applied loading conditions. Even under the worst-case scenario—where a cutting force of 1035 N is applied due to the previously outlined aggressive cutting parameters—the maximum elastic deformation the OEM tool experiences was observed at approximately 52 μm.

Since some of the TO gave way to an irregular and inhomogeneous mesh distribution near insert mating faces, surface reconstructions needed careful consideration. There were detached regions close to the insert mating faces in just one of the results, Option 3, which had the most ambitious mass reduction target. The Polygon Modeling module of Siemens NX 2412 and the PolyNURBS modeling module of Altair Inspire were used to overcome these problems; however, this revealed the necessity of FEA simulations on the new geometries. Figure 7 shows the results of FEA simulations of the reconstructed face mills that were optimized by aiming for minimum mass reductions of 15%, 25%, and 35%, respectively, for the three options. These optimizations also limited the deformation increase relative to the OEM face mill to less than 20%, 25%, and 30%. As can be seen in Figure 7, all optimization targets regarding the mass reduction were successfully met and none of the deformation constraints were violated.

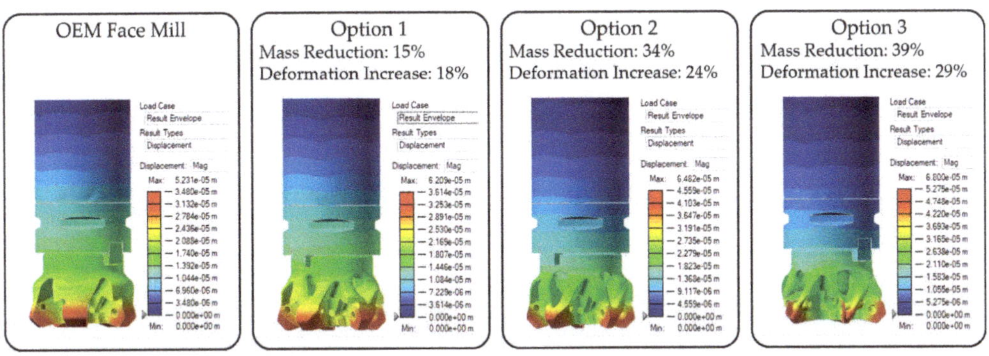

Figure 7. Benchmarking of TO results in terms of mass reduction and deformation increase.

The highest deformation increase ratio of 29%, which changed the OEM tool's 52.3 μm deformation to 68 μm for TO Option 3, still remained within elastic limits. This conclusion was confirmed by examining the stress graphs produced by FEA simulations (see Figure 8). The graph illustrates the von Mises stress levels of the tool bodies, accompanied by a common color legend on the right side. The highest stress observed in those limited areas for various TO options reached approximately 450 MPa, providing a sufficient buffer from the yield stress and preventing the risk of plastic deformations.

The reconstructed face mill geometries were produced via PBF-LB, SR and machining without any significant problems. The 3D scans of the produced face mills exhibited good dimensional accuracy in comparison to the nominal CAD model and all adhered to the tolerances specified in the Materials and Methods section. The surfaces of unoptimized AM face mills showed Sa values ranging from 0.7 μm to 1.5 μm, since they were generally machined post PBF-LB, whereas optimized face mills could only achieve this range for the flat plane between the face mill and arbor, as well as the insert seatings. The remaining freeform surfaces of all optimized face mills, particularly in the outer radial regions, showed a rougher nature, reaching Sa 14.28 µm due to the well-known stair stepping effect.

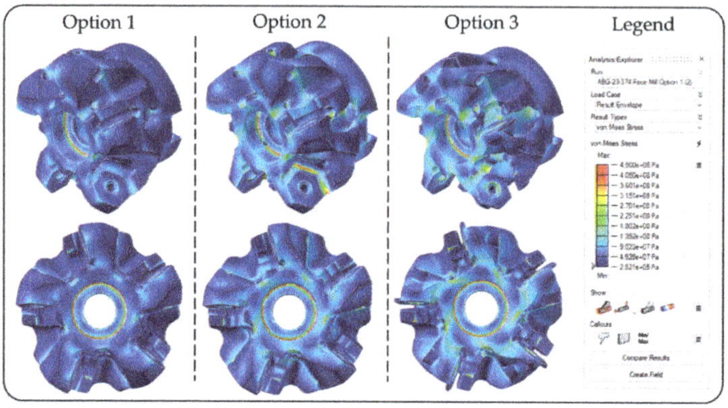

Figure 8. Benchmarking of TO results in terms of von Mises stresses.

The weight of the tools also differed due to the variations in geometries resulting from TO. As can be seen in Table 2, every optimized face mill exhibited a reduced mass when compared to the original face mill. Conversely, only the unoptimized AM face mill was heavier than the original face mill. This is because, while the volumes of both options are identical, the density of M300 maraging steel (8.1 g/cm^3) surpasses the density of the original face mill material (7.8 g/cm^3), AMS 6411 [29,38].

Table 2. Comparison of face mill masses.

	Original Face Mill	Unoptimized AM Face Mill	TO Option 1 Face Mill	TO Option 2 Face Mill	TO Option 3 Face Mill
Mass (g)	329	344	279	213	201

In addition to Figures 7 and 8, which illustrate deformations and stresses, Figure 9 compares the modal characteristics based on tap test results. Figure 9a demonstrates that tap testing solely on the extension body produced a distinct peak at approximately 1100 Hz, which was identified as the dominant frequency associated with the extension body. Upon the installation of the arbor on the extension, a new peak emerged at approximately 700 Hz, and the dominant frequency of the extension body was shifted to 1000 Hz with a significant increase in its magnitude (Figure 9b). Finally, addition of OEM face mill and inserts introduced a new peak around 500 Hz, while shifting the extension's dominant frequency to 900 Hz and increasing the magnitude of extension and arbor to around 700 Hz (Figure 9c). When the results from the OEM face mill were compared with TO Option 3, a distinct advantage of weight reduction in vibration magnitude was apparent in the 400–500 Hz frequency range (Figure 9d). After observing the advantages from the comparison of the OEM face mill and TO Option 3, the other TO options (Option 1, Option 2) and an additively manufactured version of the OEM face mill geometry were also evaluated within the same frequency range. While TO Option 1 and TO Option 2 showed benefits compared to the tool assembly with the OEM face mill, their ability to reduce vibration magnitude was less effective than that of Option 3. The unoptimized additively manufactured face mill assembly exhibited higher vibration levels than the OEM face mill assembly. This outcome was associated with the higher mass of unoptimized AM face mill when compared to the mass of the original face mill (Table 2).

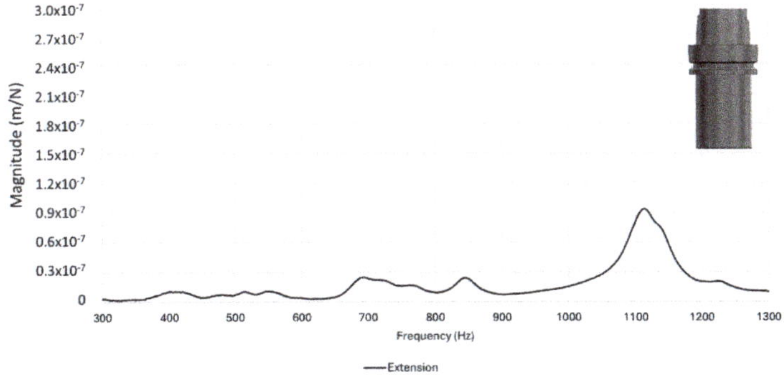

(**a**)

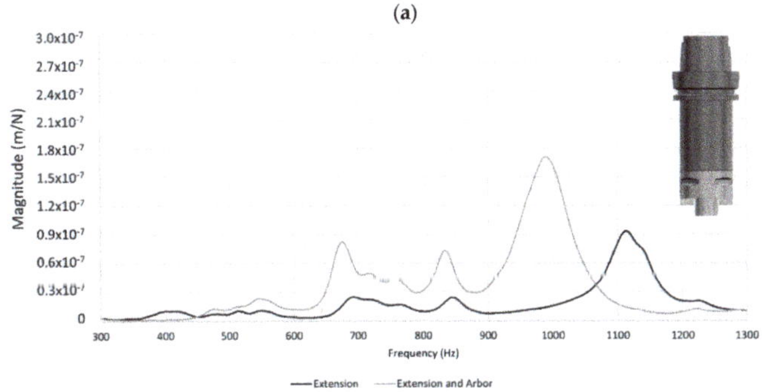

(**b**)

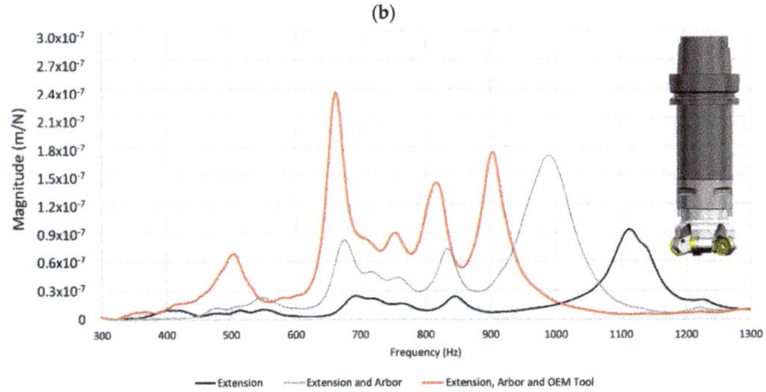

(**c**)

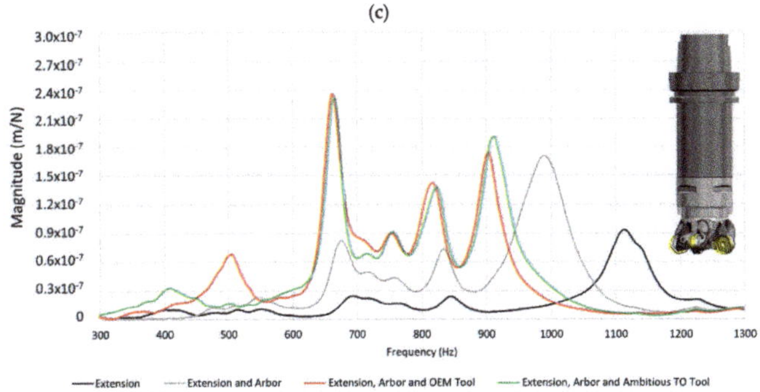

(**d**)

Figure 9. Tap testing results: (**a**) extension; (**b**) extension and arbor; (**c**) extension arbor and OEM tool; (**d**) extension, arbor, and OEM tool benchmarked with extension, arbor, and TO Option 3.

It was observed that all tools could operate for about 1 h on one cutting edge before hitting the average flank wear (VB1) limit of 0.3 mm, as shown in Figure 10, as recommended by the standard [35]. Considering the six edges on each insert and both sides, these tools could potentially be effective for up to 12 h. Since the distances of the localized flank wear (VB3) peaks to the insert tip were 2 mm in most instances, this suggests a correlation with a 2 mm cutting depth, allowing for the interpretation of notch wear (Figure 11). Nonetheless, it is acknowledged that stability problems during the cut may exacerbate notch wear.

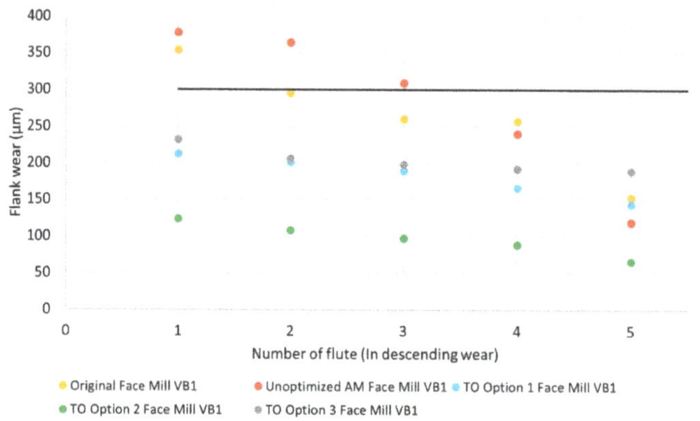

Figure 10. Average flank wear of various face mills listed according to the number of flutes in decreasing order of wear.

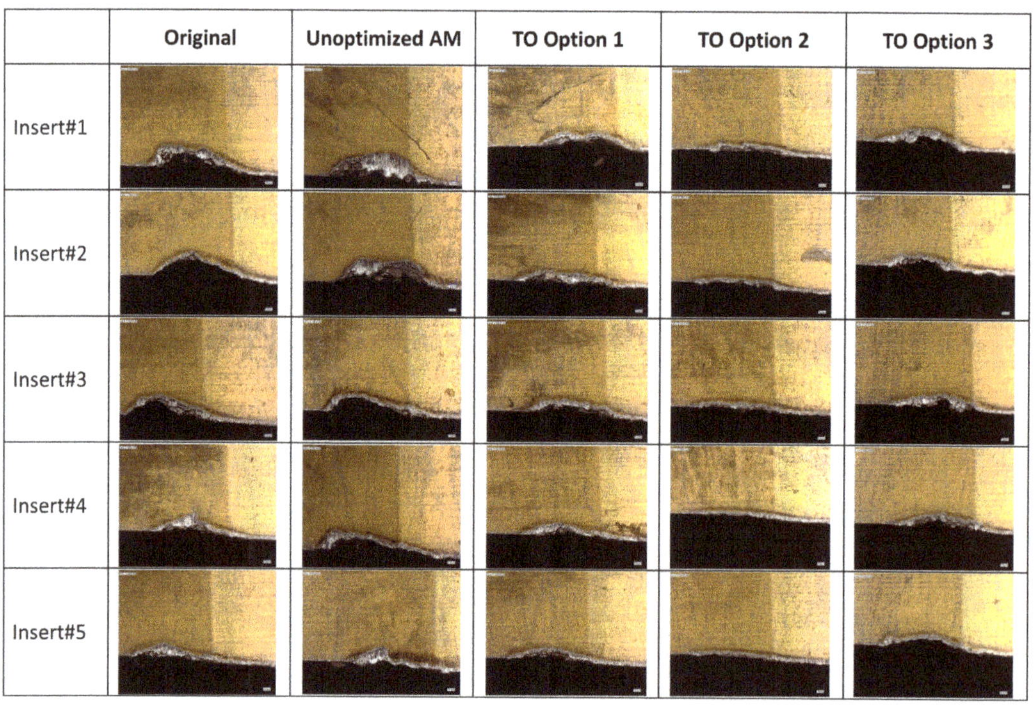

Figure 11. Benchmarking of worn edges of various face mill inserts.

The optimal outcomes were seen with TO Option 2, which had a moderate mass reduction, whereas the least favorable results were linked to the unoptimized AM tool. Of the three optimized versions, the TO Option 3 face mill had the lowest performance. Despite having more ambitious optimization goals than the other two, which results in greater reductions in mass and vibration magnitude, this observation might be linked to micro deflections in thin insert supports during the process, causing variations in chip thickness. On the other hand, the thinner sections may be less effective than the thicker ones at transferring heat to the arbor and extension body, resulting in the reduced cooling of the face mill and inserts. However, in this study, the effect of the coolant was not taken into consideration, and it is possible that this optimized geometry might have led to some unfavorable conditions regarding the cooling. Figure 11 illustrates that the inserts from TO Option 1, TO Option 2, and TO Option 3 face mills with reduced wear showed minimal variation among themselves, and the average flank wear across various inserts stayed within a steady range. The variations in average flank wear among different inserts of the original face mill and the unoptimized AM face mill ranged from 0.15 m to 0.35 mm.

Ti6Al4V plates that were machined and benchmarked with the original face mill and the topologically optimized AM face mills exhibited the same surface roughness values of Ra 0.3 μm in the cross-feed direction. On the other side, the surface roughness measurements in the feed direction fluctuated within a small range, with the original face mill demonstrating Ra 0.29 μm, whereas the optimized face mills exhibited an average of Ra 0.21 μm.

The economic viability of employing additive manufacturing technologies for producing face mills was examined once all technical evaluations were finished. Although the real cycle time computations were executed for all AM face mills taking into account PBF-LB, SR, and machining phases, an estimation was made for the conventional face mill by programming creep feed grinding of the flutes using parameters outlined in the article by Dang et al. (2022) [39]. These cycle times were subsequently multiplied by staff costs and machine charge rates. Ultimately, the costs for materials and tools were combined to determine the overall price. Since the information concerning costs is commercially confidential, the results were normalized as a percentage and are shown in Figure 12. As shown in Figure 12, the unoptimized AM face mill is the priciest tool due to its lengthy PBF-LB step, along with significant costs resulting from machining and powder material usage. Contrary to this, the original face mill, utilizing a conventional machining method that includes grinding, turning, and milling, exhibits the least production expenses. It is essential to emphasize that the material costs of the original face mill are this low and cannot be observed in the graph. Of the optimized variants, just TO Option 3 face mill can come close to the reduced expenses of original face mill, yet continues to exhibit elevated figures.

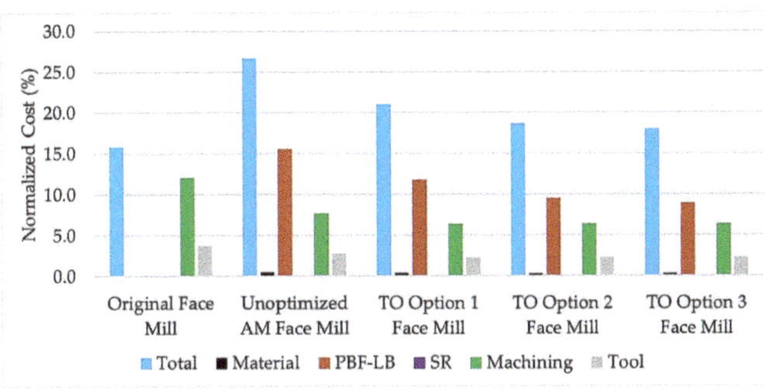

Figure 12. Normalized costs of tool productions.

4. Conclusions

This research demonstrated the application of topology optimizations and additive manufacturing of indexable face mills to improve cutting efficiency. Three weight reductions were evaluated to lower the vibration magnitudes of tool assemblies while maintaining deformation similar to traditional counterparts and within elastic limits. By outlining all essential procedures, such as surface reconstruction, laser powder bed fusion, heat treatment, and machining, it was shown that various objectives can be accomplished through the application of the suggested methodology and workflow. The advantages of utilizing optimized tools highlighted an extension of the insert tool life and successfully reduced flank wear to 100 µm after one hour of machining Ti6Al4V.

Author Contributions: Conceptualization, E.Y. and O.P.; methodology, O.P., D.T., M.E.B., A.M. and E.Y.; validation, D.T., A.M. and M.E.B., investigation, D.T., A.M., M.E.B. and O.P.; writing—original draft preparation, O.P. and E.Y.; writing—review and editing, E.Y. and J.H.; supervision, E.Y. and O.P.; project administration, J.H. and E.Y.; funding acquisition, J.H. and E.Y. All authors have read and agreed to the published version of the manuscript.

Funding: This research was funded by AMRC Membership Funding under ABG-23-374 Project entitled as "Additively Manufactured Face Mill Development FoR EnhancEd Machining (FREE)".

Data Availability Statement: The original contributions presented in this study are included in the article. Further inquiries can be directed to the corresponding author.

Acknowledgments: The authors of the study would like to acknowledge Abdul Haque for initial LPBF layout preparation and M. Bora Islier for project management related activities. Kennametal's approval to utilize their face mill as a benchmark is very much appreciated, along with Andy Boffin's and Werner Penkert's technical support and rigorous review of project tasks. Finally, the authors express their gratitude to Kennametal, Sandvik and Seco Tools for providing the images of their advanced additively manufactured face mills.

Conflicts of Interest: The authors declare no conflicts of interest.

Abbreviations

The following abbreviations are used in this manuscript:

3D	Three-dimensional
AM	Additive manufacturing
BCC	Body centered cubic
BJ	Binder jetting
CAM	Computer aided manufacturing
DED	Directed energy deposition
DfAM	Design for additive manufacturing
Dv (X)	Particle size distribution for X percentage
FEA	Finite element analysis
ME	Material extrusion
MJ	Material jetting
OEM	Original equipment manufacture
PBF	Powder bed fusion
PBF-LB	Laser-based powder bed fusion
PCD	Ploy crystalline diamond
SR	Stress relieved
stl	Standard tessellation language file format
TO	Topology optimization

VB	Flank wear
Vc	Cutting speed
WEDM	Wire electric discharge machining

References

1. Martínez-García, A.; Monzón, M.; Paz, R. Standards for Additive Manufacturing Technologies. In *Additive Manufacturing*; Elsevier: Amsterdam, The Netherlands, 2021; pp. 395–408, ISBN 9780128184110.
2. Leary, M.; Downing, D.; Lozanovski, B.; Harris, J. Design Principles. In *Fundamentals of Laser Powder Bed Fusion of Metals*; Elsevier: Amsterdam, The Netherlands, 2021; pp. 119–154, ISBN 9780128240908.
3. Guan, J.; Wang, Q. Laser Powder Bed Fusion of Dissimilar Metal Materials: A Review. *Materials* **2023**, *16*, 2757. [CrossRef] [PubMed]
4. Lupo, M.; Ajabshir, S.Z.; Sofia, D.; Barletta, D.; Poletto, M. Experimental Metrics of the Powder Layer Quality in the Selective Laser Sintering Process. *Powder Technol.* **2023**, *419*, 118346. [CrossRef]
5. Depboylu, F.N.; Yasa, E.; Poyraz, O.; Korkusuz, F. Thin-Walled Commercially Pure Titanium Structures: Laser Powder Bed Fusion Process Parameter Optimization. *Machines* **2023**, *11*, 272. [CrossRef]
6. Kundakcıoğlu, E.; Lazoglu, I.; Poyraz, Ö.; Yasa, E. Modeling of Residual Stress and Distortion in Direct Metal Laser Sintering Process: A Fast Prediction Approach. *Prod. Eng. Res. Dev.* **2022**, *16*, 769–783. [CrossRef]
7. Flores, I.; Kretzschmar, N.; Azman, A.H.; Chekurov, S.; Pedersen, D.B.; Chaudhuri, A. Implications of Lattice Structures on Economics and Productivity of Metal Powder Bed Fusion. *Addit. Manuf.* **2020**, *31*, 100947. [CrossRef]
8. Eren, O.; Kürşad Sezer, H.; Ersel Canyurt, O. Perspective Chapter: Design Considerations for Additive Manufacturing. In *Product Design—A Manufacturing Perspective*; Yasa, E., Poyraz, O., Eds.; IntechOpen: London, UK, 2024; ISBN 9781803565545.
9. Khorasani, M.; Ghasemi, A.; Rolfe, B.; Gibson, I. Additive Manufacturing a Powerful Tool for the Aerospace Industry. *Rapid Prototyp. J.* **2022**, *28*, 87–100. [CrossRef]
10. Hurtado-Pérez, A.B.; Pablo-Sotelo, A.D.J.; Ramírez-López, F.; Hernández-Gómez, J.J.; Mata-Rivera, M.F. On Topology Optimisation Methods and Additive Manufacture for Satellite Structures: A Review. *Aerospace* **2023**, *10*, 1025. [CrossRef]
11. Goldstine, H.H. *A History of the Calculus of Variations from the 17th Through the 19th Century*; Studies in the History of Mathematics and Physical Sciences; Springer: New York, NY, USA, 1980; Volume 5, ISBN 9781461381082.
12. Maxwell, J.C.I. On Reciprocal Figures, Frames, and Diagrams of Forces. *Trans. R. Soc. Edinb.* **1870**, *26*, 1–40. [CrossRef]
13. Bendsøe, M.P.; Kikuchi, N. Generating Optimal Topologies in Structural Design Using a Homogenization Method. *Comput. Methods Appl. Mech. Eng.* **1988**, *71*, 197–224. [CrossRef]
14. Altair. Topology Optimization and the Lessons of History. 2020. Available online: https://altair.com/blog/executive-insights/Topology-Optimization-and-the-Lessons-of-History#:~:text=As%20the%20world%20leader%20in,also%20saw%20the%20competitive%20advantages (accessed on 21 January 2025).
15. Liu, S.; Li, Q.; Hu, J.; Chen, W.; Zhang, Y.; Luo, Y.; Wang, Q. A Survey of Topology Optimization Methods Considering Manufacturable Structural Feature Constraints for Additive Manufacturing Structures. *Addit. Manuf. Front.* **2024**, *3*, 200143. [CrossRef]
16. Ibhadode, O.; Zhang, Z.; Sixt, J.; Nsiempba, K.M.; Orakwe, J.; Martinez-Marchese, A.; Ero, O.; Shahabad, S.I.; Bonakdar, A.; Toyserkani, E. Topology Optimization for Metal Additive Manufacturing: Current Trends, Challenges, and Future Outlook. *Virtual Phys. Prototyp.* **2023**, *18*, e2181192. [CrossRef]
17. Kelliger, T.; Meurer, M.; Bergs, T. Potentials of Additive Manufacturing for Cutting Tools: A Review of Scientific and Industrial Applications. *Metals* **2024**, *14*, 982. [CrossRef]
18. Lakner, T.; Bergs, T.; Döbbeler, B. Additively Manufactured Milling Tool with Focused Cutting Fluid Supply. *Procedia CIRP* **2019**, *81*, 464–469. [CrossRef]
19. Kugaevskii, S.; Pizhenkov, E.; Gamberg, A. The Effectiveness of Additive SLM-Technologies in the Manufacture of Cutting Tools. *Mater. Today Proc.* **2019**, *19*, 1977–1981. [CrossRef]
20. Zäh, M.; Seidel, C.; Sellmer, D. *Technologie Report 08—Additive Fertigung*; MAPAL Präzisionswerkzeuge Dr. Kress KG: Aalen, Germany, 2018.
21. Vogel, F.A.M.; Berger, S.; Özkaya, E.; Biermann, D. Vibration Suppression in Turning TiAl6V4 Using Additively Manufactured Tool Holders with Specially Structured, Particle Filled Hollow Elements. *Procedia Manuf.* **2019**, *40*, 32–37. [CrossRef]
22. Vogel, F.; Baumann, J.; Jaquet, S.; Biermann, D. Particle Damped Tool Holders Enable Higher Stability Limits When Milling EN AW-7075. *SSRN J.* **2023**, 77–84. [CrossRef]
23. Hanzl, P.; Zetek, M.; Rulc, V.; Purš, H.; Zetková, I. Finite Element Analysis of a Lightweight Milling Cutter for Metal Additive Manufacturing. *Manuf. Technol.* **2019**, *19*, 753–758. [CrossRef]
24. Hanzl, P.; Zetková, I.; Zetek, M. Comparison of Lightweight and Solid Milling Cutter Capabilities. *Manuf. Technol.* **2020**, *20*, 23–26. [CrossRef]

25. Tomasoni, D.; Giorleo, L.; Ceretti, E. Milling Tool Optimization by Topology Optimization Technique. *ESAFORM* **2021**, *2021*, 1–10. [CrossRef]
26. Kennametal. Kennametal Innovation Earns International Recognition as Recipient of R&D 100 Award. 2023. Available online: https://www.kennametal.com/de/en/news/123114.html (accessed on 23 January 2025).
27. Sandvik. How 3D-Printing and Titanium Became Game-Changers for Lightweight CoroMill® 390. 2020. Available online: https://www.metalpowder.sandvik/en/news-media/news/2020/07/additive-manufacturing-and-titanium-optimized-the-lightweight-coromill-390/ (accessed on 12 February 2025).
28. Seco. Seco Tools 3D Manufacturing Creates New Opportunities. 2023. Available online: https://www.secotools.com/article/123840?language=en (accessed on 12 February 2025).
29. Renishaw. M300 Maraging Steel Data Sheet. 2017. Available online: https://www.renishaw.com/resourcecentre/download/data-sheet-maraging-steel-m300-for-200-w-powder-for-additive-manufacturing--96325?userLanguage=en&srsltid=AfmBOoq3HAz4jS0jcXDJMppmcNjfWZyIduCVDA8sYjGEslKZRXPXXTtO (accessed on 23 January 2025).
30. Davis, J.R.; ASM International (Eds.) *Carbon and Alloy Steels*; ASM Speciality Handbook; ASM International: Materials Park, OH, USA, 1996; ISBN 9780871705570.
31. *ASM International ASM Engineered Materials Reference Book*; ASM International: Metals Park, OH, USA, 1989; ISBN 9780871703507.
32. Carpenter. PowderRange M300 Data Sheet. Available online: https://www.carpenteradditive.com/hubfs/Resources/datasheets%202023/PowderRange_M300_Datasheet.pdf (accessed on 21 February 2025).
33. *ASTM B962-23*; Standard Test Methods for Density of Compacted or Sintered Powder Metallurgy (PM) Products Using Archimedes' Principle. American Society for Testing and Materials (ASTM): West Conshohocken, PA, USA, 2023.
34. Metalweb. Titanium Ti-6Al-4V (Grade 5), STA. Available online: https://www.matweb.com/search/datasheet.aspx?MatGUID=b350a789eda946c6b86a3e4d3c577b39&ckck=1 (accessed on 25 February 2025).
35. *International Standard ISO 8688*; Tool-Life Testing in Milling. ISO: Geneva, Switzerland, 1989.
36. Liang, X.; Liu, Z.; Wang, B.; Hou, X. Modeling of Plastic Deformation Induced by Thermo-Mechanical Stresses Considering Tool Flank Wear in High-Speed Machining Ti-6Al-4V. *Int. J. Mech. Sci.* **2018**, *140*, 1–12. [CrossRef]
37. Yue, C.; Li, X.; Liu, X.; Du, J.; Liang, S.Y.; Wang, L.; Sun, Y. Wear Behavior of Tool Flank in the Side Milling of Ti6Al4V: An Analytical Model and Experimental Validation. *Proc. Inst. Mech. Eng. Part C J. Mech. Eng. Sci.* **2022**, *236*, 1631–1644. [CrossRef]
38. Dynamic Metals. AMS 6411 Alloy Steel Data. Available online: https://dynamicmetalsltd.com/products/alloy-steel/4330-alloy-steel/ (accessed on 13 March 2025).
39. Dang, J.; Zang, H.; An, Q.; Ming, W.; Chen, M. Feasibility Study of Creep Feed Grinding of 300M Steel with Zirconium Corundum Wheel. *Chin. J. Aeronaut.* **2022**, *35*, 565–578. [CrossRef]

metals

MDPI

Article

Additively Manufactured Biomedical Ti-15Mo Alloy with Triply Periodical Minimal Surfaces and Functional Surface Modification

Zerui Li [1,†], Jingyuan Xu [1,2,†], Jincheng Tang [1], Zhuo Sang [3,*] and Ming Yan [1,4,*]

1 Department of Materials Science and Engineering, Southern University of Science and Technology, Shenzhen 518055, China; 12332081@mail.sustech.edu.cn (Z.L.); jingyuan.xu@uq.net.au (J.X.); 12149007@mail.sustech.edu.cn (J.T.)
2 School of Mechanical and Mining Engineering, The University of Queensland, Brisbane 4072, Australia
3 The Eighth Affiliated Hospital, Sun Yat-sen University, Shenzhen 518033, China
4 Jiaxing Research Institute, Southern University of Science and Technology, Jiaxing 314031, China
* Correspondence: sangzh@mail.sysu.edu.cn (Z.S.); yanm@sustech.edu.cn (M.Y.)
† These authors contributed equally to this work.

Abstract: Ti and Ti alloys are being widely used as bone tissue repair materials. Progress on mechanical properties and bio-functionality is required for their applications due to the large difference in elastic modulus between bone and Ti implants and the fact that the Ti materials themselves are biologically inert. In this work, a low-modulus, β-phase Ti-15Mo alloy based on a triply periodical minimal surfaces (TPMS) structure was fabricated using a Powder Bed Fusion-Laser Beam (PBF-LB) under optimized printing parameters into implant samples with controllable porous structures. The selection of TPMS, lattice unit cell size, and relative density was based on a combination of mechanical properties and cytocompatibility. Surface modifications were used to further impart antibacterial, antioxidant, and osteogenesis properties to the implants. Broad-spectrum antibacterial Ag, antioxidant tannic acid (TA), and highly stable fluorinated hydroxyapatite ((F)HA) were applied as an advanced coating on a microporous TiO_2 surface modification layer formed by micro-arc oxidation. Ultimately, porous Ti-15Mo implant samples with a biofunctional coating were obtained with Young's modulus 15–50 GPa, a yield strength of approximately 100 MPa, and good cytocompatibility, hemocompatibility, and bactericidal effects. This study provides a systematic scheme for the preparation and surface modification of β Ti alloy implants for subsequent studies.

Keywords: PBF-LB; TPMS structure; biomedical Ti alloy; surface modification; antibacterial

1. Introduction

Titanium (Ti) alloys have excellent biomechanical properties and good biocompatibility and are commonly used as hard tissue grafting materials in medical applications. However, the widely used Ti-6Al-4V alloy and commercially pure (CP) Ti have a stress shielding effect due to the mismatch between their elastic modulus of approximately 105–110 GPa and the bone tissue [1–3], which can lead to degradation of the original bone tissue and ultimately to implant failure [4]. Moreover, Ti alloys are biologically inert materials and have difficulty forming strong biological bonds with bone tissue. β-phase Ti alloys, on the other hand, are considered to have great potential to become a new generation of Ti alloys for medical applications due to their low modulus of elasticity, high biocompatibility, and corrosion resistance [5], of which the Ti-Mo family is represented by Ti-15Mo, with an elasticity modulus of 75–80 GPa [6–8].

The elastic modulus of the biomaterial is key to the force distribution and stress shielding effect of the implant, and in this regard, porous structures are ideal candidates. Compared to dense materials, porous structures greatly reduce stiffness [9,10] and are particularly suitable for applications in β Ti alloys where the modulus still does not match the modulus of human bone (0.022 to 21.8 GPa) [11]. At the same time, porous structures with through holes also facilitate biological cell behavior, such as cell adhesion, proliferation, and bone tissue regeneration [12–15], which can lead to greater biological bonding between implant and tissue, thus enhancing implant fixation and improving implant success. Additive manufacturing techniques, such as Laser Based Powder Bed Fusion (PBF-LB) are considered to be one of the most favorable methods for the fabrication of complex porous structures for artificial bone implants, as they are capable of producing alloy components with complex internal structures [16–18]. Compared to conventional subtractive manufacturing methods, PBF-LB offers the flexibility to modulate the porosity of porous alloys and to produce complex shapes, facilitating the creation of customized porous bone implant scaffolds.

In fact, the inner surface of real bone is mostly composed of irregular surfaces, and studies have shown that the surface curvature of the scaffold microstructure has an important impact on the regeneration process of bone tissue [19,20]. Bidan et al. [21,22] proposed that the average curvature of the structural surface in trabecular bone is approximately 0, which has drawn attention to the submicroscopic structure of the bone and the method of constructing porous bone implant scaffolds with minimal curved surfaces. Unlike conventional microstructures in porous scaffold designs, since the curvature on triply periodical minimal surfaces (TPMS) is 0 everywhere, the local surface energy of TPMS bone scaffolds is minimal [23], and the thermodynamic state is stable [20]. An implant of TPMS lattice has good connectivity, relatively high specific surface area, and relatively tortuous internal channels. Studies have shown that when the porosity of the implant of TPMS lattice is similar to that of trabecular bone, the structure has a higher yield stress and a lower elastic modulus [24]. In addition, functional gradient-type scaffolds with morphological features such as variable density can be generated by adjusting the surface equations [25,26]. The best pore size, however, has been determined differently by various studies. For instance, in a biomaterial structure, pores between 100 and 350 μm were deemed to be appropriate for cell colonization and vascularization and aiding bone tissue penetration [27]. Capillary growth will be encouraged in the bone scaffold by pores larger than 140 μm [28], and cell proliferation will be aided by pores larger than 300 μm [14]. According to research by Murphy et al. [29], bone production peaked at 325 μm pore size. When the pore size was between 300 and 1200 μm, bone production did not differ noticeably, as demonstrated by Hollister et al. [30]. Moreover, rats with femoral defects experienced a positive healing effect through a cylindrical bone scaffold of 80% porosity with pore sizes between 400 and 600 μm [31].

In addition to modification of the substrate of Ti alloy implants, surface modification and loading with various functional physiochemical coatings to improve osseointegration has been the focus of research [3,32,33]. Ti surface roughening and micro-nano treatment can effectively improve the bonding performance with bone tissue [34,35]. Micro-arc Oxidation (MAO) is a new modification technique based on the anodic oxidation of valve metals (Al, Ti, Ag, Nb, etc.) to produce a layer of porous TiO_2 ceramics grown in situ on the surface [35–37]. It uses a high working voltage and introduces the working area into the high voltage discharge zone, instead of the Faraday zone of anodic oxidation [38]. The TiO_2 ceramic layer, which is firmly bonded to the substrate, can be modified by adjusting the power supply parameters and electrolyte formulation to change the hardness, wear resistance, and roughness of the Ti alloy surface, thus improving the bonding properties to the bone tissue [39,40].

To prevent bacterial adhesion, antibacterial bioactive coatings can be prepared on the surface of Ti implants. Nano Ag has broad-spectrum antimicrobial properties, but it has a dose-dependent cytotoxicity, so the appropriate concentration is important for striking a balance between biocompatibility and antimicrobial properties. Hydroxyapatite (HA) has good bioactivity and osteogenic properties and can be compounded with various antimicrobial agents. Studies have shown that the doping of HA with F can improve its stability [41–44], and the corrosion resistance of fluoride-hydroxyapatite ((F)HA) increases with increasing F content [45]. Cellular experiments have shown that F-doping can improve the biological properties of HA [46,47], and a small amount of F ions can improve osteoblast activity [48] and promote cell proliferation [45]. Coating Ag nanoparticles with HA reduces the risk to human tissues while maintaining the good antimicrobial properties of Ag. Hence, in order to further improve the osteogenicity on a cell adhesion scale and make the bonding of HA and Ti stronger, microporous TiO_2 prepared by MAO is designed as an intermediate layer between Ti alloy and coating. In the presence of inflammation or certain pathologies (e.g., diabetes), excessive production and accumulation of reactive oxygen species (ROS) can trigger an oxidative stress response that can affect the biological performance of surrounding osteoblasts, which in turn can lead to deterioration and failure of the implant. Tannic acid (TA) with many pyrogallol groups as part of a composite coating can increase the antibacterial and antioxidant capacity of the specimen, which has been used in the dental field [49–51].

In this paper, pre-alloyed Ti-15Mo powder was used, and TPMS lattices were designed and selected with full consideration of the mechanical properties and biocompatibility. After PBF-LB preparation and annealing, the microstructure and mechanical properties of the samples were characterized. Samples with superior properties were selected for surface modification to enhance bio-functionality. In order to impart antioxidant properties to Ti alloy implants, this study prepared a layer of TA-Ca complex on the surface of MAO produced microporous TiO_2 as a calcium site for (F)HA formation using strongly reducing TA as an antioxidant, and then induced the formation of (F)HA coating by a hydrothermal method; the prepared coating is expected to have good antioxidant activity and biological properties. Considering the antimicrobial and anti-infective capacity, Ag nanoparticles are covered by (F)HA in the composite coating to create a slow release effect.

2. Materials and Methods

2.1. Raw Materials

The Ti-15Mo powder (Figure 1) used in this experiment was purchased from Northwest Institute for Non-ferrous Metal Research (Xi'an, China) and produced by the gas atomization method. A laser particle size analyzer was used to assess the particle size, and the results revealed that the particle size was Dv (50) = 38.4 μm. The as-cast sample used for comparison was taken from the powder supplier (Keruikerui Co., Ltd., Wuxi, China).

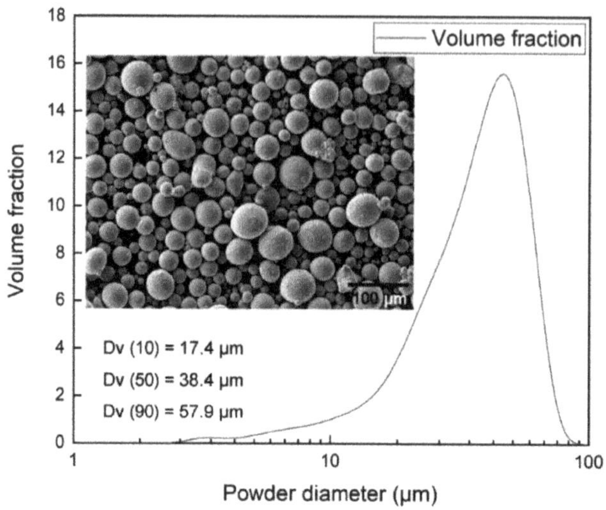

Figure 1. Powder size distribution of pre-alloyed Ti-15Mo.

2.2. Lattices Design and Samples Design

Uniform TPMS structures were designed by MSLattice 1.0 [52], a software based on MATLAB R2020a, applying the mathematical description of those surfaces to produce 3D models. The nodal coordinates that represent a minimal surface have been described using a number of different mathematical techniques. The level-set approximation strategy, the most straightforward and popular technique, uses a set of trigonometric functions satisfying F(x,y,z) = const (constant for an iso-surface). The following are the level-set equations describing the minimal surfaces chosen in this work. Figure 2a represents the design and preparation process of the porous Ti-15Mo lattices. Both solid and sheet lattices derived from the surfaces are shown. Gyroid (G), sheet IWP (ISH), and solid IWP (ISO) structures were chosen because studies [53,54] have shown that they use less material and have better mechanical properties under the same unit cell size.

$$\text{Gyroid} \qquad \sin X \cos Y + \sin Y \cos Z + \sin Z \cos X = const \tag{1}$$

$$\text{IWP} \quad 2(\cos X \cos Y + \cos Y \cos Z + \cos Z \cos X) - (\cos 2X + \cos 2Y + \cos 2Z) = const \tag{2}$$

where $X = 2\alpha\pi x$, $Y = 2\beta\pi y$, $Z = 2\pi\gamma z$, in which α, β, and γ are constants associated with the lattice unit cell parameters in the x, y, and z directions, respectively.

When the above unit cell size was set to 1.5 mm and the relative density (RD) was set to 30–40% (Figure 3a), the lattice derived using MSLattice presented a suitable layer thickness and pore size. As introduced before, these lattices facilitate cell growth with the desired strength at lower densities and lower moduli, while being within the limits of the MSLattice design and PBF-LB fabrication. Samples of various shapes for testing mechanical properties and performing in vitro experiments were designed after determining the lattices. Figure 3b,c show two sets of the designs of samples. The samples are named by lattices and densities for G30, G35, G40; ISH30, ISH35, ISH40; ISO30, ISO35, and ISO40, respectively. It needs to be mentioned that based on the NumPy library, which supports complex computations in multiple dimensions, some programs were written in Python 3.8.8 to parse .stl files to calculate the surface area of the .stl models by summing up all polygonal regions of the mesh.

(**a**)

(**b**)

(**c**)

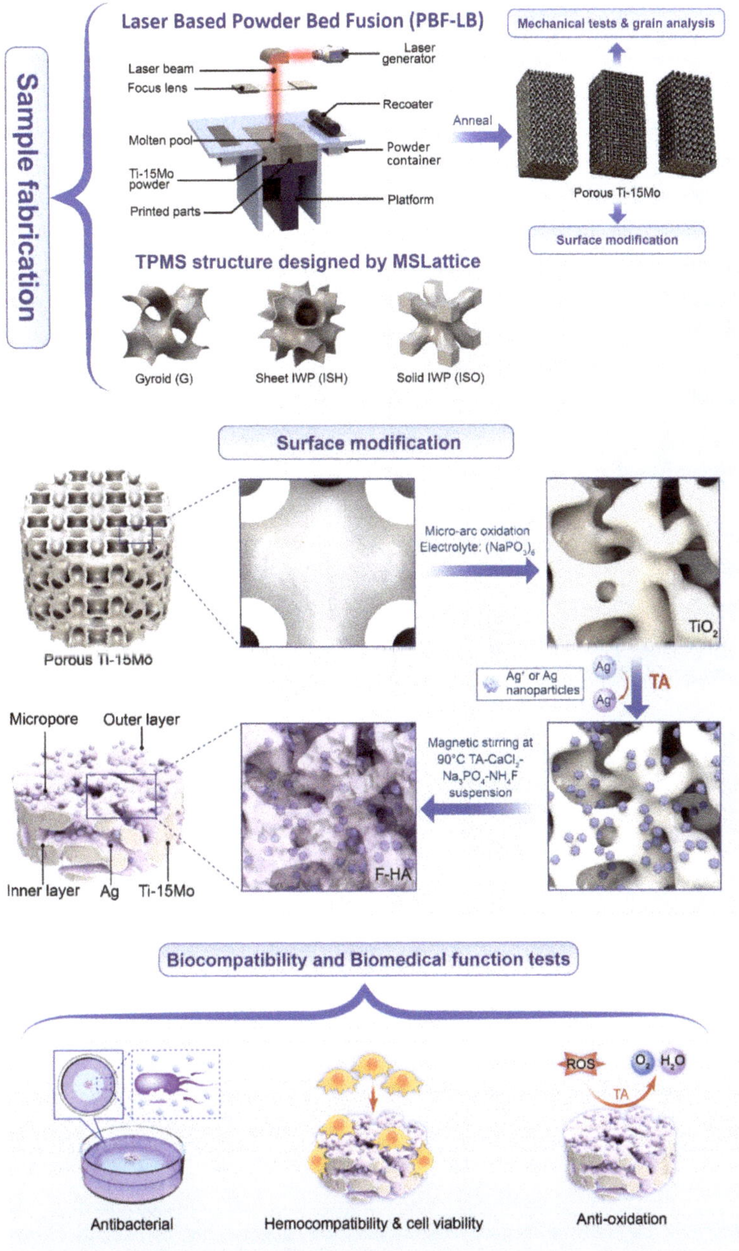

Figure 2. Scheme diagram of (**a**) sample fabrication process; (**b**) the surface modification process; (**c**) the in vitro biocompatibility and biomedical function tests.

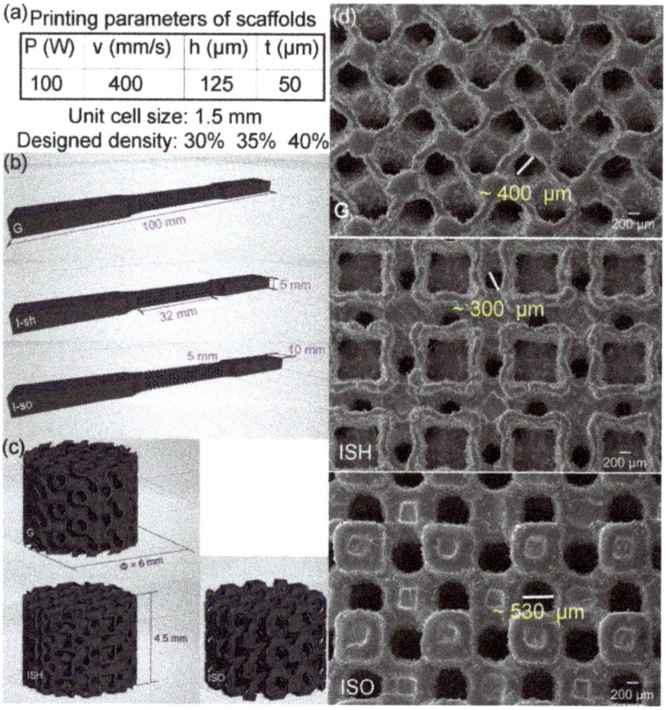

Figure 3. The models designed by MSLattice and the formability of printed samples: (**a**) the optimized printing and designed parameters; (**b**) the models of tensile bars; (**c**) the cylindric samples for in vitro tests are shown with dimensions; (**d**) the forming accuracy of as-printed G, ISH and ISO samples.

2.3. PBF-LB Processing and Process Optimization

The PBF-LB printer used in the experiments was an EOS M290 3D printer (EOS GmbH Electro Optical Systems, Munich, Germany) equipped with a 400 W fiber laser (1064 nm). Its spot diameter is about 70 μm, and it can prepare materials in the range of 250 × 250 × 325 mm^3 in a high-purity argon atmosphere. The interlayer rotation angle was 67°, and the scanning strategy was Chess Scan. The volume energy density (VED) was used for understanding the laser energy input. Equation (3) defines the calculation of this parameter [55]:

$$VED = \frac{P_L}{v_s \times h_s \times t_l} \quad (3)$$

where P_L is laser power, v_s is scanning speed, h_s is scanning pitch, and t_l is single-layer layer thickness.

The layer thickness was set to 50 μm according to the average powder particle size. After multiple trials, the optimum sample relative density of 99.3% was determined when printing a dense bulk sample utilizing the drainage method. At this moment, the corresponding VED = 100 J/mm^3, and the lattice's VED was designed for 40% of it. The optimized printing parameters are shown in Figure 3a.

All as-printed samples were annealed in a vacuum heat treatment furnace (MH(BR)-669-HV, FULCRUM, Shanghai, China) at a rate of 5 °C/min to 500 °C, held for 3 h [56,57]. After the samples were naturally cooled to 300 °C in the furnace, Ar was passed through the furnace to reduce to room temperature.

2.4. Microstructural Analysis

An optical metallographic microscope (OM) (Zeiss, Oberkochen, Germany) was used for the metallographic observation, which also required polishing and etching before observation. Afterwards, the alloy surface was etched using Kroll solution (3 mL HF:5 mL HNO_3:100 mL H_2O). The samples were subsequently ultrasonically cleaned with anhydrous alcohol and dried completely before further study by microscopy.

For the X-ray diffractometer (XRD, SmartLab, Rigaku, Tokyo, Japan) analysis, the 2θ angle range was set to 30–80°, the step size was set to 0.02°, and the test speed was set to 6°/min. The grazing incidence small-angle X-ray scattering (GISAXS) patterns were obtained using the same X-ray diffractometer and 2 h scanning from 10–30°. The morphology of the unit cells of the printed samples and sample microstructures was examined using the secondary electron mode of a TESCAN MIRA3 (TESCAN, Brno, Czech Republic), HITACHI SU8230 (Hitachi High-Tech Corporation, Tokyo, Japan), and Zeiss Merlin (Carl Zeiss AG, Oberkochen, Germany) scanning electron microscope (SEM, accelerating voltage 5 kV, operating current 100 pA). The energy-dispersive X-ray spectroscopy (EDS) mode (accelerating voltage 25 kV, operating current 2 nA) was used to analyze the surface chemical composition.

The electron backscatter diffraction (EBSD) accessory in the Zeiss Merlin SEM characterizes the phase distribution, grain size, and grain boundary information at a 20 kV voltage, 5000 pA current, and 15 mm working distance. The test data were analyzed by the OIM Analysis 7 software 7.3.1.1.

2.5. Mechanical Property Testing

With the use of the HX-1000TM/LCD automatic turret microhardness tester (Shanghai optical instrument factory, Shanghai, China), the specimens' hardness was determined. In the studies, 100 gf of load and 15 s of loading time were the parameters employed. To lessen the experimental error, each sample was measured three times. To mitigate the impact of surface contaminants and scratches on the measurement results, the sample's surface was ground, polished, and ultrasonically cleaned with anhydrous ethanol before the hardness test was conducted.

A universal mechanical testing equipment (Instron 68TM-30, Norwood, MA, USA) with a non-contacting video extensometer was used to conduct a room temperature tensile and compression test. Figure 3b depicts the form of the tensile specimens. The elongation gauge of the tensile and compression sample used in the experiments has a scale of about 30 mm, and the set strain rate for the experiments was 0.002/s. Triplicate specimens were prepared for each distinct structure. The tensile bars' fractography images were detected by SEM.

2.6. Surface Modification and Characterization

2.6.1. Micro-Arc Oxidation (MAO) Process

Small cylindric samples measuring 6 mm in diameter and 4.5 mm in height (Figure 3c) with 30% RD were used for surface modification. Prior to treatment, each sample was ultrasonically cleaned for 10 min in acetone, alcohol, and deionized (DO) water. Temperatures were kept below 40 °C throughout the MAO process using the porous Ti-15Mo samples, which served as the cathode and a stainless-steel container as the anode. In this experiment, the MAO was carried out in a 10 g/L $(NaPO_3)_6$ (Sinopharm Chemical 5Reagent, Shanghai, China) basic electrolyte solution [58] with the following parameters: 450–500 V voltage, 10% duty ratio, 500 Hz pulse frequency, and 12 min oxidation time. The oxidized samples (denoted G(M) group, ISH(M) group and ISO(M) group) were then cleaned using ultrasonic vibration and distilled water for 15 min, followed by an overnight dry at 60 °C.

2.6.2. Manufacturing of Multi-Biofunctional Coating

When the introduction of bio-functionality was considered, lattice structures with better mechanical properties were selected. The antimicrobial properties were first introduced into the MAO-treated porous Ti-15Mo scaffolds. Deionized (DI) water was used for the preparation of the solution and for cleaning thoroughly. The samples were immersed in 0.05 M, 0.1 M, and 0.2 M high-concentration $AgNO_3$ solutions [59,60], protected from light, and shaken with a magnetic stirrer for 1 h. The samples were then immersed in 10 g/L aqueous TA solution and shaken for 1 h. The samples were then removed and rinsed with DI water and sonicated for 20 min to remove the loose Ag nanoparticles and dried overnight in an oven at 55 °C.

A mixture of TA at a concentration of 30 g/L, $CaCl_2$ at 0.5 M, and NH_4F at (1/18) M [61] was prepared and heated to 90 °C. The corresponding amount of Na_3PO_4 at a concentration of (1/3) M was quickly added to the mixture solution and magnetically stirred. Each sample with reduced Ag nanoparticles on the surface was suspended and soaked in the middle of the solution with a thin wire for 10 h of constant temperature cross-linking reaction. The samples were then removed and rinsed with deionized water, sonicated, and dried overnight in an oven at 55 °C.

A summary of the experimental methods of surface modification is shown in Figure 2b. Since the Ag loading was the only variable in the coating for all the scaffolds, the subsequent sets of samples were named accordingly, e.g., ISH0.05, ISH0.1, and ISH0.2. The names of samples with a complete coating remained unchanged.

2.6.3. Surface Characterization

The morphology of the multi-biofunctional coating was observed by SEM, and the elemental composition and distribution were detected and analyzed by SEM-EDS. By using an auto scratch coating tester (WS-2005, Zhongke Kaihua, Dongguan, China), the coatings underwent scratch testing, and their adhesion force to substrates was assessed. The load applied by the tester was set from 0.2 N to 16 N at a loading speed of 3 N/min and a scratching speed of 3 mm/min [61,62].

2.7. Bio-Property Test: In Vitro Test

2.7.1. Cytotoxicity Test and Ion Precipitation Analysis

The samples were immersed in acetone, alcohol with a 95% alcohol concentration, and DI water for 10 min ultrasonically. The samples were dried before being sanitized for later use at a high temperature and pressure.

Preparation of leachate: Each sterilized sample was immersed in 2 mL of serum-containing cell culture medium with sodium alginate spheres. The samples were extracted at 37 °C for 24, 48, and 72 h. The leaching solutions were obtained by removing the samples.

Cytotoxicity assay: MG-63 cells were prepared into a cell suspension at a concentration of 5×10^4 cells/mL, inoculated into 96-well culture plates at 200 μL per well, and cultured in an incubator. When the cells grew to 80% confluence, different sample leachates (experimental group), sodium alginate sphere leachates (negative control), 6.4% phenol (positive control), and normal culture medium (blank control) were added sequentially. 5 replicates were set up for each group and incubated for 24 h. The number of cells in each well was determined by absorbance (OD) on a microplate reader at 450 nm. Cell viability was characterized by cell relative growth rate (*RGR*, Equation (4) [63]).

$$RGR\% = \frac{OD_{sample} - OD_{blank}}{OD_{control} - OD_{blank}} \times 100\% \tag{4}$$

Ion precipitation analysis: Quantitative determination of ion precipitation by inductively coupled plasma mass spectrometry (ICP-MS). The 7700 Series ICP-MS (Agilent, Lexington, MA, USA) was turned on with Ar gas. After tuning, the mixed standard solutions were drawn, and the standard curves were made with the elements Ti, Mo, and Ag in the standard sample. The nitrified leaching solutions and blank control solution were drawn one by one, and the content of each element was determined by the full quantitative method. Finally, the element concentrations of Ti, Mo, and Ag in each leachate were calculated according to the standard curves.

2.7.2. Hemocompatibility Tests

Each experimental group of G0.05-ISH0.2, the negative control group without a sample, and the positive control group using distilled water were kept in a 24-well plate at a constant temperature at 37 °C. A fresh mouse eyeball blood solution containing 3.8% sodium citrate diluted with normal saline at a ratio of 1:5 was added to each well and incubated at 37 °C for 1 h. After the liquid was centrifuged, the *OD* of the supernatant was detected at a wavelength of 545 nm. *Hemocompatibility* was characterized as Equation (5) [64]:

$$Hemocompatibility\ \% = \frac{OD_{sample} - OD_{negative\ control}}{OD_{positive\ control} - OD_{negative\ control}} \times 100\% \tag{5}$$

2.7.3. Total Antioxidant Capacity Test

Due to the presence of TA, the antioxidative capacity of the coated samples was determined by a total antioxidant capacity assay kit (BEYOTIME, Haimen, China) with the ferric reducing ability of plasma (FRAP) method [65]. The kit is based on the principle that under acidic conditions, the antioxidants reduce Fe^{3+}-tripyridyltriazine (TPTZ) to produce a blue Fe^{2+}-TPTZ, which is then measured at 593 nm using a microplate reader to obtain the total antioxidant capacity of the sample. The antioxidant capacity of Trolox, a positive control with similar antioxidant capacity to that of vitamin E, was used as a benchmark for calibration. Each sample was incubated with FRAP reagent for 15–90 min at 37 °C.

2.7.4. Antibacterial Test

As a semi-quantitative assay, the inhibition zone test was utilized to assess the antimicrobial effect of the multi-coatings against *Escherichia coli* (*E. coli*) and *Staphylococcus aureus* (*S. aureus*). Both the experimental group containing the coating and the control group treated with MAO only were autoclaved. Chilled dormant *E. coli* and *S. aureus* were activated to obtain bacterial suspensions, and 100 µL of each was pipetted onto plates poured with cooled sterile solid LB medium. The bacterial suspensions were spread evenly with a spreader. Each sample was sterilized again with an alcohol lamp and pressed firmly to embed in the agar in the center of one plate. A camera was used to capture digital pictures of the samples after the plates had been removed from an incubator where they had been incubated for 24 h at 37 °C. The size of the zone was then measured.

A summary of the experimental methods of in vitro biocompatibility and the biomedical function tests is shown in Figure 2c.

2.8. Statistical Analysis

All data were analyzed by one-way ANOVA using SPSS 17.0 software (SPSS, Chicago, IL, USA) and presented as average ± standard deviation. Experiments were performed in triplicate with statistical values indicated in relevant analyses. Statistical significance was set at $p < 0.05$.

3. Results and Discussion

3.1. Results and Discussion of the As-Printed Dense and Porous Ti-15Mo

Under the parameters of $VED = 100\ J/mm^3$, $P_L = 200$ W, $v_s = 400$ mm/s, $h_s = 100$ μm, and $t_l = 50$ μm, dense Ti-15Mo alloy was printed with a result of 99.29% RD. Figure 4a,b gives an example of the microstructure morphology of dense samples printed with pre-alloyed Ti-15Mo powder by OM. The average melt channel width of 117.0 μm can be measured from the channels showing the full width in Figure 4a. The defects in the fusion channels and micro poles resulting from the laser scan are visible in the figure, as are the grain boundaries in the martensite and β phases after deep etching.

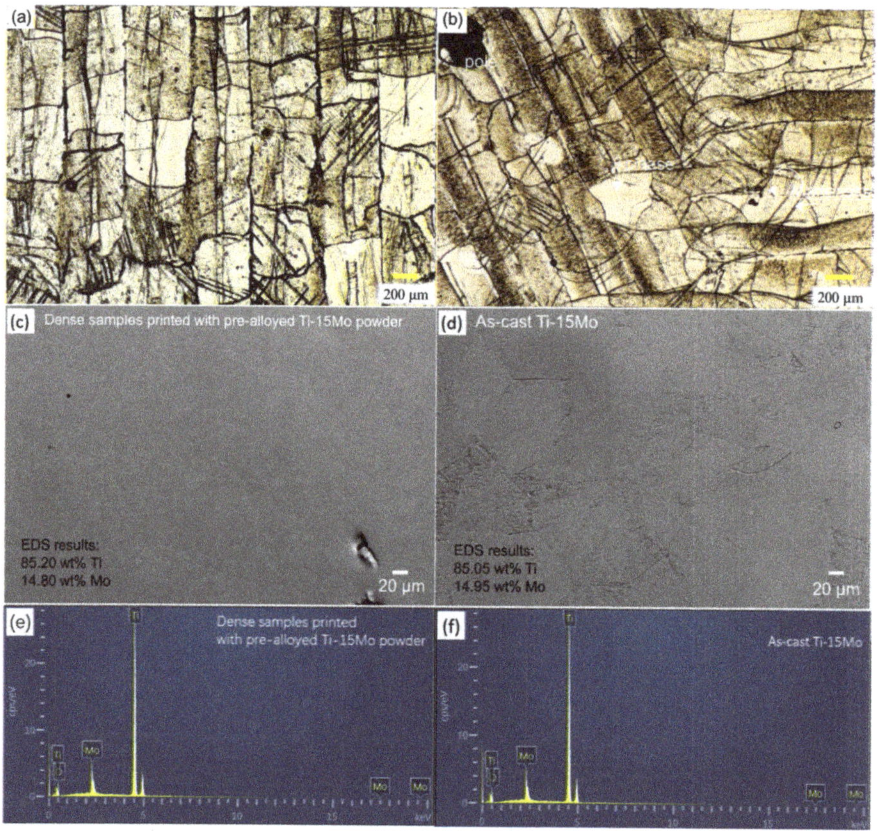

Figure 4. Images of a polished and etched dense Ti-15Mo sample by OM. (**a**) Powder melting/laser scan path shown as uniform vertical streaks; (**b**) Most of the β-phase with very little of the needle-like martensite present; images of the SE mode of the SEM and results of EDS of the etched (**c**) as-printed and (**d**) as-casted dense Ti-15Mo samples; EDS spectrum of (**e**) as-printed and (**f**) as-casted dense Ti-15Mo samples, respectively.

Figure 4c,d give an example of the surface morphology and compositions of a dense sample printed with pre-alloyed Ti-15Mo powder and an as-casted sample as a control group by the SEM SE mode. The large grain boundaries of the as-cast sample are very evident when comparing the moderately etched as-printed and as-cast samples under the SE of the SEM. The EDS results show that the elemental composition of the sample is indeed 85 wt% Ti and 15 wt% Mo.

For the dense samples, tensile tests were carried out, and the results are shown in Table 1 along with others. The PBF-LB laser gave a large increase in hardness to the alloy and maintained a high level of strength while keeping the modulus low and the elongation ideal. The numbers in parentheses are standard deviations.

Table 1. Mechanical properties of dense Ti-15Mo with and without heat treatment. compared with the literature.

Samples	$\sigma_{0.2}$ (MPa)	UTS (MPa)	Elongation (%)	E (GPa)	Hardness (Hv)	Ref.
As-printed, dense	816 ± 4.5	910 ± 9.3	19 ± 3.0	82.1 ± 2.5	391.9 ± 7.0	This article
As-cast	802 ± 3.3	905 ± 8.7	15 ± 5.0	78.3 ± 1.7	319.8 ± 7.9	This article
Ti-15Mo (in situ alloyed by PBF-LB, 1100 °C treated)	1170 ± 17	1180 ± 22	1.6 ± 0.2	86.3 ± 1.9	379.3 ± 13	This article
Ti-15Mo	584	934	14	/	321	[66,67]
Ti-15Mo (cast)	745 ± 13	921 ± 19	25 ± 2	84 ± 1	310 ± 10	[68]

Regarding the porous samples, their SEM images are shown in Figure 3d. The obtained 300–550 μm pores are favorable for cell growth. A major problem with the preparation of porous structures from metal powders by PBF-LB is that partially molten powder is on the lattice surface, making the scaffold morphology less smooth, and a morphology similar to spheroidization appears. For the purpose of in vivo bone defect applications, physical treatment such as sandblasting cannot be used to treat the surface at a cell size of 1.5 mm. The reason for using only prolonged ultrasound to remove the loose powder and not acid etching or electrolytic polishing to give the scaffold a smooth appearance is that our previous studies [69] have demonstrated to a certain extent that the metal powder in this case does not have a significant negative effect in vivo. Methods such as acid etching are more suitable for porous structures with larger cell sizes, i.e., thicker struts and larger pores. Therefore, on balance, a smaller unit cell with a simpler initial surface treatment is a more desirable combination.

The results of the XRD analysis of powder stock, as-printed scaffolds, and as-cast alloy are shown in Figure 5a. Only the β phase was detected in the interval 30–80°, with the highest peaks occurring on the β {110} crystal plane. This indicates that the martensite content is below the XRD detection limit, which can demonstrate the phase composition of the as-printed scaffolds. In the presence of a large amount of β-phase stabilizing element Mo, the transformation of the β-phase into the α-phase is suppressed in the melting pool with a fast cooling rate, and only a small amount of the β-phase is transformed into a supersaturated solid solution (martensite) of the same composition but with a different crystal structure. This martensitic phase transformation is a diffusion-free phase transformation in which no atomic diffusion takes place; only lattice reconstruction occurs [70].

The results of the EBSD analysis of the as-printed scaffolds and as-cast alloy are shown in Figure 5b–f. As can be seen from Figure 5b, the as-printed scaffolds exhibited fine polycrystalline microstructures with grain sizes significantly smaller than those of the as-cast alloy. The average particle size of the former was about 40 μm, while that of the latter was about 100 μm, which was also consistent with the mapping results of Figure 5c,d. This phenomenon is due to the extremely short existence time of each melting pool of the PBF-LB technique, which effectively refines the Ti-15Mo alloy grains. The kernel average misorientation (KAM) results are shown in Figure 5e,f. The method is based on the deviation of the orientation gradients between pixels and therefore corresponds to the geometric dislocation density and local residual stresses for assessing the stress state inside a metallic material. As the results showed very low orientation gradients, only 40% of the standard color scale bar in the range 0–5 was used here for display purposes. The as-cast sample had almost no dislocations, while the as-printed Ti-15Mo showed yellow-red aggregations at the low-angle grain boundaries, indicating relatively high residual stresses in these areas.

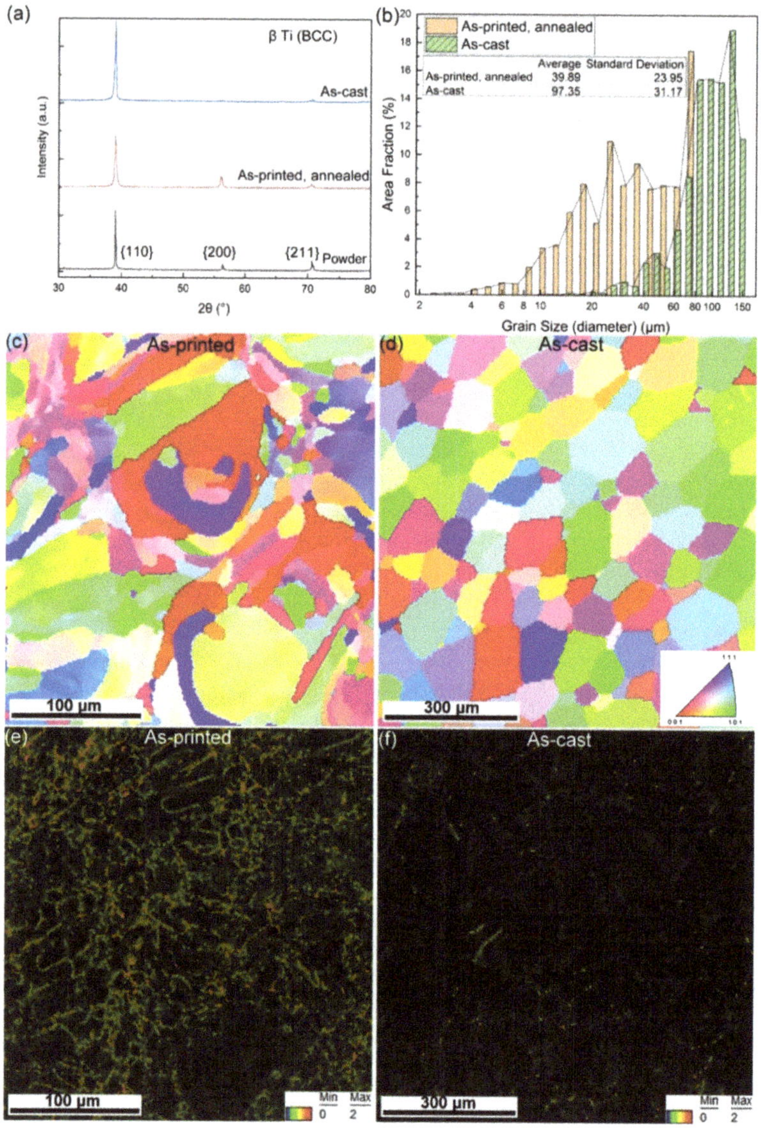

Figure 5. (**a**) XRD and (**b–f**) EBSD results of as-printed and as-cast Ti-15Mo. (**a**) Indicates the grain size distribution. (**c**,**d**) EBSD mappings illustrate the as-printed samples have smaller grains. (**e**,**f**) EBSD KAM figures demonstrate very low internal stress found in the as-printed sample, though higher than that of the as-cast sample.

The results of the compression and tensile tests on the as-printed porous Ti-15Mo samples are summarized in Tables 2 and 3. Figure 6 shows the results of the compression tests on the three TPMS lattices with different RD values. The yield strength $\sigma_{0.2}$ ranged from 78 to 130 MPa, and Young's modulus ranged from 16 to 48 GPa for all experimental samples. Figure 6d–f show that the yield strength and Young's modulus of TPMS lattices increase with increasing RD. Comparing the three groups with the same structural parameters and different TPMS, such as G35, ISH35, and ISO35, the Young's modulus and yield strength are different, indicating that the different structures have a significant effect on the mechanical properties, given the same structural parameters.

Table 2. Tensile tests results of as-printed Ti-15Mo of G, ISH, and ISO structures.

Structures	Tensile Tests Results				
	RD	**YS (MPa)**	**UTS (MPa)**	**ε (%)**	**E (GPa)**
G	30%	158.84 ± 2.62	160.10 ± 3.02	1.10 ± 0.12	18.16 ± 0.32
	35%	216.50 ± 4.17	217.98 ± 1.69	1.15 ± 0.17	23.24 ± 0.38
	40%	300.11 ± 5.18	313.95 ± 2.66	1.15 ± 0.09	33.53 ± 0.45
ISH	30%	139.26 ± 3.31	143.04 ± 1.02	1.07 ± 0.07	15.80 ± 0.26
	35%	167.62 ± 1.44	168.73 ± 5.71	1.00 ± 0.04	21.10 ± 0.08
	40%	170.04 ± 4.75	173.34 ± 2.54	0.61 ± 0.07	35.62 ± 0.17
ISO	30%	131.12 ± 5.29	131.21 ± 4.63	1.24 ± 0.10	12.65 ± 0.13
	35%	154.75 ± 3.85	158.96 ± 2.08	1.08 ± 0.06	18.38 ± 0.61
	40%	151.85 ± 5.58	159.27 ± 0.94	1.02 ±0.11	22.84 ± 0.73

Table 3. Compression tests results of as-printed Ti-15Mo of G, ISH, and ISO structures (with 30%, 35%, and 40% RD).

Structures	Compression Tests Results				
	RD	**$\sigma_{0.2}$ (MPa)**	**Final Strength (MPa)**	**ε (%)**	**E (GPa)**
G	30%	90.99 ± 2.11	167.85 ±3.24	20.2 ± 0.2	31.59 ± 0.26
	35%	117.40 ± 4.54	192.20 ± 4.72	20.1 ± 0.1	36.60 ± 0.07
	40%	129.91 ± 2.86	213.62 ± 3.39	20.0 ± 0.2	40.37 ± 0.85
ISH	30%	78.82 ± 1.15	125.52 ± 2.13	39.3 ± 0.1	31.10 ± 0.06
	35%	107.10 ± 4.63	157.62 ± 2.41	39.3 ± 0.1	43.00 ± 0.73
	40%	118.51 ± 3.75	183.83 ± 6.87	39.1 ± 0.3	47.3 ± 0.55
ISO	30%	82.31 ± 5.97	113.51 ± 2.24	39.2 ± 0.4	16.80 ± 0.34
	35%	98.92 ± 2.42	147.64 ± 4. 79	39.0 ± 0.2	19.90 ± 0.31
	40%	120.10 ± 1.86	178.08 ± 4.58	39.0 ± 0.3	24.20 ± 0.17

Furthermore, the data in Tables 2 and 3 show that the elongation of the porous structures remained relatively constant in the tensile and compression tests, but there was some degree of difference in Young's modulus and yield strength. Considering that porous implants are mainly subjected to compressive stresses in vivo, the relatively low-strength structure of ISO will be removed from subsequent in vitro experiments and used as a control group. The mechanical properties of the G structure are slightly stronger than those of the ISH. The strut thicknesses of 0.15 mm and 0.11 mm for G30 and ISH30, respectively, are also consistent with the results of the mechanical properties analyzed in the literature [71,72], i.e., the mechanical properties are stronger with a wider support structure in the load direction, all other things being equal.

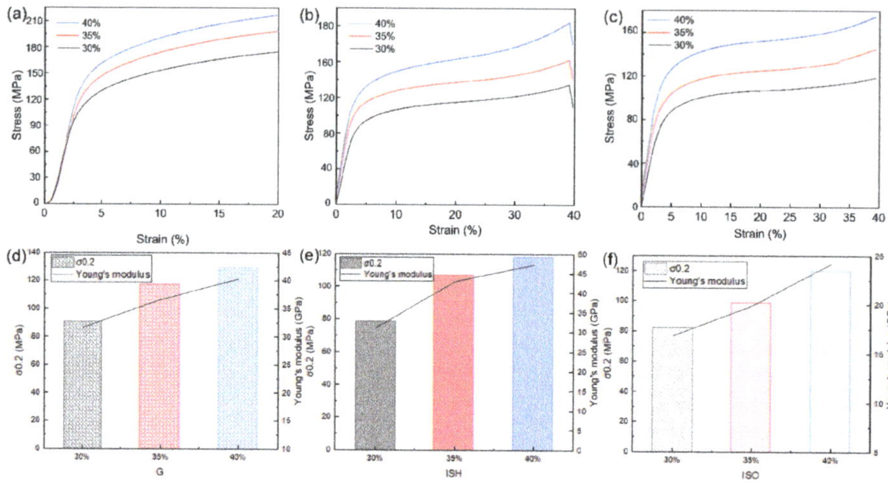

Figure 6. Compression stress–strain curve, yield strength, and corresponding Young's modulus of (**a**,**d**) G, (**b**,**e**), ISH, and (**c**,**f**), ISO structures.

Figure 7 shows the tensile test results for the three TPMS lattices with different RD values. All flat tensile specimens show similar mechanical behavior trends to the compression specimens, with yield strengths ranging from 131 to 300 MPa and Young's modulus ranging from 12 to 36 GPa. Figure 7g–i show the fracture analysis of three sets of tensile specimens, where the joint appearance of the cleavage planes and dimples demonstrates that all three tensile failure mechanisms were a mixture of brittle and ductile fracture.

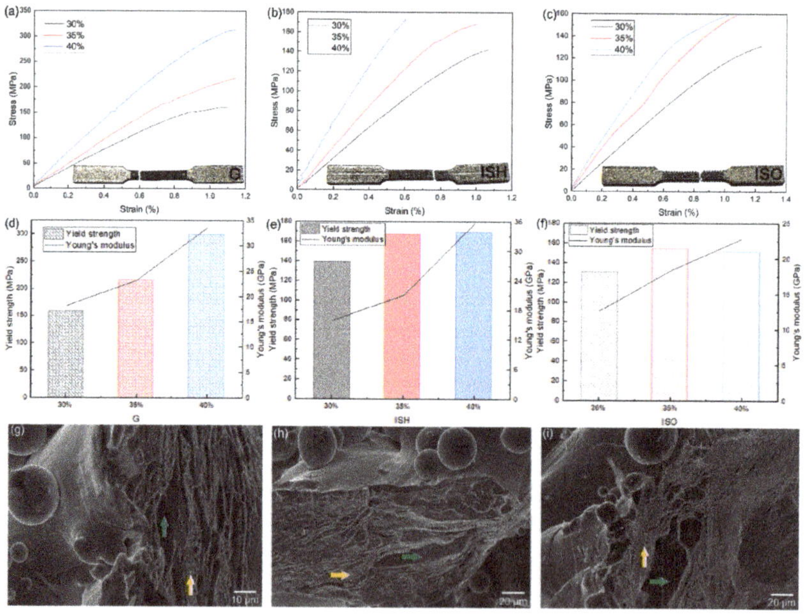

Figure 7. Tensile stress–strain curve, yield strength, and the corresponding Young's modulus corresponding fractography images of (**a**,**d**,**g**) G, (**b**,**e**,**h**), ISH, and (**c**,**f**,**i**), ISO structures. The figures inserted are the corresponding tensile bars. (In the fractography images, green arrows point to cleavage planes of brittle fractures, and yellow arrows point to dimples of ductile fractures).

3.2. Results and Discussion of the Surface Modification of Ti-15Mo Lattices

Figure 8 represents the initial stages of surface modification. All samples designed for in vitro tests were treated by MAO and named as G(M), ISH(M), and ISO(M), respectively. The 0.1–1 μm holes were evenly distributed and tightly covered on the substrate scaffold. The composite structure of the 100-micron-scale macropore substrate scaffold and the micron-scale ceramic coating laid a foundation for the subsequent attachment of multi-biofunctional coatings with stable thermodynamic properties, larger surface area, and easier adhesion. Based on the experimental results of the mechanical properties, G(M) and ISH(M) were selected as the experimental group, and ISO(M) was henceforth set as the control group. The SEM image of the reduced Ag particles is shown in Figure 8a. The smallest individual Ag particle size was about 20 nm, with some Ag particles agglomerated up to 400 nm. Nano Ag particles were distributed evenly inside and outside the micron-scale TiO_2 pores. Figure 8c,d show the EDS results for the corresponding color star symbols in Figure 8a, where the Cl peak was derived from the electrolyte of MAO. The bright white particles in the SEM were thus confirmed to be all nano Ag and attached to the microporous TiO_2.

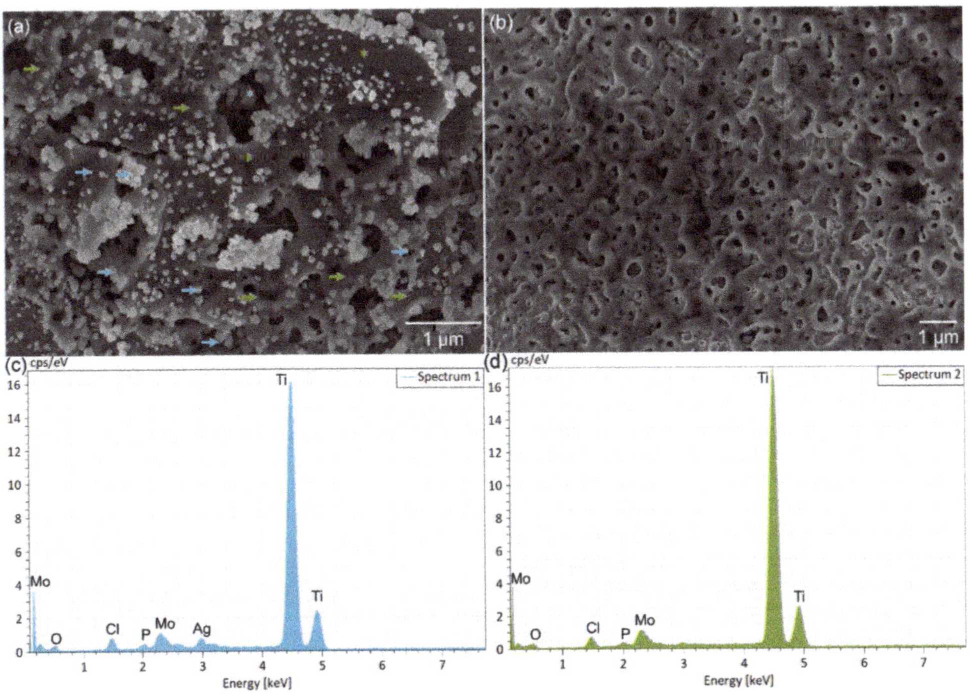

Figure 8. (**a**) The reduced Ag nanoparticles are attached to the (**b**) TiO_2 porous surface by MAO; (**c**,**d**) show the EDS results of the Ag nanoparticle and TiO_2 coating from (**a**) respectively. The blue markings in (**a**) represent nano Ag. and the green ones represent the microporous TiO_2 ceramic layer. The blue asterisks and arrows in (**a**) represent nano Ag, and the green asterisks and arrows represent the microporous TiO_2 ceramic layer.

Figure 9 shows the (F)HA-TA-Ag-TiO_2 coating after completion of the surface modification steps. Figure 9a shows flake HA covering the outside and inside of micron-scale TiO_2 pores, while Figure 9b shows another kind of rough surface spherical HA aggregated morphology, both with a tiny amount of uncovered nano Ag. Figure 9c,d add the Ca and P peaks representing HA and the weak peak of the reactant Na based on the previous EDS. Figure 9e–g show the evidence of all the results of surface modification. TiO_2, nano Ag, and HA have been distributed as the corresponding element mappings. According to Figure S1, the MAO coating was composed of an amorphous phase, whose amorphous bulge was at 20–30° in the results of GISAXS. An XRD peak corresponded to the hydroxyapatite (HA) phase at $2\theta = 25.36°$. These results align with the SEM observations and theoretical expectations for hydrothermal crystallization behavior [58].

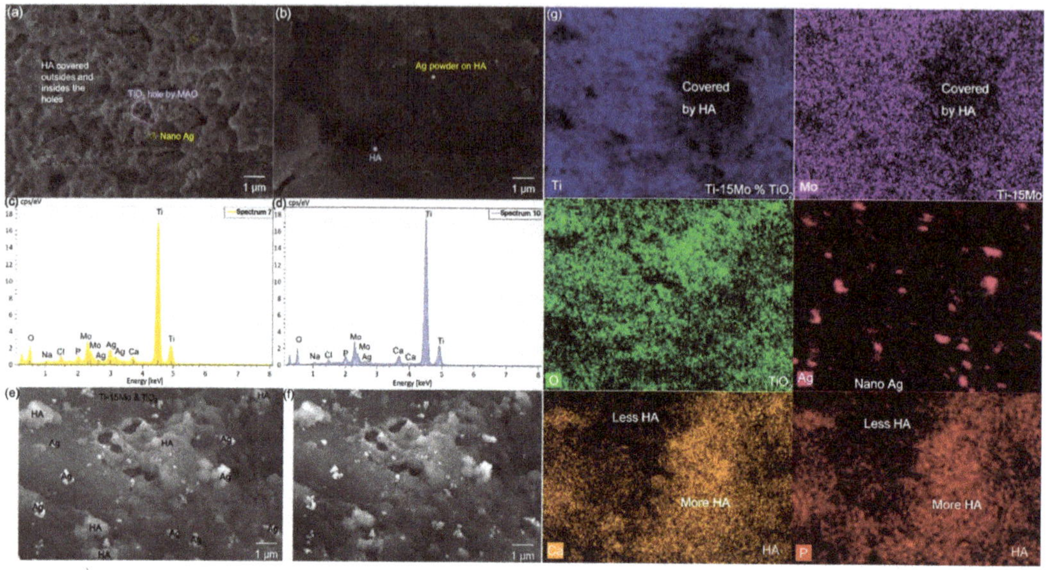

Figure 9. (**a**) Morphology of final (F)HA-TA-Ag-TiO_2 coatings with HA (multi)crystals covered. (**c**,**d**) Show the curves of the EDS results at the corresponding marked positions in the SEM image of (**b**), respectively. (**e**) SEM image of rare one area with all three materials giving evidence of all the results of surface modification. (**g**) Is an EDS mapping of the SEM image of (**f**) based on the detection of Ti, Mo, O, Ag, Ca, and P elements.

The change of the tangential friction force was measured by the acoustic emission method to determine the bonding force between the coating and the substrate. The indenter gradually increased the pressure on the surface. When the relative displacement of the coating and the substrate occurred, an acoustic signal would be generated, as shown in Figure 10. When the coating failed, there would be acoustic feedback, and the corresponding normal critical load (Lc) was used to characterize the bonding force between the film and the substrate. The larger the critical load, the stronger the bonding between the coating and the substrate. The average binding force (Lc) of TiO_2 prepared by MAO was ~30 N, and the average bonding force between the complete coating and the substrate was 9.98 N, which is comparable to the conclusions of some existing literature [61]. The data indicate that MAO can indeed prepare a well-bonded ceramic layer; the bonding force of the composite coating is greater than the literature reference value of ~6.5 N for HA, also indicating the advantage of F-doping on the stability of the outermost HA layer.

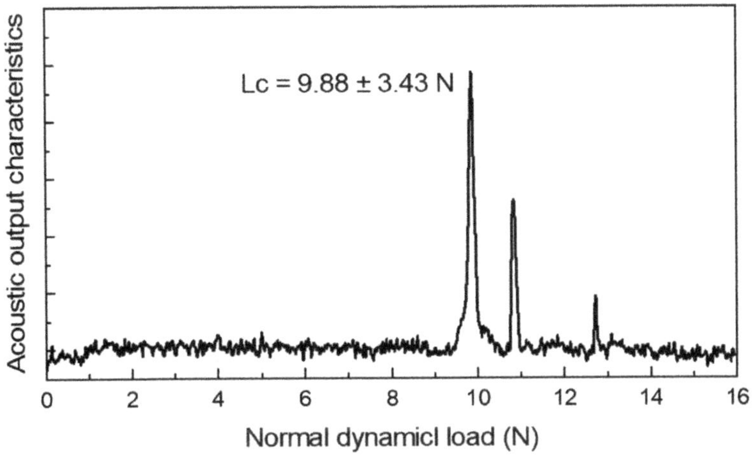

Figure 10. The results of bonding strength between functional (F)HA-TA-Ag-TiO_2 coatings and the substrate were recorded by the auto scratch coating tester. Lc = 9.98 ± 3.43 N.

3.3. *Results and Discussion of the In Vitro Analysis of Coated Ti-15Mo Scaffolds*

When testing the biologically relevant properties of coated porous Ti-15Mo, the experimental groups involved were G and ISH samples treated with all surface modification steps (i.e., G0.05 to ISH0.2), and the control group was mainly the MAO-treated ISO(M) only. The ion release amounts of Ti, Mo, and Ag from the leachates for 0–24 h, 24–48 h, and 48–72 h for each group of samples were calculated and are presented in the histograms in Figure 11. The ionic release of Ti and Mo was minimal for all groups of samples, and no Ag was detected for ISO(M), as no Ag was added. For each experimental group, the line graph in Figure 11c systematically reflects the overall precipitation of Ag from the coating over the three days, with a steady increase in the amount released. This demonstrates not only the feasibility of releasing the effective ions from the coating in the incubation environment, but also the slow release effect. The minimum total Ag release was about 13.5 μg, and the G0.05 and ISH0.05 groups released about 18.5 μg and 17 μg, respectively, which is high compared to the literature results of several micrograms [59,73].

Figure 12 shows the results of cell and blood compatibility. As Ti-15Mo and TiO_2 have not been shown to be antibacterial and cytotoxic, the 94.2% RGR result of the control group ISO(M) allowed for a rough estimate of the cell proliferative effect of MAO-treated as-printed Ti-15Mo lattices. The cytotoxicity results showed that the RGR of each experimental group was higher than 94.2%, indicating that the high Ag ion release concentration and the TiO_2 ceramic layer of the composite coating were non-toxic to cells, and the overall cytocompatibility of the composite coating was better than that of microporous TiO_2. In the hemolysis assay, the sample was in direct contact with the blood cells, and hemolysis is related to osmotic pressure and hemoglobin degeneration caused by ions. As can be seen from Figure 12b, G0.1 had the highest hemolysis rate, followed by G0.2 and ISH0.1, all of which were higher than the 5% hemolysis safety line. ISH0.2 had a relatively low hemolysis rate, but still reached 3.498% and should not be preferred. G0.05 and ISH0.05 reflected ideal safety.

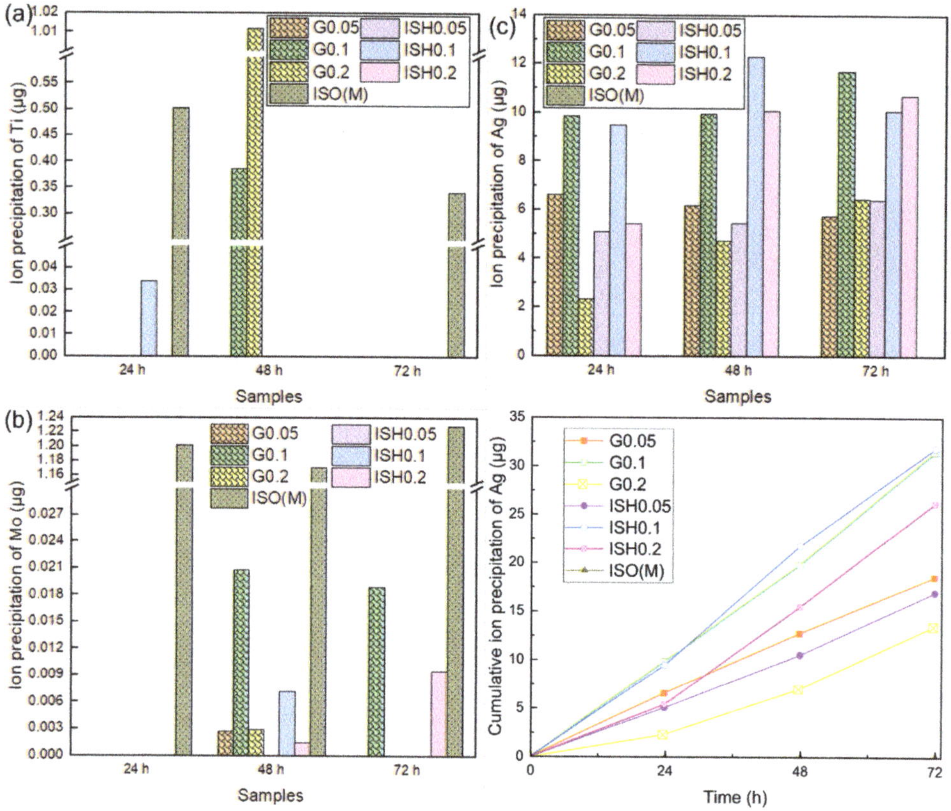

Figure 11. (**a**) Ti, (**b**) Mo, and (**c**) Ag ion precipitation results of the leaching solution obtained after co-incubation with each group of samples for 24, 48, and 72 h, respectively, measured by ICP-MS. The line graph of cumulative ion precipitation of Ag shows the continuous release of Ag from the coating over 72 h.

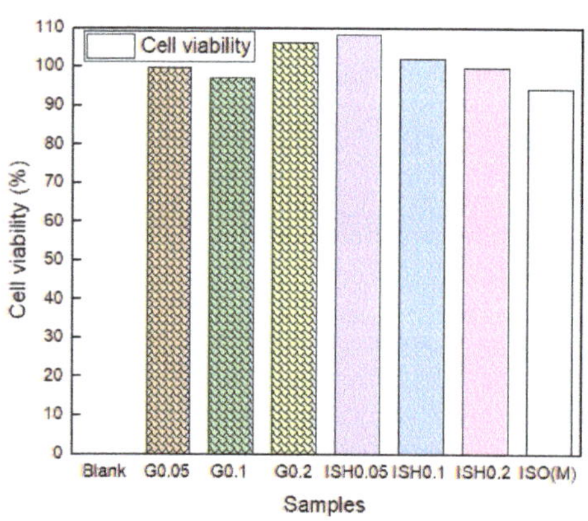

(**a**)

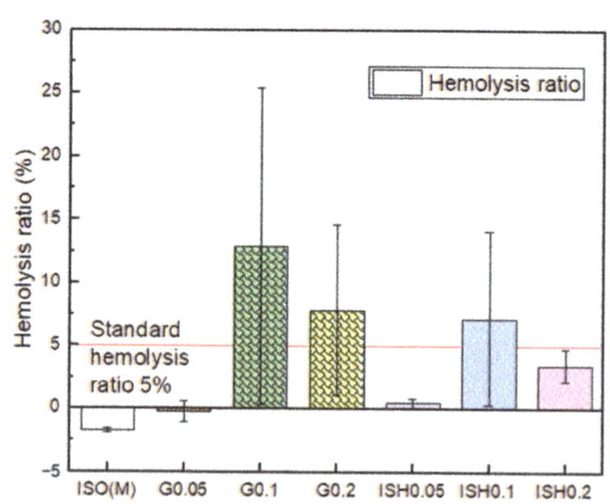

(**b**)

Figure 12. (**a**) Cell viability (RGR%) at 72 h and (**b**) the hemolysis ratio of the samples.

After the comprehensive ion precipitation, cytocompatibility, and hemocompatibility tests results, G0.05 and ISH0.05 were selected for superiority. The two groups of samples with $AgNO_3$ concentration of 0.1 M at the time of preparation showed the highest Ag concentration in both ion release and hemocompatibility assays, while G0.2 and ISH0.2 did not show the most significant toxicity. The reason for such results may be that at high Ag levels, the Ag ion in the leaching solution reaches an appropriate amount to bind to proteins such as serum in the medium, and hemoglobin is directly bound by the excess Ag in the sample, in order to form a stable precipitate without exerting significant toxicity.

The results of antioxidant properties are shown in Figure 13. Figure 13a shows that the absorbance values used to measure the total antioxidant capacity to ROS of the (F)HA-TA-Ag-TiO_2 coating in G0.05 and ISH0.05 gradually increase with incubation time, due to the slow sustained release of TA from the coating. Therefore, the prepared (F)HA-TA-Ag-TiO_2 coating has long-term anti-oxidation properties. The antioxidant capacity of the coatings with different TPMS structures was found to be similar, further indicating that the antioxidant properties of the composite coatings are conferred by the TA.

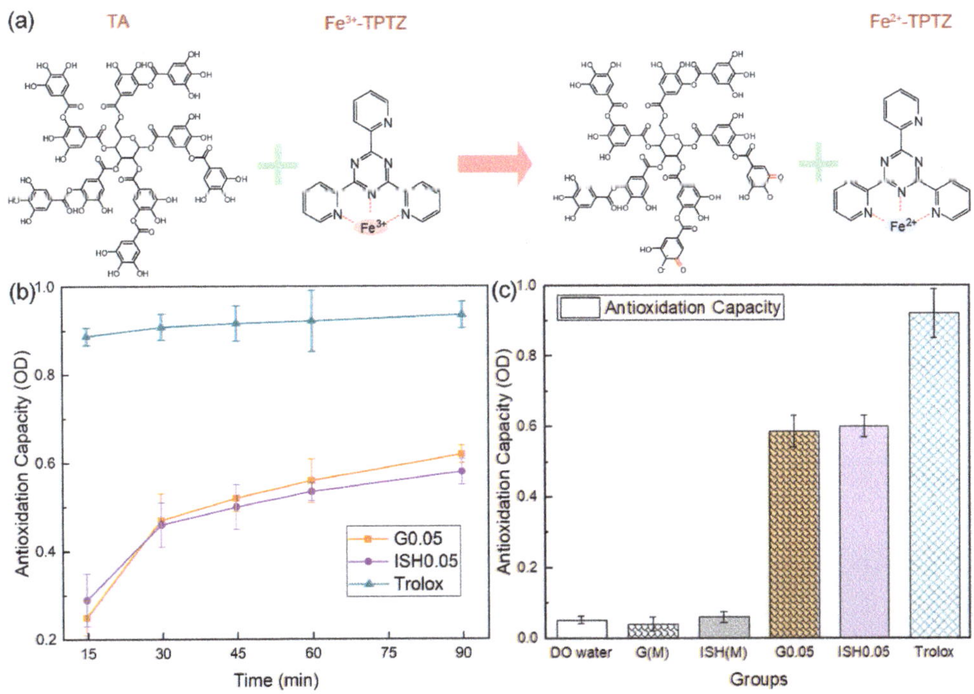

Figure 13. (**a**) Diagram of the antioxidant mechanism of TA [74,75]. (**b**) The total antioxidant capacity of G0.05 and ISH0.05 compared with Trolox for 15–90 min. (**c**) The total antioxidant capacity of G0.05 and ISH0.05 compared with control groups at 90 min.

The antibacterial test results of the samples are shown in Figure 14. As a semi-quantitative test, the size of the inhibition zone is summarized in Table 4. The inhibition zone pattern of the ISH0.05 group was complete, the size was relatively larger, and the higher silver ion release radius showed better antibacterial performance. In contrast, the G0.05 group also had a good performance of the inhibition zone, but the edge of the zone was not smooth, and the size was relatively small. The control group had no bactericidal properties because it lacked Ag. The results reflect that different TPMS structures affect the amount of Ag released, with almost a twofold difference in the radial length of the corresponding inhibition circles.

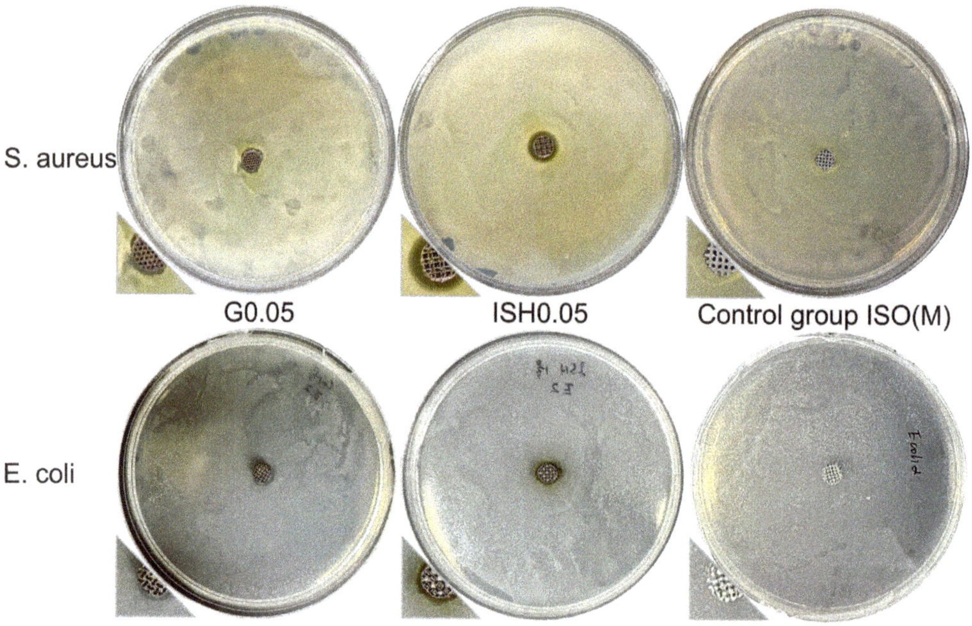

Figure 14. G0.05, ISH0.05, and control ISO(M) results for *S. aureus* and *E. coli* inhibition circles. The lower left triangular division of each Petri dish photograph shows the clearer inhibition effect after the removal of the corresponding sample at the end of the experiment.

Table 4. Results of the inhibition zones (expressed in millimeters).

Inhibition Zone/mm	G0.05	ISH0.05	ISO(M)
S. aureus	0.75 ± 0.02	1.50 ± 0.01	0
E. coli	0.55 ± 0.01	1.00 ± 0.02	0

When the structural parameters are the same, it means that the same amount of pre-alloyed powder is used. At this time, it can be assumed that the larger the surface area, the smaller/thinner the struts of the lattice. In order to explain the gap of doubling the antimicrobial capacity of ISH0.05 versus G0.05, the surface area of two CAD models with the same structural parameters, G30 and ISH30, was analyzed using Python 3.8.8 to obtain a result of 542.98 mm^2 for G30 and 629.33 mm^2 for ISH30, consistent with the previous results for the thickness of struts. With similar ion precipitation results, a difference in surface area of approximately 16% may not be sufficient to determine the antimicrobial efficiency gap alone, so the pore structure of the ISH structure may be more favorable to the coating of the active ingredient for bactericidal applications.

4. Conclusions

In this work, porous β-phase Ti-15Mo TPMS alloys based on Gyroid, I-WP Sheet, and I-WP solid lattices were fabricated by PBF-LB. The unit cell size was 1.5 mm, the relative density (RD) was 30–40%, the compressive modulus was 16–48 GPa, and the yield strength $\sigma_{0.2}$ ranged from 78 to 130 MPa, which was effectively close to the mechanical parameters of human bone. After using MAO to prepare functional modification layers, the antibacterial, antioxidative and osteopromoting composite coatings were successfully prepared by using broad-spectrum antibacterial Ag, antioxidative tannic acid and (F)HA with high stability, respectively. It was demonstrated that the G and ISH groups, with an Ag ion release of about 6 μg/D and an RD of 30% possessed better antioxidant and biocompatible properties, with the ISH group having better antibacterial ability. For the same structural parameters, ISH lattices have somewhat weaker mechanical properties than G lattices and are more biofunctional with the multi-coatings.

Supplementary Materials: The following supporting information can be downloaded at: https://www.mdpi.com/article/0.3390/met15040355/s1, Figure S1: (**a**) Grazing incidence small-angle X-ray scattering patterns of final (F)HA-TA-Ag-TiO_2 coatings. (**b**) XRD results of (F) HA-TA-Ag-TiO_2 coatings.

Author Contributions: Conceptualization, Z.S.; Methodology, Z.L. and J.X.; Software, J.X.; Validation, Z.L. and J.T.; Investigation, Z.L. and J.X.; Resources, J.T. and Z.S.; Data curation, Z.L. and J.T.; Writing—original draft, Z.L. and J.X.; Writing—review & editing, J.T. and M.Y.; Visualization, J.X.; Supervision, Z.S. and M.Y.; Project administration, M.Y. All authors have read and agreed to the published version of the manuscript.

Funding: This work is supported by the Key Research and Development Program of Jiangsu Province (K22251901), the National Natural Science Foundation of China (52271032), and the Shenzhen Science and Technology Innovation Commission (JCYJ20220818100612027).

Data Availability Statement: The original contributions presented in the study are included in the article/Supplementary Materials, further inquiries can be directed to the corresponding authors.

Acknowledgments: The authors would also like to acknowledge the technical support from SUSTech CRF.

Conflicts of Interest: The authors declare no conflict of interest.

References

1. *ASTM F67-13*; Standard Specification for Unalloyed Titanium for Surgical Implant Applications. American Society for Testing and Materials: Philadelphia, PA, USA, 2017.
2. Li, Y.; Wang, F. Review of 3D-Printed Titanium-Based Implants: Materials and Post-Processing. *ChemBioEng Rev.* **2024**, *11*, e202400032. [CrossRef]
3. Xu, J.; Zhang, J.; Shi, Y.; Tang, J.; Huang, D.; Yan, M.; Dargusch, M.S. Surface Modification of Biomedical Ti and Ti Alloys: A Review on Current Advances. *Materials* **2022**, *15*, 1749. [CrossRef] [PubMed]
4. Mesa-Restrepo, A.; Byers, E.; Brown, J.L.; Ramirez, J.; Allain, J.P.; Posada, V.M. Osteointegration of Ti Bone Implants: A Study on How Surface Parameters Control the Foreign Body Response. *ACS Biomater. Sci. Eng.* **2024**, *10*, 4662–4681. [CrossRef] [PubMed]
5. Raja, P.; Babu, K.B.; Kannan, R.; Shanmugam, V.; Balaji, N.S.; Sahani, R.; Behera, L.; Pugazhenthi, A.; Thansekhar, M.R. Progress in the Optimization of Compositional Design and Thermomechanical Processing of Metastable β Ti Alloys for Biomedical Applications. *ACS Biomater. Sci. Eng.* **2024**, *10*, 3528–3547. [CrossRef]
6. Yan, M.; Qian, M.; Kong, C.; Dargusch, M. Impacts of trace carbon on the microstructure of as-sintered biomedical Ti-15Mo alloy and reassessment of the maximum carbon limit. *Acta Biomater.* **2014**, *10*, 1014–1023. [CrossRef]
7. Geetha, M.; Singh, A.K.; Asokamani, R.; Gogia, A.K. Ti based biomaterials, the ultimate choice for orthopaedic implants—A review. *Prog. Mater. Sci.* **2009**, *54*, 397–425. [CrossRef]
8. dos Santos, M.L.; Riccardi, C.d.S.; Filho, E.d.A.; Guastaldi, A.C. Calcium phosphates of biological importance based coatings deposited on Ti-15Mo alloy modified by laser beam irradiation for dental and orthopedic applications. *Ceram. Int.* **2018**, *44*, 22432–22438. [CrossRef]
9. Pronina, Y.; Narykova, M.; Kachanov, M. Relating Stiffness Changes in Porous Materials to the Evolution of Pore Space. *Mech. Mater.* **2025**, *202*, 105236. [CrossRef]
10. Li, Z.; Mo, H.; Tian, J.; Li, J.; Xia, S.; Jia, X.; Zhou, L.; Lu, Y. Compressive Properties and Fracture Behaviours of Ti/Al Interpenetrating Phase Composites with Additive-Manufactured Triply Periodic Minimal Surface Porous Structures. *Met. Mater. Int.* **2024**, *31*, 955–970. [CrossRef]

11. Bram, M.; Stiller, C.; Buchkremer, H.P.; Stöver, D.; Baur, H. High-porosity titanium, stainless steel, and superalloy parts. *Adv. Eng. Mater.* **2000**, *2*, 196–199. [CrossRef]
12. Guo, C.; Ding, T.; Cheng, Y.; Zheng, J.; Fang, X.; Feng, Z. The Rational Design, Biofunctionalization and Biological Properties of Orthopedic Porous Titanium Implants: A Review. *Front. Bioeng. Biotechnol.* **2025**, *13*, 1548675. [CrossRef]
13. Chang, S.Y.; Kang, D.H.; Cho, S.K. Innovative Developments in Lumbar Interbody Cage Materials and Design: A Comprehensive Narrative Review. *Asian Spine J.* **2024**, *18*, 444–457. [CrossRef] [PubMed]
14. Karageorgiou, V.; Kaplan, D. Porosity of 3D biomaterial scaffolds and osteogenesis. *Biomaterials* **2005**, *26*, 5474–5491. [CrossRef]
15. Balla, V.K.; Bodhak, S.; Bose, S.; Bandyopadhyay, A. Porous tantalum structures for bone implants: Fabrication, mechanical and in vitro biological properties. *Acta Biomater.* **2010**, *6*, 3349–3359. [CrossRef]
16. Cwieka, K.; Wysocki, B.; Skibinski, J.; Chmielewska, A.; Swieszkowski, W. Numerical Design of Open-Porous Titanium Scaffolds for Powder Bed Fusion Using Laser Beam (PBF-LB). *J. Mech. Behav. Biomed. Mater.* **2023**, *151*, 106359. [CrossRef]
17. Mohsan, A.U.H.; Wei, D. Advancements in Additive Manufacturing of Tantalum via the Laser Powder Bed Fusion (PBF-LB/M): A Comprehensive Review. *Materials* **2023**, *16*, 6419. [CrossRef]
18. Tang, J.; Li, S.; Liu, Y.; Pei, G.; Yan, M. Significantly Enhanced Biocompatibility and Performance of 3D-Printed Porous 316L Stainless Steel via a Simple and Efficient Surface Polishing Approach. *Mater. Des.* **2025**, *251*, 113640. [CrossRef]
19. Liu, H.; Chen, H.; Sun, B.; Fan, D.; Zhang, A.; Liu, H.; Wei, H.; Yang, W.; Li, Y.; Xia, P.; et al. Enhancing Angiogenesis and Osseointegration through a Double Gyroid Ti6Al4V Scaffold with Triply Periodic Minimal Surface. *Bio. Design Manuf.* **2024**, *8*, 36–54. [CrossRef]
20. Lai, R.; Jiang, J.; Huo, Y.; Wang, H.; Bosiakov, S.; Lyu, Y.; Li, L. Design of Novel Graded Bone Scaffolds Based on Triply Periodic Minimal Surfaces with Multi-Functional Pores. *Front. Bioeng. Biotechnol.* **2025**, *13*, 1503582. [CrossRef]
21. Bidan, C.M.; Wang, F.M.; Dunlop, J.W. Dunlop, A three-dimensional model for tissue deposition on complex surfaces. *Comput. Methods Biomech.* **2013**, *16*, 1056–1070. [CrossRef]
22. Jinnai, H.; Watashiba, H.; Kajihara, T.; Nishikawa, Y.; Takahashi, M.; Ito, M. Surface curvatures of trabecular bone microarchitecture. *Bone* **2002**, *30*, 191–194. [CrossRef]
23. Pinkall, U.; Polthier, K. Computing Discrete Minimal Surfaces and Their Conjugates. *Exp. Math.* **1993**, *2*, 15–36. [CrossRef]
24. Çalışkan, C.İ.; Şahin, M.; Koca, A.; Tarakçı, G.; Okbaz, A.; Dalkılıç, A.S. Mechanical Performance and ANN-Based Prediction of Co-Cr Dental Alloys with Gyroid Cellular Structures Produced by LPBF Technology. *Proc. Inst. Mech. Eng. Part C J. Mech. Eng. Sci.* **2025**, 1–19. [CrossRef]
25. Zou, S.; Gong, H.; Gao, J. Additively Manufactured Multilevel Voronoi-Lattice Scaffolds with Bonelike Mechanical Properties. *ACS Biomater. Sci. Eng.* **2022**, *8*, 3022–3037. [CrossRef]
26. Afshar, M.; Anaraki, A.P.; Montazerian, H.; Kadkhodapour, J. Additive manufacturing and mechanical characterization of graded porosity scaffolds designed based on triply periodic minimal surface architectures. *J. Mech. Behav. Biomed. Mater.* **2016**, *62*, 481–494. [CrossRef]
27. Simske, S.; Ayers, R.; Bateman, T. Porous materials for bone engineering. *Mater. Sci. Forum* **1997**, *250*, 151–182. [CrossRef]
28. Klenke, F.M.; Liu, Y.; Yuan, H.; Hunziker, E.B.; Siebenrock, K.A.; Hofstetter, W. Impact of pore size on the vascularization and osseointegration of ceramic bone substitutes in vivo. *J. Biomed. Mater. Res. Part A* **2008**, *85A*, 777–786. [CrossRef]
29. Murphy, C.M.; Haugh, M.G.; O'Brien, F.J. The effect of mean pore size on cell attachment, proliferation and migration in collagen-glycosaminoglycan scaffolds for bone tissue engineering. *Biomaterials* **2010**, *31*, 461–466. [CrossRef]
30. Hollister, S.; Lin, C.; Saito, E.; Schek, R.; Taboas, J.; Williams, J.; Partee, B.; Flanagan, C.; Diggs, A.; Wilke, E.; et al. Engineering craniofacial scaffolds. *Orthod. Craniofacial Res.* **2005**, *8*, 162–173. [CrossRef] [PubMed]
31. Kumar, A.; Webster, T.J.; Biswas, K.; Basu, B. Flow cytometry analysis of human fetal osteoblast fate processes on spark plasma sintered hydroxyapatite-titanium biocomposites. *J. Biomed. Mater. Res. Part A* **2013**, *101*, 2925–2938. [CrossRef]
32. Gnilitskyi, I.; Dolgov, L.; Tamm, A.; Ferraria, A.M.; Diedkova, K.; Kopanchuk, S.; Tsekhmister, Y.; Veiksina, S.; Polewczyk, V.; Pogorielov, M. Enhanced Osteointegration and Osteogenesis of Osteoblast Cells by Laser-Induced Surface Modification of Ti Implants. *Nanomed. Nanotechnol. Biol. Med.* **2024**, *62*, 102785. [CrossRef]
33. Sun, X.D.; Liu, T.T.; Wang, Q.Q.; Zhang, J.; Cao, M.S. Surface Modification and Functionalities for Titanium Dental Implants. *ACS Biomater. Sci. Eng.* **2023**, *9*, 4442–4461. [CrossRef]
34. Costa Filho, P.M.D.; Marcantonio, C.C.; de Oliveira, D.P.; Lopes, M.E.S.; Puetate, J.C.S.; Faria, L.V.; Carvalho, L.d.F.; de Molon, R.S.; Junior, I.R.G.; Nogueira, A.V.B.; et al. Titanium Micro-Nano Textured Surface with Strontium Incorporation Improves Osseointegration: An In Vivo and In Vitro Study. *J. Appl. Oral Sci.* **2024**, *32*, e20240144. [CrossRef]
35. Robertson, S.F.; Bandyopadhyay, A.; Bose, S. Titania nanotube interface to increase adhesion strength of hydroxyapatite sol-gel coatings on Ti-6Al-4V for orthopedic applications. *Surf. Coatings Technol.* **2019**, *372*, 140–147. [CrossRef]
36. Li, G.; Ma, F.; Liu, P.; Qi, S.; Li, W.; Zhang, K.; Chen, X. Review of Micro-Arc Oxidation of Titanium Alloys: Mechanism, Properties and Applications. *J. Alloys Compd.* **2023**, *948*, 169773. [CrossRef]

37. Verdier, S.; Boinet, M.; Maximovitch, S.; Dalard, F. Formation, structure and composition of anodic films on AM60 magnesium alloy obtained by DC plasma anodizing. *Corros. Sci.* **2005**, *47*, 1429–1444. [CrossRef]
38. Dittrich, K.; Krysmann, W.; Kurze, P.; Schneider, H.G. Structure and Properties of Anof Layers. *Cryst. Res. Technol.* **1984**, *19*, 93–99. [CrossRef]
39. Liu, X.Y.; Chu, P.K.; Ding, C.X. Surface modification of titanium, titanium alloys, and related materials for biomedical applications. *Mater. Sci. Eng. R Rep.* **2004**, *47*, 49–121. [CrossRef]
40. Ma, C.F.; Li, D.M.; Jiang, B.L.; Li, H.J. Study on surface bio-modification of titanium implant using microarc oxidation technology. *J. Fourth Mil. Med. Univ.* **2004**, *25*, 4–7.
41. Gross, K.A.; Rodríguez-Lorenzo, L.M. Sintered hydroxyfluorapatites. Part 1: Sintering ability of precipitated solid solution powders. *Biomaterials* **2004**, *25*, 1375–1384. [CrossRef]
42. Gross, K.A.; Rodríguez-Lorenzo, L.M. Sintered hydroxyfluorapatites. Part II: Mechanical properties of solid solutions determined by microindentation. *Biomaterials* **2004**, *25*, 1385–1394. [CrossRef]
43. Li, B.; Zhao, P.; Liu, S.; Wang, D.; Li, H.; Liang, C.; Wang, H. Preparation and Properties of Hydroxyapatite/Chitosan Coatings on Ti Surface by Different Electrochemical Methods. *J. Mater. Eng. Perform.* **2024**, 1–9. [CrossRef]
44. Boanini, E.; Torricelli, P.; Gazzano, M.; Della Bella, E.; Fini, M.; Bigi, A. Combined effect of strontium and zoledronate on hydroxyapatite structure and bone cell responses. *Biomaterials* **2014**, *35*, 5619–5626. [CrossRef]
45. Qu, H.; Wei, M. Effect of fluorine content on mechanical properties of sintered fluoridated hydroxyapatite. *Mater. Sci. Eng. C* **2006**, *26*, 46–53. [CrossRef]
46. Qu, H.; Wei, M. The effect of fluoride contents in fluoridated hydroxyapatite on osteoblast behavior. *Acta Biomater.* **2006**, *2*, 113–119. [CrossRef]
47. LeGeros, R.Z. Properties of osteoconductive biomaterials: Calcium phosphates. *Clin. Orthop. Relat. Res.* **2002**, *395*, 81–98. [CrossRef]
48. Ratha, I.; Datta, P.; Reger, N.C.; Das, H.; Balla, V.K.; Devi, K.B.; Roy, M.; Nandi, S.K.; Kundu, B. In Vivo Osteogenesis of Plasma Sprayed Ternary-Ion Doped Hydroxyapatite Coatings on Ti6Al4V for Orthopaedic Applications. *Ceram. Int.* **2022**, *48*, 11475–11488. [CrossRef]
49. Kaczmarek-Szczepańska, B.; Zasada, L.; Michalska-Sionkowska, M.; Vishnu, J.; Manivasagam, G. The Modification of Titanium Surface by Decomposition of Tannic Acid Coating. *Appl. Sci.* **2023**, *13*, 5204. [CrossRef]
50. DOh, D.X.; Prajatelistia, E.; Ju, S.-W.; Kim, H.J.; Baek, S.-J.; Cha, H.J.; Jun, S.H.; Ahn, J.-S.; Hwang, D.S. A rapid, efficient, and facile solution for dental hypersensitivity: The tannin-iron complex. *Sci. Rep.* **2015**, *5*, srep10884. [CrossRef]
51. Huang, D.; Lin, Q.; Yin, M.; Wei, Y.; Du, J.; Hu, Y.; Zhao, L.; Lian, X.; Chen, W. Ag nanoparticles decorated electrospinning carbon nanotubes/polyamide nanofibers. *J. Biomater. Sci. Polym. Ed.* **2019**, *30*, 1744–1755. [CrossRef]
52. Al-Ketan, O.; Abu Al-Rub, R.K. MSLattice: A free software for generating uniform and graded lattices based on triply periodic minimal surfaces. *Mater. Des. Process. Commun.* **2021**, *3*, e205. [CrossRef]
53. Al-Ketan, O.; Assad, M.A.; Abu Al-Rub, R.K. Mechanical properties of periodic interpenetrating phase composites with novel architected microstructures. *Compos. Struct.* **2017**, *176*, 9–19. [CrossRef]
54. Al-Ketan, O.; Rowshan, R.; Abu Al-Rub, R.K. Topology-mechanical property relationship of 3D printed strut, skeletal, and sheet based periodic metallic cellular materials. *Addit. Manuf.* **2018**, *19*, 167–183. [CrossRef]
55. Zhu, W.; Moumni, Z.; Zhu, J.; Zhang, Y.; Li, S.; Zhang, W. A Multi-Scale Experimental Investigation for Fatigue Limit and Fatigue Crack Initiation Behavior of Powder Bed Fusion-Laser Beam 316L Stainless Steel. *Mater. Sci. Eng. A* **2023**, *866*, 144692. [CrossRef]
56. Schuh, B.; Völker, B.; Todt, J.; Schell, N.; Perrière, L.; Li, J.; Couzinié, J.; Hohenwarter, A. Thermodynamic instability of a nanocrystalline, single-phase TiZrNbHfTa alloy and its impact on the mechanical properties. *Acta Mater.* **2018**, *142*, 201–212. [CrossRef]
57. Li, C.-L.; Wang, C.-S.; Narayana, P.; Hong, J.-K.; Choi, S.-W.; Kim, J.H.; Lee, S.W.; Park, C.H.; Yeom, J.-T.; Mei, Q. Formation of equiaxed grains in selective laser melted pure titanium during annealing. *J. Mater. Res. Technol.* **2021**, *11*, 301–311. [CrossRef]
58. Tang, J.; Wu, Z.; Yao, X.; Zhou, Y.; Xiong, Y.; Li, Y.; Xu, J.; Dargusch, M.S.; Yan, M. From bio-inertness to osseointegration and antibacterial activity: A one-step micro-arc oxidation approach for multifunctional Ti implants fabricated by additive manufacturing. *Mater. Des.* **2022**, *221*, 110962. [CrossRef]
59. Jia, Z.; Xiu, P.; Li, M.; Xu, X.; Shi, Y.; Cheng, Y.; Wei, S.; Zheng, Y.; Xi, T.; Cai, H.; et al. Bioinspired anchoring AgNPs onto micro-nanoporous TiO_2 orthopedic coatings: Trap-killing of bacteria, surface-regulated osteoblast functions and host responses. *Biomaterials* **2016**, *75*, 203–222. [CrossRef]
60. Huang, D.; Lin, Q.X.; Zhou, Y.J.; Li, J.X.; Wei, Y.; Hu, Y.C.; Lian, X.J.; Chen, S.; Chen, W.Y. Ag nanoparticles incorporated tannic acid/nanoapatite composite coating on Ti implant surfaces for enhancement of antibacterial and antioxidant properties. *Surf. Coatings Technol.* **2020**, *399*, 126169. [CrossRef]

61. He, G.; Guo, B.; Wang, H.; Liang, C.; Ye, L.; Lin, Y.; Cai, X. Surface characterization and osteoblast response to a functionally graded hydroxyapatite/fluoro-hydroxyapatite/titanium oxide coating on titanium surface by sol-gel method. *Cell Prolif.* **2014**, *47*, 258–266. [CrossRef]
62. Xu, S.; Long, J.; Sim, L.; Diong, C.H.; Ostrikov, K. RF plasma sputtering deposition of hydroxyapatite bioceramics: Synthesis, performance, and biocompatibility. *Plasma Process. Polym.* **2005**, *2*, 373–390. [CrossRef]
63. Shi, R.; Huang, Y.; Zhang, J.; Wu, C.; Gong, M.; Tian, W.; Zhang, L. Effective delivery of mitomycin-C and meloxicam by double-layer electrospun membranes for the prevention of epidural adhesions. *J. Biomed. Mater. Res. Part B Appl. Biomater.* **2020**, *108*, 353–366. [CrossRef]
64. Wei, Z.; Tian, P.; Liu, X.; Zhou, B. Hemocompatibility and selective cell fate of polydopamine-assisted heparinized PEO/PLLA composite coating on biodegradable AZ31 alloy. *Colloids Surfaces B Biointerfaces* **2014**, *121*, 451–460. [CrossRef]
65. Benzie, I.F.F.; Strain, J.J. The ferric reducing ability of plasma (FRAP) as a measure of "antioxidant power": The FRAP assay. *Anal. Biochem.* **1996**, *239*, 70–76. [CrossRef]
66. Sugano, M.; Tsuchida, Y.; Satake, T.; Ikeda, M. A microstructural study of fatigue fracture in titanium-molybdenum alloys. *Mater. Sci. Eng. A Struct. Mater. Prop. Microstruct. Process.* **1998**, *243*, 163–168.
67. Júnior, J.R.S.M.; Nogueira, R.A.; de Araújo, R.O.; Donato, T.A.G.; Arana-Chavez, V.E.; Claro, A.P.R.A.; Moraes, J.C.S.; Buzalaf, M.A.R.; Grandini, C.R. Preparation and Characterization of Ti-15Mo Alloy used as Biomaterial. *Mater. Res.* **2011**, *14*, 107–112. [CrossRef]
68. Ho, W.-F. A comparison of tensile properties and corrosion behavior of cast Ti-7.5Mo with c.p. Ti, Ti-15Mo and Ti-6Al-4V alloys. *J. Alloys Compd.* **2008**, *464*, 580–583. [CrossRef]
69. Tang, J.; Luo, J.; Huang, Y.; Sun, J.; Zhu, Z.; Xu, J.; Dargusch, M.; Yan, M. Immunological response triggered by metallic 3D printing powders. *Addit. Manuf.* **2020**, *35*, 101392. [CrossRef]
70. Wang, C.H.; Yang, C.D.; Liu, M.; Li, X.; Hu, P.F.; Russell, A.M.; Cao, G.H. Martensitic microstructures and mechanical properties of as-quenched metastable β-type Ti–Mo alloys. *J. Mater. Sci.* **2016**, *51*, 6886–6896. [CrossRef]
71. Luo, J.; Huang, Y.; Xu, J.; Sun, J.; Dargusch, M.; Hou, C.; Ren, L.; Wang, R.; Ebel, T.; Yan, M. Additively manufactured biomedical Ti-Nb-Ta-Zr lattices with tunable Young's modulus: Mechanical property, biocompatibility, and proteomics analysis. *Mater. Sci. Eng. C* **2020**, *114*, 110903. [CrossRef]
72. Cheng, H. Additive Manufacturing and Mechanical Properties of Functionally Graded Medical Ti–35Nb–7Zr–5Ta Porous Scaffolds. *J. Physics Conf. Ser.* **2021**, *1986*, 012036. [CrossRef]
73. Jamuna-Thevi, K.; Bakar, S.; Ibrahim, S.; Shahab, N.; Toff, M. Quantification of silver ion release, in vitro cytotoxicity and antibacterial properties of nanostuctured Ag doped TiO_2 coatings on stainless steel deposited by RF magnetron sputtering. *Vacuum* **2011**, *86*, 235–241. [CrossRef]
74. Labieniec, M.; Gabryelak, T.; Falcioni, G. Antioxidant and pro-oxidant effects of tannins in digestive cells of the freshwater mussel Unio tumidus. *Mutat. Res. Toxicol. Environ. Mutagen.* **2003**, *539*, 19–28. [CrossRef]
75. Parashar, U.K.; Kumar, V.; Bera, T.; Saxena, P.S.; Nath, G.; Srivastava, S.K.; Giri, R.; Srivastava, A. Study of mechanism of enhanced antibacterial activity by green synthesis of silver nanoparticles. *Nanotechnology* **2011**, *22*, 415104. [CrossRef]

metals

MDPI

Article

Enhanced Compressive Properties of Additively Manufactured Ti-6Al-4V Gradient Lattice Structures

Zhengjie Guo [1], Yuting Ma [1], Tayyeb Ali [2], Yi Yang [1,*], Juan Hou [1], Shujun Li [2] and Hao Wang [2,*]

[1] School of Materials and Chemistry, University of Shanghai for Science and Technology, Shanghai 200093, China; 222203099@st.usst.edu.cn (Z.G.); 223353166@st.usst.edu.cn (Y.M.); hou18217727686@163.com (J.H.)

[2] Institute of Metal Research, Chinese Academy of Sciences, Shenyang 110016, China; tayyebali@imr.ac.cn (T.A.); shjli@imr.ac.cn (S.L.)

* Correspondence: yangyi@usst.edu.cn (Y.Y.); haowang@imr.ac.cn (H.W.)

Abstract: Lattice structures are widely used in the aerospace and biomedical fields, due to their lightweight, high specific strength, large specific surface area, good biocompatibility, etc. However, the balancing of the weight and the mechanical properties remains a challenge in designing lattice structures. Combining experiments and simulations, the present work first designs and evaluates the mechanical properties of uniform and gradient topology-optimized Ti-6Al-4V lattices with the same overall porosity of 84.27%, employing finite element simulations. Then, laser powder bed fusion technology is used to fabricate the uniform and gradient Ti-6Al-4V lattices, and their compressive performance is tested. The results indicate that, under longitudinal compression, the gradient lattice structure exhibits good layer-by-layer collapse deformation behavior, achieving better comprehensive performance than the uniform lattice structure. While under horizontal compression, the deformation behavior of the gradient lattice structure is similar to that of the uniform lattice structure, and the deformation is mostly randomly distributed. The cumulative energy absorption of the gradient lattice structure increased by approximately 20% compared with that of the uniform lattice structure. The results provide a technical basis for the integrated design of structural and functional components for aerospace applications.

Keywords: lattice structure; Ti-6Al-4V; laser powder bed fusion; mechanical property; energy absorption

1. Introduction

Titanium and its alloys have been widely used in both aerospace and biomedical applications due to their excellent properties such as high strength, excellent fatigue resistance, corrosion resistance, and outstanding biocompatibility [1–3]. Compared with traditional energy-absorbing materials, the specific energy absorption of titanium alloys is relatively inadequate [4,5]. By adopting a lattice matrix structure, the energy absorption effect can be effectively improved to achieve a specific energy absorption effect that is substantially better than that of steel materials [6].

Lattice material is a new type of engineering material that combines functional and structural properties. It has the characteristics of light weight, large specific surface area, high specific strength, low thermal conductivity, and impact resistance and is widely used in fields such as aerospace, biomedicine, and petrochemicals [7–9]. However, balancing the weight, thermal insulation, and mechanical properties of lattice materials is still a challenge when designing lattice materials (structures) for some special applications. A gradient lattice structure refers to a lattice structure in which the pore size and arrangement change according to a certain pattern with depth. It has many advantages such as improved residual stress distribution, enhanced thermal properties, improved fracture toughness, and reduced stress intensity factor [10]. Its excellent structural characteristics can effectively improve the energy absorption rate and mechanical properties of titanium alloys. Topology optimization is an advanced computational technique that optimizes material layout within a predefined design domain to meet specific constraints and achieve performance optimization, especially suitable for the design of complex porous lattice structures [11]. Topology optimization systematically analyzes and adjusts the number, size, shape, and location of holes in a structure in order to maximize the overall performance of the structure while achieving a lightweight design [12–14]. Within a certain range, the volume fraction of topology-optimized structures is proportional to their moduli, and, thus, structures with a higher volume fraction exhibit greater strength and superior fatigue life [15,16]. In addition, topology-optimized structures can effectively reduce the stress concentration phenomenon and, thus, inhibit the emergence of fatigue cracks, which significantly improves the reliability and durability of the structures in long-term use [13,17].

Laser powder bed fusion (LPBF) technology, as a popular additive manufacturing process, is based on three-dimensional digital models [18–21]. These models are sliced into two-dimensional plane coordinates with high-precision parameters. Using a high-intensity laser as the energy source, the laser beam melts the powder layer by layer through heat (near-infrared light with a wavelength of approximately 1064–1080 nm), achieving rapid processing. The material then cools and solidifies, accumulating to form three-dimensional (3D) parts that meet the dimensional requirements [22–24]. Liu et al. [25] used LPBF to fabricate topologically optimized porous titanium structures, which exhibited high plasticity and excellent fatigue resistance, significantly reducing the fatigue crack propagation rate and extending their fatigue life. Chen et al. [26] investigated the effect of microstructure on the fatigue behavior of graded porous titanium alloys formed by LPBF, gaining insights into the relationship between fatigue performance and microstructure. They found that differences in gradient density distribution had a significant impact on fatigue performance. Hu et al. [27] used LPBF to investigate the microstructure and mechanical properties of porous titanium implants. It was found that the performance along the scanning direction was slightly better than that along the forming direction, and that the lattice materials with a porosity of 60%~80% had superior compressive strength and elastic modulus. Ren et al. [28] used LPBF to fabricate a hexagonal-body-centered (HBC) lattice structure, and the experiments showed that the HBC lattice structures produced by LPBF not only had good formability but also demonstrated excellent mechanical properties and energy absorption capacity.

Despite the previous intensive investigations on uniform lattice structures, gradient lattice structures have been much less covered. Indeed, both new deformation mechanisms and mechanical properties are present in gradient lattice structures. In this work, we focus on the enhanced compressive property of gradient lattices compared with uniform lattices with the same overall porosity, primarily aiming at the applications of complex lattice structures in the aerospace field in which the mechanical properties require further improvement. Moreover, we employed as-print samples because, in certain circumstances, subsequent hot isostatic pressing or heat treatment is not applicable.

2. Experimental Procedures and Programs

2.1. Experimental Materials

Spherical gas-atomized pre-alloyed Ti-6Al-4V (TC4) powder (provided by Aerospace Seahawk (Harbin) Titanium Industry Co., Ltd., Harbin, China) having an average size ranging from 15 to 53 μm was used in this study (Figure 1a). The specific parameters of the

Ti-6Al-4V powder are shown in Table 1, and the average sphericity of the Ti-6Al-4V powder is about 0.96 (Sphericity is the ratio of the surface area of a sphere with the same volume as the particle to the actual surface area of the particle. It is a dimensionless quantity), as detected by the National Nonferrous Metals and Electronic Materials Analysis and Testing Center of National Standard (Beijing) Inspection and Certification Co. (Beijing, China). Figure 1 shows the typical morphology of TC4 powder used for developing the lattice structures by using an LPBF printer (AmPro Innovations Technology Co., Ltd., Suzhou, China). Furthermore, the spherical shape of the powder particles with a smooth surface and their size range (15 to 53 micrometers) both met the LPBF printer's feed requirements.

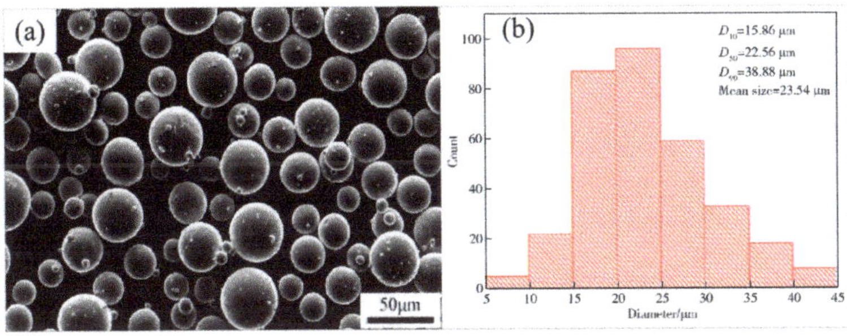

Figure 1. (**a**) SEM image of initial TC4 powder; (**b**) powder particle size distribution chart.

Table 1. Chemical composition (wt%) of TC4 powder metallurgy titanium alloy.

Al	V	Fe	C	N	O	H	Ti
6.42	4.22	0.044	0.052	0.012	0.097	<0.0020	Bal.

2.2. *Lattice Structure Design and Fabrication*

Currently, the topologically optimized BCC lattice structure has been well investigated and exhibits superior mechanical properties [17,29,30]. Our research focuses on exploring the performance and behavior of lattice structures during the LPBF forming process. The experiments utilized a combination of Solidworks (Dassault Systèmes (Shanghai) Co., Ltd., Shanghai, China) and Rhino Grasshopper (Robert McNeel & Associates, Seattle, WA, USA) to design three topology optimization-based BCC uniform lattice structures and gradient lattice structures all with an average porosity of 84.32%, as shown in Figure 2. Solidworks (https://www.solidworks.com/, accessed on 12 September 2024) is used to construct lattice unit cells and generate uniform lattice structures through linear arrays. Due to the varying interlayer densities in the gradient lattice structure, Rhino Grasshopper (https://www.rhino3d.com/learn/?query=kind:%20grasshopper, accessed on 12 September 2024) is used to generate the gradient lattice structure by defining the number of layers, along with the porosity of the top and bottom layers, using a linear relationship. Lattice unit cells with different densities are obtained by adjusting the rod width, in which the smallest unit cell size is 3.3 mm × 3.3 mm × 3.3 mm, the maximum rod width is 0.80 mm, and the minimum width is 0.25 mm. Consequently, the porosity of the unit cell structures ranges from a minimum of 80.63% to a maximum of 88.01%. The gradient lattice with varying porosity gradients is generated using arrays and combinations, with the porosity gradient increasing from 1.05% at G-A to 3.69% at G-C. The experiment employed the SP101 (AmPro Innovations, AmPro Innovations Technology Co., Ltd., Suzhou, China) LPBF system (Figure 3a). Using this equipment, Ti-6Al-4V lattice structure specimens were prepared. The process flow for this preparation is illustrated in Figure 3b. A circular TC4 substrate with a diameter of Φ100 mm was selected for the test, and a stripe scanning strategy was adopted. The laser path rotation angle was 66.67°; the stripe width was 10 mm; and the stripe overlap was 0.08 mm (Figure 3c). Titanium alloy has an extremely strong oxidation capability. To prevent oxidation during the experiment, the entire forming process was conducted in an inert gas environment. Argon gas was selected as the protective gas, and the oxygen content was controlled to be below 900 ppm to reduce unnecessary contamination. After printing was completed, wire cutting was used to remove the sample from the substrate. To remove any unmelted powder from the samples, all the specimens were cleaned by ultrasonic oscillation in an ultrasonic cleaning machine with anhydrous ethanol for 15 min. After cleaning and drying, the samples were stored in sealed bags.

The laser energy input into the powder bed in LPBF is controlled by the laser power P (W), scanning speed v (mm/s), scanning spacing s (mm), and layer thickness t (mm). It is typically expressed as a function of laser energy density Ed (J/mm^3) and used for parameter optimization [31,32], but there are certain limitations.

$$Ed = \frac{P}{v\,s\,t} \tag{1}$$

The processing parameters were obtained after the parameter optimization by using the Doehlert matrix method [33,34], with a laser power of 180 W, scanning speed of 1000 mm/s, energy density range of 66.67 J/mm^3, laser spacing of 0.09 mm, and layer thickness of 30 μm. The samples fabricated using the optimal process parameters are shown in Figure 4. The microstructure of the as-fabricated samples is mainly composed of α grains and acicular metastable α' martensites (for details, please refer to our previous works [35,36]).

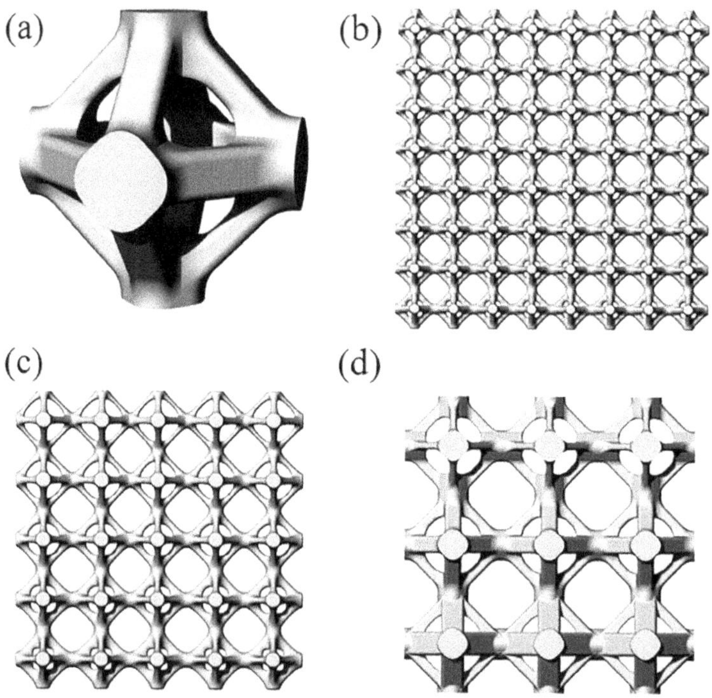

Figure 2. (**a**) Topology-optimized structural single-cell model; (**b**) 8 × 8 × 8 gradient lattice structure G-A; (**c**) 5 × 5 × 5 gradient lattice structure G-B; (**d**) 3 × 3 × 3 gradient lattice structure G-C.

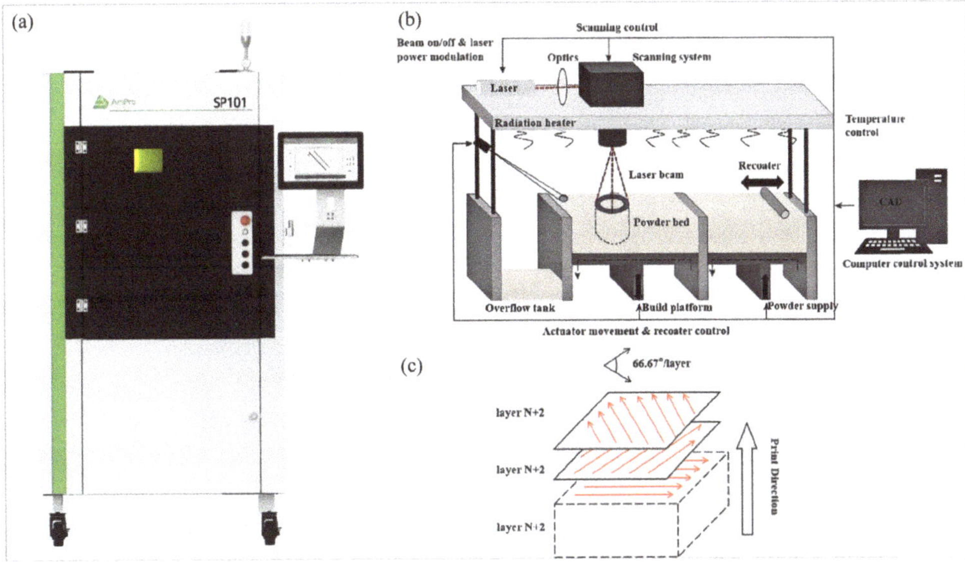

Figure 3. (**a**) Physical diagram of the device; (**b**) workflow diagram; (**c**) scanning strategy.

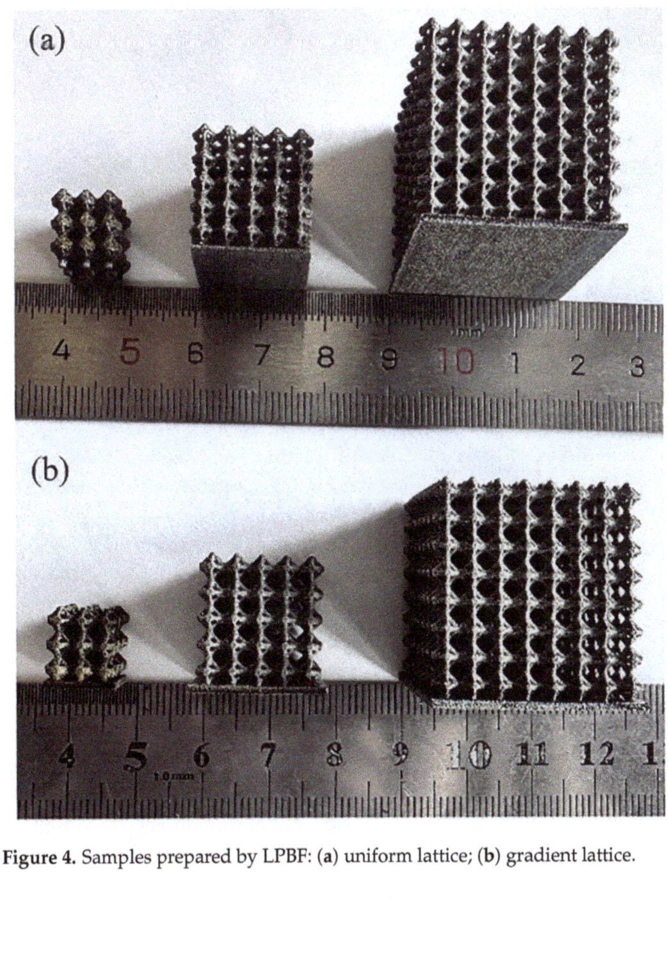

Figure 4. Samples prepared by LPBF: (**a**) uniform lattice; (**b**) gradient lattice.

2.3. Quasi-Static Compression

Based on the ISO 13314-2011 standard [37,38], the samples were compressed using a Zwick/Roell Z100 universal testing machine from Ulm, Germany at a speed of 1.0 mm/min, with a maximum experimental force of 100 kN. The principle of the compression test is shown in Figure 5. Compression tests were conducted on different uniform lattice structures and gradient lattice structures, with at least five compression tests for each sample. The section morphology was observed using a scanning electron microscope (SEM, Zeiss SUPRA55, Oberkochen, Germany) under the secondary-electron-imaging mode (SEI), and its specific deformation mechanism and fracture surface microstructure were analyzed. The experimental stress–strain curves were averaged from multiple tests. The energy absorption was then calculated according to the stress–strain curves.

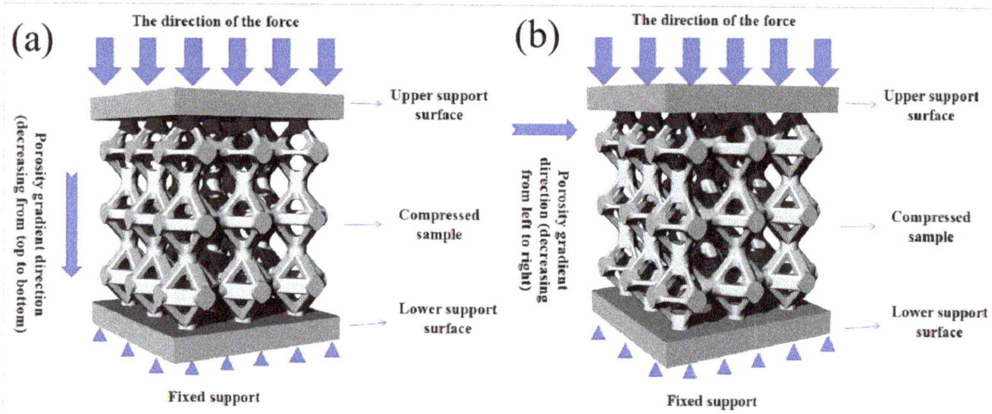

Figure 5. Schematic diagram of compression test: (**a**) longitudinal compression; (**b**) transverse compression.

2.4. Finite Element Analysis

Before conducting the static compression experiment, numerical simulation analysis was carried out to predict the mechanical properties of a Ti-6Al-4V lattice structure designed based on topology optimization. The compression model mainly consists of three parts, with the middle part being the lattice structure, and the upper and lower sides being rigid planes, respectively. The lower rigid plane is fixed with constraints, and the compression process of the pressing plate is simulated in the explicit dynamics module, in which the upper plate is pressed down at a speed of 1.0 mm/min. The structural model employs Cartesian grid elements to mesh the complex lattice structure, with a grid size of 0.3 mm [39,40]. The dynamic friction coefficient between the structure and the surfaces of the lower and upper supports is 0.15, and the static friction coefficient is 0.25. The elastic properties of the material are set with an elastic modulus of 110 GPa and a Poisson's ratio of 0.33 [41]. An isotropic hardening constitutive model is adopted in this study, and the Johnson–Cook failure model is used for the fracture failure model [42,43].

3. Results and Discussion

3.1. Finite Element Simulations

To validate the feasibility of the experiment, this study used finite element simulation to predict the deformation within the compressed samples. Due to the lack of an anisotropic failure model for the present system, an isotropic failure model was employed to estimate the stress and strain distribution. As shown in Figure 6a,b, the gradient lattices exhibit a distinct stress–strain behavior compared with the uniform lattices. The layer-by-layer failure behavior results in an overall uphill stress–strain relation. During the compression process after the first failure, the stress rises again due to the densification of the first layer. In contrast, for the uniform lattices, the stress decreases after the peak and only reaches a higher value when the sample is finally completely densified. The corresponding stress distribution is shown in Figure 6c. The simulation results demonstrate that the gradient lattices possess better compression characteristics than the uniform lattices.

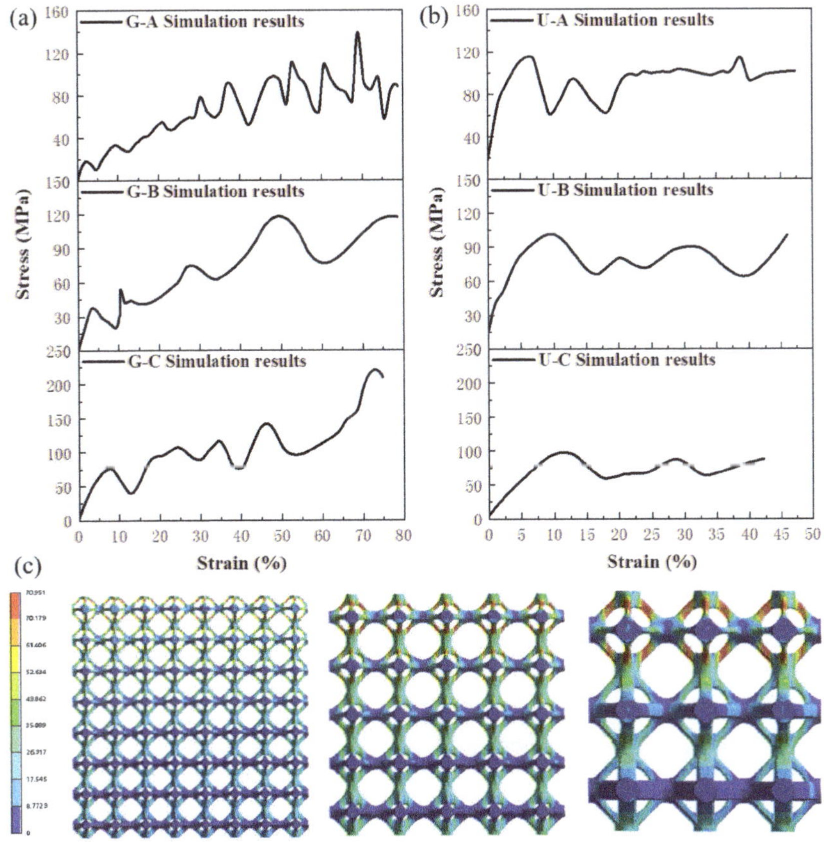

Figure 6. Finite-element-simulated compressive stress–strain curves and stress distribution: (**a**) gradient lattice longitudinal compressive load; (**b**) uniform lattice compressive load; (**c**) stress distribution.

3.2. Deformation Behavior and Failure Mechanisms

During the longitudinal compression process, the layer with lower relative density will preferentially withstand higher stress and, thus, undergo collapse deformation behavior, as shown in Figure 7a (the percentage on the left side represents the degree of deformation). As the relative density increases, the stress value gradually increases. Other layers gradually collapse as deformation progresses, and the overall structure exhibits a clear layer-by-layer collapse behavior. Compared to the layer-by-layer collapse behavior during longitudinal compression, the deformation of the uniform lattice structure did not change in sequence, as shown in Figure 7b. The deformation of the uniform lattice structure exhibited a random distribution, starting from any layer and then randomly spreading to another layer. The overall structure gradually compressed, expanded, and flattened, ultimately leading to the complete failure of the sample. Since there is no difference between transverse and longitudinal loading for a uniform lattice, it was also found that the deformation behavior of the gradient lattice structure was similar to that of the uniform lattice structure during compression under transverse loading (as shown in Figure 8, the percentage on the left side represents the degree of deformation). This deformation mechanism is consistent with the failure mechanism of gyroid structures reported by Polley et al. [44] and Li et al. [45]. Based on these conclusions, we can obtain the desired properties by applying forces in different loading directions. For example, when subjected to longitudinal loads, the lattice structure can enhance the energy absorption of the structure and, thus, provide greater strain protection; when subjected to transverse loads, the lattice structure can be used for lightweight structures or to simulate tissue density in the medical field.

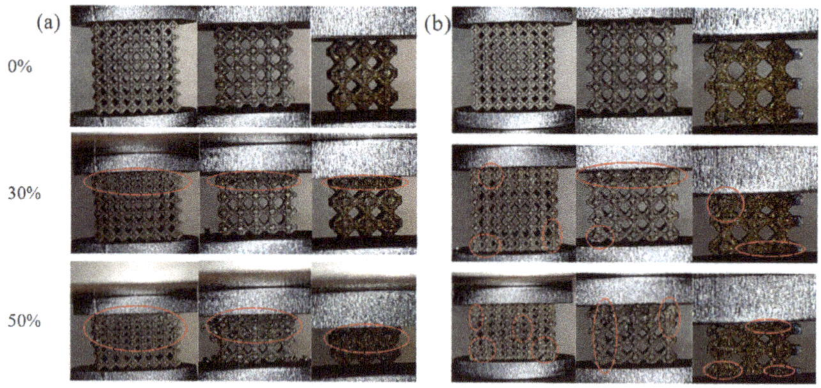

Figure 7. Longitudinal compression deformation of different lattice structures: (**a**) gradient lattice structure; (**b**) uniform lattice structure.

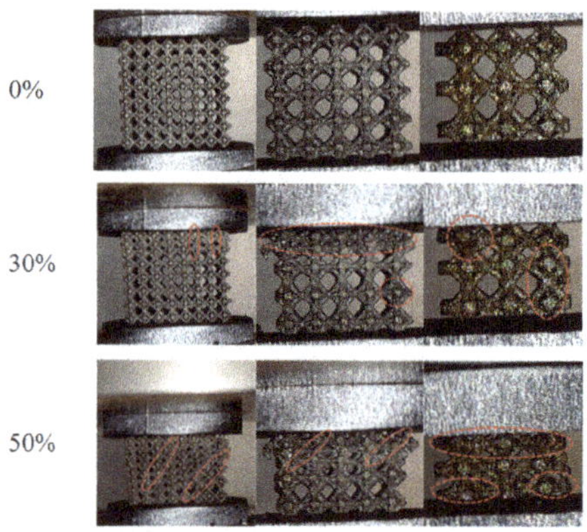

Figure 8. Transverse compression deformation of gradient lattice.

To further investigate the failure mechanism of the Ti6Al4V alloy lattice structure, scanning electron microscopy (SEM) was used to observe the fracture surface morphology of the samples after different compression load tests, as shown in Figure 9. It was found that the fracture surface morphology consisted of smooth planar surfaces and a few large-sized tear ridges, both under longitudinal and transverse compressive loading conditions. This indicates that the primary fracture mechanism of the samples is brittle fracture, with a localized post-yield response occurring before damage. Therefore, the compression fracture mechanism of the Ti6Al4V lattice structure fabricated by LPBF is a quasi-cleavage fracture, which is also consistent with the results of Sun et al. [46] and Zhang et al. [47].

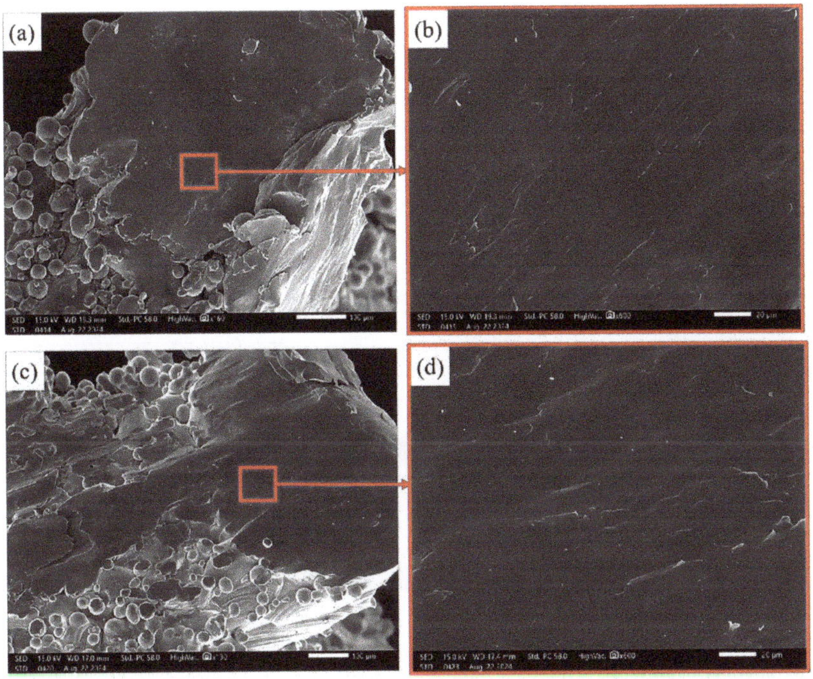

Figure 9. SEM images of compression fracture surface morphology of Ti6Al4V lattice structure: (**a**) overall fracture morphology of longitudinal compression; (**b**) enlarged image of the red-marked area in (**a**); (**c**) overall fracture morphology of transverse compression; (**d**) enlarged image of the red-marked area in (**c**).

3.3. Compression Deformation Characteristics

Figure 10 demonstrates the relationship between stress and strain in compression along different loading directions for gradient lattice and uniform lattice structures with equal average porosity. It can be found that, at the initial stage of loading, both gradient lattice and uniform lattice structures under the two loading conditions show obvious elastic behaviors. As the deformation increases, they gradually enter the plastic plateau and densification stages. The samples exhibit significant post-yield strain hardening under longitudinal compressive loading as shown by the curves in Figure 10a–c. Although a significant post-yield stress drop occurs for all three structures, as compression proceeds, the stress values again reach the initial peak stress and gradually rise above the peak stress. When the average porosity is constant, the average compressive strength of the LPBF gradient lattice structure increases with the increase in the porosity gradient. As the porosity gradient increases from 1.05% (Figure 10a) to 3.69% (Figure 10c), the average compressive strength rises from 108.78 MPa to 166.82 MPa, which is 16.40%, 20.63%, and 50.87% higher than that of the corresponding uniform lattice structures, respectively. These experimental results are also consistent with those of Choy et al. [48] and Cardeña et al. [49]. The average compressive strength of the same structure is about 12%–15% higher than that of Sancho's BCC lattice structure. This phenomenon can be attributed to the result of the increase in diameter of the single-cell pillars and the increase in the porosity gradient, leading to an increase in the yield strength of the LPBF gradient lattice structure.

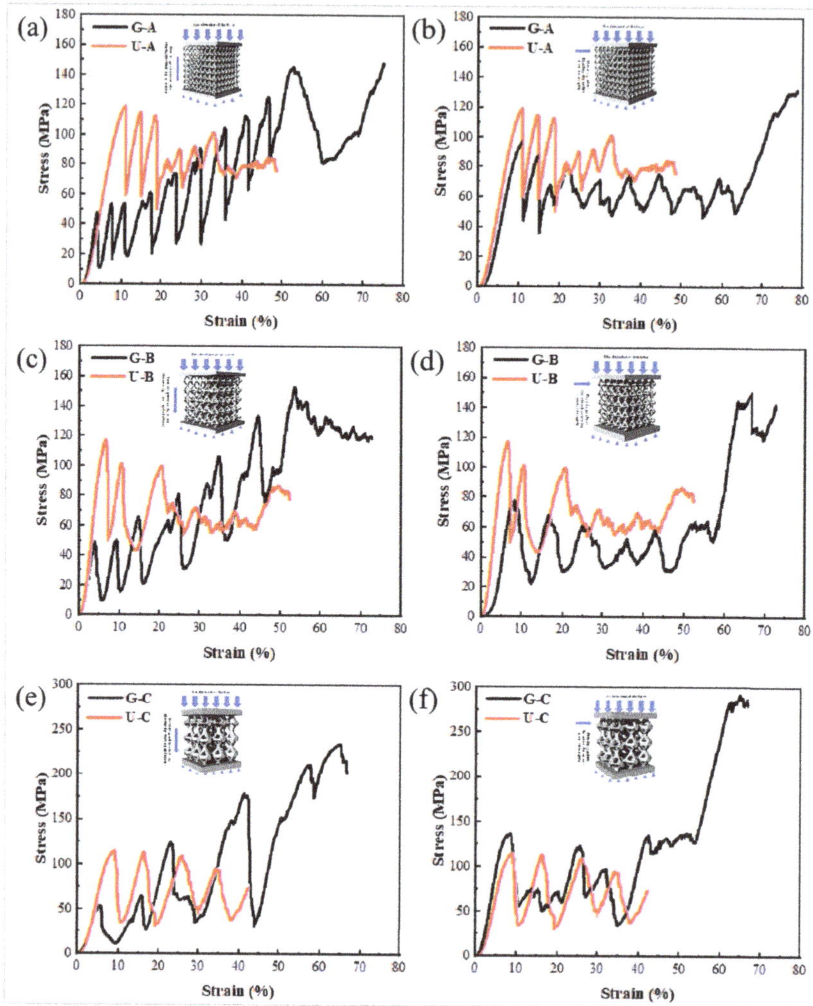

Figure 10. Compressive stress–strain curves of different gradient lattice and uniform lattice structures with equal average porosity: (**a–c**) longitudinal loading; (**d–f**) compression under transverse loading.

As shown in the curves of Figure 10d–f, the deformation of the samples under transverse compressive loading exhibits a strain-softening behavior similar to that of a uniform lattice structure. After the stress reaches the first peak value, the stress remains below the peak stress before the sample undergoes densification. This is because the overall structure of the sample has been locally destroyed (Figure 8). The compressive strength is significantly reduced, which leads to a lower loading capacity of the sample after plastic failure. It was observed that during the plastic plateau stage, the stress value of G-A showed the largest change compared to the maximum compressive strength, decreasing by approximately 21.43%. It exhibited low-amplitude stress variations, demonstrating a more uniform plastic plateau region. The large increase in the stress values of the samples with increasing deformation is due to the gradual densification of the samples as compression proceeds.

3.4. Cumulative Energy Absorption and Energy Absorption Efficiency Under Compressive Deformation Behavior

Cumulative energy absorption (W) is usually an important measure of the ductility of lattice structures and can be used to determine the total amount of energy absorbed by the material or system under external pressure. The calculation formula is given in Equation (2) [50]. Maximum stress is an important factor in energy absorption applications. The energy absorption efficiency (η) is used to evaluate the energy absorption characteristics of lattice structures and is defined as the energy absorbed at any strain divided by the peak stress. The calculation formula is given in Equation (3) [51].

$$W = \int_0^\varepsilon \sigma(\varepsilon) d\varepsilon \tag{2}$$

$$\eta(\varepsilon) = \frac{1}{\sigma(\varepsilon)} \int_0^\varepsilon \sigma(\varepsilon) d\varepsilon \tag{3}$$

In the equation, σ and ε are the stress and strain during the compression test. W is the cumulative energy absorption value (MJ/mm^3), which can be calculated from the compressive stress–strain curve. $\sigma(\varepsilon)$ is the corresponding peak stress at a certain ε, and $\eta(\varepsilon)$ is the energy absorption efficiency.

The relationships between cumulative energy absorption and energy absorption efficiency and strain were experimentally investigated for gradient lattice and uniform lattice structures with equal porosity under longitudinal loading conditions, as shown in Figure 11. The results show that the cumulative energy absorption of both the gradient lattice structure and the uniform lattice structure increased with the increase in deformation, almost linearly (see Figure 11a–c). In the elastic deformation stage, the cumulative energy absorption of the uniform lattice structure is higher than that of the gradient lattice structure. As the deformation increases, the energy absorption of the gradient lattice structure gradually increases. The cumulative energy absorption of the gradient lattice structure appears to be higher than that of the uniform lattice structure in the plateau region. This is because, with the increase in deformation, the energy absorption efficiency of the gradient lattice structure gradually surpasses that of the uniform lattice structure (e.g., Figure 11d–f). Meanwhile, with the increase in the porosity gradient, the gradient lattice structure can absorb more energy with a smaller deformation. The cumulative energy absorption of G-B and G-C increased by 20.43% and 7.69%, respectively, compared to that of the uniform lattice structure. The energy absorption efficiency was approximately 32% and 47% higher than that of the corresponding uniform lattice structures. The reason that the cumulative energy and energy absorption efficiency of the G-A structure approximates that of the uniform dot matrix structure may be because the layer-by-layer porosity gradient of the G-C structure is only 1%. The overall structure tends to be more like that of the uniform dot matrix structure. It is observed that the gradual decrease in deformation corresponding to the critical point from Figure 11a–c is due to the increase in porosity gradient. In the subsequent deformation process, the energy absorption of the gradient lattice structure is greater than that of the uniform lattice structure and continues to increase. The characteristics of the energy absorption efficiency and strain relationship curves in Figure 11d–f also confirm this result effectively.

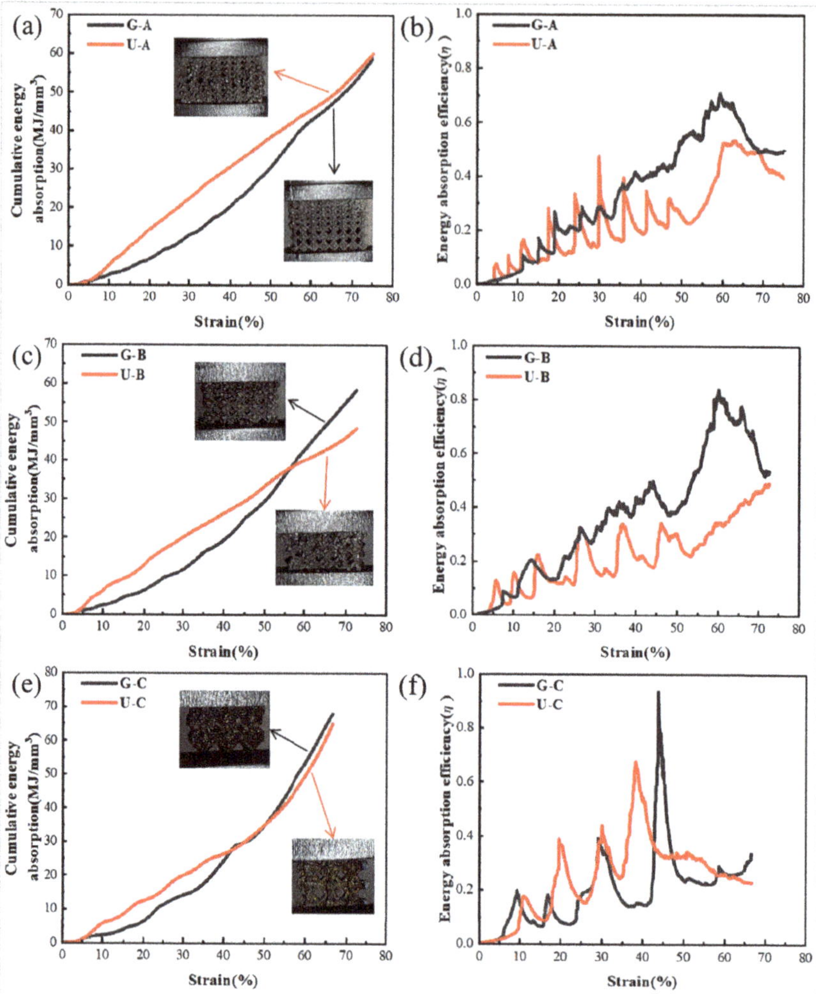

Figure 11. Uniform lattice structure and gradient lattice structure: (**a**–**c**) cumulative energy absorption–strain curve; (**d**–**f**) cumulative absorption efficiency–strain curve.

It is crucial to compare the average compressive strength and cumulative energy absorption of this study with the results from the existing literature. However, research on gradient lattice structures with energy absorption characteristics manufactured by LPBF is very limited. Choy et al. [48] reported that, for gradient lattice structures of the same material and type, the average compressive strength reached 113 MPa, and the cumulative energy absorption was 95 MJ/mm^3. In comparison, the corresponding uniform lattice structure had an even higher average compressive strength of 172 MPa. Compared to our study, the higher average compressive strength and cumulative energy absorption in their research can be attributed to the fact that their lattice structure had a porosity of only 71%, which is about 14% lower than the porosity of the lattice structure in our experiment. As a result, their overall structure was heavier, allowing them to bear larger loads. Although it is not entirely fair to directly compare the experimental results due to the differences in porosity and porosity gradient, the impact of varying porosity and porosity gradients on compressive strength and cumulative energy absorption is worth further discussion.

4. Conclusions

In this work, uniform and gradient lattice structures with the same porosity based on topology-optimized design were fabricated by an LPBF process, and their compression behavior was investigated. The conclusions are as follows:

(1) With the same microstructure and average porosity, the gradient lattice exhibits better compression performance, with a cumulative energy absorption capacity significantly superior to that of the uniform lattice structure.
(2) This superior performance is attributed to the fact that, under compressive loading, the gradient lattice exhibits layer-by-layer shear failure, whereas the uniform lattice experiences overall random failure.
(3) The gradient structure exhibits a greater total deformation before failure than the uniform lattice structure. However, as the deformation increases, the energy absorption efficiency slightly decreases after the deformation exceeds 50%. This indicates that the densification of the structure reduces its energy absorption efficiency.

Author Contributions: Conceptualization, H.W.; Methodology, Z.G. and Y.M.; Software, Z.G. and Y.M.; Validation, T.A., Y.Y., J.H., S.L. and H.W.; Formal analysis, Z.G. and Y.M.; Investigation, Z.G., Y.M. and T.A.; Resources, Y.Y., J.H., S.L. and H.W.; Writing—original draft, Z.G. and Y.M.; Writing—review and editing, Z.G., T.A., Y.Y., J.H., S.L. and H.W.; Visualization, Z.G. and Y.M.; Supervision, Y.Y., J.H., S.L. and H.W.; Project administration, H.W.; Funding acquisition, H.W. All authors have read and agreed to the published version of the manuscript.

Funding: This work is financially supported by the National Natural Science Foundation of China (U2241245 and 52321001), the Aeronautical Science Foundation of China (2022Z053092001), the Natural Science Foundation of Shenyang (23-503-6-05), the Shanghai Science and Technology Development Funds (22QB1406500), and the Liaoning Provincial Science and Technology Major Project (2024JH1/11700028).

Data Availability Statement: The original contributions presented in the study are included in the article, further inquiries can be directed to the corresponding authors.

Conflicts of Interest: The authors declare no conflict of interest.

References

1. Hao, Y.-L.; Li, S.-J.; Yang, R. Biomedical titanium alloys and their additive manufacturing. *Rare Met.* **2016**, *35*, 661–671. [CrossRef]
2. Qian, M.; Bourell, D.L. Additive Manufacturing of Titanium Alloys. *JOM* **2017**, *69*, 2677–2678. [CrossRef]
3. Wang, M.; Lin, X.; Huang, W. Laser additive manufacture of titanium alloys. *Mater. Technol.* **2016**, *31*, 90–97. [CrossRef]
4. Zhang, J.; Liu, Y.; Babamiri, B.B.; Zhou, Y.; Dargusch, M.; Hazeli, K.; Zhang, M.-X. Enhancing specific energy absorption of additively manufactured titanium lattice structures through simultaneous manipulation of architecture and constituent material. *Addit. Manuf.* **2022**, *55*, 102887. [CrossRef]
5. Sun, G.; Chen, D.; Zhu, G.; Li, Q. Lightweight hybrid materials and structures for energy absorption: A state-of-the-art review and outlook. *Thin-Walled Struct.* **2022**, *172*, 108760. [CrossRef]
6. Pan, X.; Qian, G.; Wu, S.; Fu, Y.; Hong, Y. Internal crack characteristics in very-high-cycle fatigue of a gradient structured titanium alloy. *Sci. Rep.* **2020**, *10*, 4742. [CrossRef] [PubMed]
7. Zhan, Z.; Li, H.; Lam, K. Development of a novel fatigue damage model with AM effects for life prediction of commonly-used alloys in aerospace. *Int. J. Mech. Sci.* **2019**, *155*, 110–124. [CrossRef]
8. He, S.; Zhu, J.; Jing, Y.; Long, S.; Tang, L.; Cheng, L.; Shi, Z. Effect of 3D-Printed Porous Titanium Alloy Pore Structure on Bone Regeneration: A Review. *Coatings* **2024**, *14*, 253. [CrossRef]
9. Song, C.; Liu, L.; Deng, Z.; Lei, H.; Yuan, F.; Yang, Y.; Li, Y.; Yu, J. Research progress on the design and performance of porous titanium alloy bone implants. *J. Mater. Res. Technol.* **2023**, *23*, 2626–2641. [CrossRef]
10. Bian, X.; Yuan, F.; Zhu, Y.; Wu, X. Gradient structure produces superior dynamic shear properties. *Mater. Res. Lett.* **2017**, *5*, 501–507. [CrossRef]
11. Xiao, Z.; Yang, Y.; Xiao, R.; Bai, Y.; Song, C.; Wang, D. Evaluation of topology-optimized lattice structures manufactured via selective laser melting. *Mater. Des.* **2018**, *143*, 27–37. [CrossRef]

12. Chen, L.; Che, J.; Liang, S.; Wang, Y. Multiscale topology optimization of gradient lattice structure based on volume parametric modeling. *Compos. Struct.* **2023**, *328*, 117746. [CrossRef]
13. Li, J.; Huang, Z.; Liu, G.; An, Q.; Chen, M. Topology optimization design and research of lightweight biomimetic three-dimensional lattice structures based on laser powder bed fusion. *J. Manuf. Process.* **2021**, *74*, 220–232. [CrossRef]
14. Li, T.; Sun, B.; Gan, N. Lattice infill structure design of topology optimization considering size effect. *Math. Mech. Solids* **2022**, *28*, 873–890. [CrossRef]
15. Belda, R.; Megías, R.; Marco, M.; Vercher-Martínez, A.; Giner, E. Numerical analysis of the influence of triply periodic minimal surface structures morphometry on the mechanical response. *Comput. Methods Programs Biomed.* **2023**, *230*, 107342. [CrossRef] [PubMed]
16. Zhao, M.; Liu, F.; Fu, G.; Zhang, D.Z.; Zhang, T.; Zhou, H. Improved Mechanical Properties and Energy Absorption of BCC Lattice Structures with Triply Periodic Minimal Surfaces Fabricated by SLM. *Materials* **2018**, *11*, 2411. [CrossRef] [PubMed]
17. Bai, L.; Zhang, J.; Chen, X.; Yi, C.; Chen, R.; Zhang, Z. Configuration Optimization Design of Ti6Al4V Lattice Structure Formed by SLM. *Materials* **2018**, *11*, 1856. [CrossRef]
18. Bahnini, I.; Rivette, M.; Rechia, A.; Siadat, A.; Elmesbahi, A. Additive manufacturing technology: The status, applications, and prospects. *Int. J. Adv. Manuf. Technol.* **2018**, *97*, 147–161. [CrossRef]
19. Trevisan, F.; Calignano, F.; Lorusso, M.; Pakkanen, J.; Aversa, A.; Ambrosio, E.P.; Lombardi, M.; Fino, P.; Manfredi, D. On the Selective Laser Melting (SLM) of the AlSi10Mg Alloy: Process, Microstructure, and Mechanical Properties. *Materials* **2017**, *10*, 76. [CrossRef] [PubMed]
20. Razavykia, A.; Brusa, E.; Delprete, C.; Yavari, R. An Overview of Additive Manufacturing Technologies—A Review to Technical Synthesis in Numerical Study of Selective Laser Melting. *Materials* **2020**, *13*, 3895. [CrossRef] [PubMed]
21. Oropeza, D.; Seager, T.; Firdosy, S.; Guerra, J.; Billings, K.; Jones, J.-P.; Hofmann, D.C.; Roberts, S. Porosity control of copper-based alloys via powder bed fusion additive manufacturing for spacecraft applications. *J. Porous Mater.* **2024**, *31*, 779–791. [CrossRef]
22. Yao, L.; Ramesh, A.; Xiao, Z.; Chen, Y.; Zhuang, Q. Multimetal Research in Powder Bed Fusion: A Review. *Materials* **2023**, *16*, 4287. [CrossRef] [PubMed]
23. Hayat, M.D.; Singh, H.; He, Z.; Cao, P. Titanium metal matrix composites: An overview. *Compos. Part A Appl. Sci. Manuf.* **2019**, *121*, 418–438. [CrossRef]
24. Gao, B.; Zhao, H.; Peng, L.; Sun, Z. A Review of Research Progress in Selective Laser Melting (SLM). *Micromachines* **2022**, *14*, 57. [CrossRef] [PubMed]
25. Liu, Y.; Ren, D.; Li, S.; Wang, H.; Zhang, L.; Sercombe, T. Enhanced fatigue characteristics of a topology-optimized porous titanium structure produced by selective laser melting. *Addit. Manuf.* **2020**, *32*, 101060. [CrossRef]
26. Chen, J.; Fan, M.; Zhou, L.; Chen, W.; Ren, Y.; Li, W.; Huang, W.; Niu, Y.; Li, Z.; Li, C. The effect of microstructure on the fatigue behavior of titanium alloy graded porous structures fabricated by selective laser melting. *J. Mater. Res. Technol.* **2023**, *27*, 4290–4304. [CrossRef]
27. Hu, Y.; Chen, H.; Liang, X.; Jia, M.; Lei, J. Microstructure and Biomechanical Properties in Selective Laser Melting of Porous Metal Implants. *3D Print. Addit. Manuf.* **2021**, *10*, 1003–1014. [CrossRef] [PubMed]
28. Ren, Y.; Ran, W.; Nie, Y.; Liu, Z.; Lou, C.; Li, J.; Chen, W. Excellent Mechanical Properties and Energy Absorption of Hexagonal-Body-Centered Lattice Structure Fabricated by Selective Laser Melting. *Adv. Eng. Mater.* **2023**, *25*, 2300431. [CrossRef]
29. Li, H.; Yang, G.; Lv, S.; Zhou, Q.; Fan, L.; Liu, R.; Yao, J.; Tong, P. Effects of Unit Cell Topology on the Mechanical Properties of Porous Tantalum Structures via Laser Powder Bed Fusion. *Adv. Eng. Mater.* **2022**, *25*, 2201431. [CrossRef]
30. Wang, Z.; Jiang, X.; Yang, G.; Song, B.; Sha, H. Design and mechanical performance analysis of T-BCC lattice structures. *J. Mater. Res. Technol.* **2024**, *32*, 1538–1551. [CrossRef]
31. Pauly, S.; Schricker, C.; Scudino, S.; Deng, L.; Kühn, U. Processing a glass-forming Zr-based alloy by selective laser melting. *Mater. Des.* **2017**, *135*, 133–141. [CrossRef]
32. Li, X.; Roberts, M.; O'Keeffe, S.; Sercombe, T. Selective laser melting of Zr-based bulk metallic glasses: Processing, microstructure and mechanical properties. *Mater. Des.* **2016**, *112*, 217–226. [CrossRef]
33. Chen, Z.; Wu, X.; Davies, C.H. Process variation in Laser Powder Bed Fusion of Ti-6Al-4V. *Addit. Manuf.* **2021**, *41*, 101987. [CrossRef]
34. Chen, Z.; Wu, X.; Tomus, D.; Davies, C.H. Surface roughness of Selective Laser Melted Ti-6Al-4V alloy components. *Addit. Manuf.* **2018**, *21*, 91–103. [CrossRef]
35. Liu, J.; Zhang, K.; Liu, J.; Wang, H.; Yang, Y.; Yan, L.; Tian, X.; Zhu, Y.; Huang, A. Investigation of fatigue behavior of laser powder bed fusion Ti-6Al-4V: Roles of heat treatment and microstructure. *Int. J. Fatigue* **2023**, *176*, 107839. [CrossRef]
36. Liu, J.; Zhang, K.; Liu, J.; Xu, Y.; Zhang, R.; Zeng, Z.; Zhu, Y.; Huang, A. In-situ investigation into the deformation behavior of Ti-6Al-4V processed by laser powder bed fusion. *Mater. Charact.* **2022**, *194*, 112434. [CrossRef]
37. Aslan, B.; Yıldız, A.R. Optimum design of automobile components using lattice structures for additive manufacturing. *Mater. Test.* **2020**, *62*, 633–639. [CrossRef]

38. *ISO 13314-2011*; Mechanical Testing of Metals—Ductility Testing—Compression Test for Porous and Cellular Metals. ISO: Geneva, Switzerland, 2011.
39. Sugaya, K.; Imamura, T. Turbulent flow simulations of the common research model on Cartesian grids using recursive fitting approach. *J. Comput. Phys.* **2022**, *467*, 111460. [CrossRef]
40. Hu, C.; Liu, C. Development of Cartesian grid method for simulation of violent ship-wave interactions. *J. Hydrodyn.* **2016**, *28*, 1003–1010. [CrossRef]
41. Marker, C.; Shang, S.-L.; Zhao, J.-C.; Liu, Z.-K. Effects of alloying elements on the elastic properties of bcc Ti-X alloys from first-principles calculations. *Comput. Mater. Sci.* **2018**, *142*, 215–226. [CrossRef]
42. Ren, Y.; Ma, Y.; Pan, Q. Effect of Microstructure on the Johnson-Cook Constitutive Model Parameters of Ti-6Al-4V Alloy. *J. Phys. Conf. Ser.* **2020**, *1637*, 012018. [CrossRef]
43. Löschner, P.; Gupta, M.K.; Niesłony, P.; Korkmaz, M.E.; Jamil, M. Determination and Verification of the Johnson–Cook Constitutive Model Parameters in the Precision Machining of Ti6Al4V Alloy. *Machines* **2024**, *12*, 709. [CrossRef]
44. Polley, C.; Radlof, W.; Hauschulz, F.; Benz, C.; Sander, M.; Seitz, H. Morphological and mechanical characterisation of three-dimensional gyroid structures fabricated by electron beam melting for the use as a porous biomaterial. *J. Mech. Behav. Biomed. Mater.* **2022**, *125*, 104882. [CrossRef] [PubMed]
45. Li, X.; Xiao, L.; Song, W. Compressive behavior of selective laser melting printed Gyroid structures under dynamic loading. *Addit. Manuf.* **2021**, *46*, 102054. [CrossRef]
46. Sun, Q.; Sun, J.; Guo, K.; Wang, L. Compressive mechanical properties and energy absorption characteristics of SLM fabricated Ti6Al4V triply periodic minimal surface cellular structures. *Mech. Mater.* **2022**, *166*, 104241. [CrossRef]
47. Zhang, M.; Yang, Y.; Wang, D.; Xiao, Z.; Song, C.; Weng, C. Effect of heat treatment on the microstructure and mechanical properties of Ti6Al4V gradient structures manufactured by selective laser melting. *Mater. Sci. Eng. A* **2018**, *736*, 288–297. [CrossRef]
48. Choy, S.Y.; Sun, C.-N.; Leong, K.F.; Wei, J. Compressive properties of functionally graded lattice structures manufactured by selective laser melting. *Mater. Des.* **2017**, *131*, 112–120. [CrossRef]
49. Cardeña, A., Sancho, R.; Barba, D.; Gálvez, F. Dynamic behaviour of additively manufactured Ti6Al4V BCC lattice-based structures. *Mater. Lett.* **2023**, *354*, 135286. [CrossRef]
50. Tan, Q.; Liu, P.; Du, C.; Wu, L.; He, G. Mechanical behaviors of quasi-ordered entangled aluminum alloy wire material. *Mater. Sci. Eng. A* **2009**, *527*, 38–44. [CrossRef]
51. Wu, F.; Zhou, Z.; Yao, B.; Xiao, Z. Anisotropic Compressive Properties and Energy Absorption Efficiency of Porous Twisted Short Fiber Materials. *Steel Res. Int.* **2016**, *87*, 1534–1542. [CrossRef]